D0591118

Coastal and Estuarine Studies

Series Editors:
Malcolm J. Bowman   Christopher N.K. Mooers

Coastal
and Estuarine Studies

51

George A. Maul (Ed.)

Small Islands:
Marine Science and
Sustainable Development

American Geophysical Union

Washington, DC

Series Editors

Malcolm J. Bowman
Marine Sciences Research Center, State University of New York
Stony Brook, N.Y. 11794, USA

Christopher N.K. Mooers
Division of Applied Marine Physics
RSMAS/University of Miami
4600 Rickenbacker Cswy.
Miami, FL 33149-1098, USA

Editor

George A. Maul
Florida Institute of Technology
Melbourne, Florida 32901-6988

Small Islands : marine science and sustainable development / George A.
    Maul (ed.).
            p.   cm. -- (Coastal and estuarine studies ; 51)
        Papers derived from a meeting held in Martinique, French Antilles,
    Nov. 8-10, 1993
        ISBN 0-87590-265-0
        1. Islands--Congresses. 2. Oceanography--Congresses. I. Maul,
    George A. II. American Geophysical Union. III. Series.
    GB461.S63 1996
    551.4'2--dc20   96-29237
                                                                       CIP

ISSN 0733-9569

ISBN 0-87590-265-0

CONTENTS

PREFACE
    George A. Maul ix

PART I: SUMMARY
1   Ocean Science in Relation to Sustainable Economic Development
and Coastal Area Management of Small Islands
    George A. Maul 1

PART II: PHYSICAL SCIENCES
2   Observations and Modeling of Mesoscale Ocean Circulation Near
a Small Island
    Malcolm J. Bowman, David E. Dietrich, and Charles A. Lin 18

3   Descriptive Physical Oceanography of the Caribbean Sea
    Artemio Gallegos 36

4   Oceanic Features Influencing Small Island Circulation Patterns:
Case Studies
    Roy A. Watlington and Maria C. Donoso 56

5   Subtidal Circulation in Fort-De-France Bay
    Pascal Lazure, Jean Claude Salomon, and Maruerite Breton 71

6   Sea Level, Tides, and Tsunamis
    George A. Maul, Malcolm D. Hendry, and Paolo A. Pirazzoli 83

7   Practical Aspects of Physical Oceanography for Small Island States
    Judith Wolf 120

8   Design Values of Extreme Winds in Small Island States
    Calvin R. Gray 132

9   A Real-Time System for Forecasting Hurricane Storm Surges
Over the French Antilles
    Pierre Daniel 146

PART III: GEOLOGY AND ECOLOGY
10   Geography of Small Tropical Islands: Implications for Sustainable
Development in a Changing World
    Orman E. Granger 157

11   Small Island Geology: An Overview
    Georges Vernette 188

12 The Geological Legacy of Small Islands at the Caribbean-Atlantic
    Boundary
        Malcolm D. Hendry 205

13 Geology and Development Facilities of Small Islands Belonging
    to the Atlantic Margin of Africa and Europe
        André Klingebiel and Georges Vernette 225

14 Beach Erosion and Mitigation: The Case of Varadero Beach, Cuba
        Guillermo García Montero and José L. Juanes Martí 238

15 Surficial Geology on the Insular Shelf of Martinique (French
    West Indies)
        Françoise Durand, Claude Augris, and Patrice Castaing 250

16 Littoral Ecological Stability and Economic Development in
    Small Island States: The Need for an Equilibrium
        Vance P. Vicente 266

17 Microbial Water Quality on a Caribbean Island (Martinique)
        Monique Pommepuy, Annick Derrien, Françoise Le Guyader,
        Dominique Menard, Marie-Paule Caprais, Eric Dubois,
        Elizabeth Dupray, and Michele Gourmelon 284

18 Fisheries of Small Island States and Their Oceanographic
    Research and Information Needs
        Robin Mahon 298

PART IV: MANAGEMENT
19 Towards Integrated Coastal Zone Management in Small
    Island States
        Gillian Cambers 323

20 Water Supply and Sewerage in a Small Island Environment:
    The Bahamian Experience
        Richard V. Cant 329

21 Coastal and Marine Environments of Pacific Islands: Ecosystem
    Classification, Ecological Assessment, and Traditional Knowledge
    for Coastal Management
        Paul F. Holthus 341

22 Coastal Management, Oceanography, and Sustainability of
    Small Island Developing States
        Paul H. Templet 366

23   **Sustainable Development and Small Island States of the Caribbean**
    Erik Blommestein, Barbara Boland, Trevor Harker,
    Swinburne Lestrade, and Judith Towle 385

24   **Socio-Economic Databases in the Caribbean: Status and Desiderata**
    Lancelot A. Busby 420

25   **Numerical Modeling of Small Island Socio-Economics to Achieve Sustainable Development**
    Guy Engelen, Roger White, Inge Uljee, and Serge Wargnies 437

LIST OF CONTRIBUTORS 464

PREFACE

Small tropical and subtropical islands are particularly attuned to the effects of the surrounding ocean and atmosphere. Frequently, these islands are densely populated and rely on the sea for sustenance and economic viability. Oftentimes too, individual island nations do not have the resources to address issues in oceanography and marine meteorology that are central to their future well-being. United Nations agencies, government and non-government organizations, and universities are dimly aware of their role to enhance collaboration between natural scientists and resource managers, particularly for Small Island Developing States (SIDS). Accordingly, a meeting entitled "Small Island Oceanography in Relation to Sustainable Economic Development and Related Coastal Area Management" was held in Martinique, French Antilles, November 8–10, 1993.

This collection of 25 integrated papers derived from the meeting are grouped into three major sections: physical sciences, geology and ecology, and management. The volume provides a summary of state-of-the-art knowledge in applying physics, chemistry, biology, and geology to managing natural resources in the context of sustaining economic growth and stability. Particular attention is given to the application of numerical models to quantitative decision making. The goal is to apply science to societal needs.

This volume is an outgrowth of the IOCARIBE/UNEP interdisciplinary scientific effort, which included equally marine scientists and decision-makers with environmental responsibilities. It addresses the need for developing quantitative management tools, paying particular attention to the effects of the ocean and atmosphere on small tropical and subtropical islands and on their economic sustainability. Since the biology, chemistry, geology, and physics of the ocean and the lower atmosphere are intertwined with the economics and health of SIDS, an integrated approach to management is required for sustainable development. The development of quantitative models in which the ocean's physical variability is coupled to biogeochemical and socioeconomic effects is a primary goal. Envisioned are a series of nested numerical models each providing the boundary conditions from a coarse-resolution regional ocean/atmosphere coupled circulation model to a fine-resolution island-scale circulation model to a cellular automata socioeconomic decision support model. Some of these models are PC based, and thus are amenable to the available resources on most developing island states.

The meeting on which this volume is based was suggested by Gunnar Kullenberg, Secretary-General of the Intergovernmental Oceanographic Commission (IOC) of UNESCO. It was organized by the IOC Subcommission for the Caribbean and Adjacent Regions (IOCARIBE) with support from the Caribbean Environmental Programme of UNEP, the University of Puerto Rico, and the U.S. National Oceanic and Atmospheric Administration. The Scientific Organizing Committee was George A. Maul (USA), Chairman; Avril Suing-Chang (Trinidad and Tobago), Biology and Fisheries; Artemio Gallegos (Mexico), Physics and Chemistry; Calvin R. Gray (Jamaica), Climate and Meteorology; and Georges Vernette

(France), Coastal Management and Geology. The meeting was hosted by Alain Soulan, Director METEO-FRANCE/French Antilles and Guyane, with organizational arrangements coordinated by Fernando L. E. Robles, IOC Senior Assistant Secretary for IOCARIBE. The workshop proceedings were edited as IOC Workshop Report 97, which was made available to the subsequent Global Conference on the Sustainable Economic Development of Small Island States, Barbados, 1994 (UNCED: Agenda 21, Chapter 17 G).

I express my appreciation to each of these persons and to Christopher N. K. Mooers, University of Miami, for guidance in compiling the volume. The unheralded efforts of the 50-plus peer reviewers is gratefully acknowledged. Word processing by Gail M. Derr, NOAA Atlantic Oceanographic and Meteorological Laboratory, and page layout by Susan Heyse and Damian Hite, Florida Institute of Technology, is also acknowledged.

George A. Maul
Melbourne, Florida
July 1996

1

Ocean Science in Relation to Sustainable Economic Development and Coastal Area Management of Small Islands

George A. Maul

Abstract

Small island developing states (SIDS) were recognized at the 1992 United Nations Conference on Environment and Development [UNCED] as requiring special attention by the scientific community in order to address problems associated with sustainable economic development and coastal area management. The Intergovernmental Oceanographic Commission of UNESCO convened a meeting on small island oceanography (Martinique, 8-10 November 1993) that brought together specialists from marine science, geography, hydrology, ecology, fisheries, management, economics, and sociology. Thirty participants from 15 countries and SIDS in the Atlantic, Pacific, and Indian Oceans approached the challenge from UNCED with cross-cutting lectures, debates, and problem-solving exercises focusing on the three "c's" of successful interdisciplinary endeavor: communication, collaboration, and coordination.

Oceanographic science information to support coastal zone management was deemed essential to the following most critical and related SIDS issue areas: (1) water quality; (2) natural hazards prediction and response planning; (3) harvest of living marine resources; (4) shoreline coastal dynamics; and (5) habitat conservation. Cross-cutting solutions to these five issue areas emphasize the need for numerical modeling, enhanced monitoring and inter-calibration, education and training, data sharing, regional coordination and networking, and appropriate methodology development for decision-making. In particular, quantitative data from coastal physical oceanography was identified as of the highest priority in an issues/solutions matrix, and the expanded use of PC or workstation computer models that integrate the physical environment with

Small Islands: Marine Science and Sustainable Development
Coastal and Estuarine Studies, Volume 51, Pages 1–17

socio-economic response such as climate change, sea level rise, and population growth was recommended.

Introduction

At UNCED, the United Nations Conference on Environment and Development (Rio, June 1992), the special concerns of small island developing states (SIDS) were singled out. In Agenda 21, Chapter 17 of UNCED, U.N. agencies were especially requested to address the needs of SIDS within their special area of expertise. Accordingly, the Inter-governmental Oceanographic Commission (IOC) of UNESCO decided to convene a meeting on the oceanography of small islands in relation to sustained economic development and coastal area management.

At the annual meeting in February 1992, officers of the IOC Subcommission for the Caribbean and Adjacent Regions (IOCARIBE), Dr. Gunnar Kullenberg, IOC Secretary-General, requested that IOCARIBE undertake to organize and conduct a workshop that specifically addressed SIDS issues. The Subcommission Officers agreed and Dr. George A. Maul (USA), IOCARIBE Vice Chairman, volunteered to chair the effort. He recruited four eminent scientists from the region to assist in the program development: Dr. Artemio Gallegos (Mexico), Mr. Calvin Gray (Jamaica), Dr. Avril Suing-Chang (Trinidad and Tobago), and Dr. Georges Vernette (France). Although the Workshop was to be conducted on a small Caribbean island, it was understood that a global view was to be taken.

A proposal was prepared for presentation at the Subcommission IOCARIBE Fourth Meeting (SC-IOCARIBE-IV) to the Member States in December 1992. After much discussion and with the collaboration of the United Nations Environment Programme's Caribbean Regional Coordinating Unit (UNEP/RCU), SC-IOCARIBE-IV voted to conduct the Workshop within the general guidelines outlined in Appendix I. The representative of France (Mr. Alain Soulan) agreed to request the support of his government and to conduct the Workshop on the Caribbean island of Martinique, French Antilles.

Financial support for the Workshop was solicited from a variety of organizations, and eventually acquired from the U.S. National Oceanic and Atmospheric Administration, the University of Puerto Rico, UNEP's Caribbean Environment Programme, the IOC, and METEO-FRANCE, the Workshop's host [IOC, 1993. Funding for the meeting totaled about U.S. $35,000, and support in kind for salaries, travel, transportation, and facilities was generously provided by the attendees' organizations.

The meeting's plan was to have an interdisciplinary group from varied backgrounds, but all of whom were interested in the SIDS issue. Appendix II is the meeting agenda, and it reflects membership from amongst oceanographers, meteorologists, geologists, geographers, economists, coastal managers, hydrologists, social modelers, fisheries scientists, environmentalists, and administrators. A series of lectures each followed by discussion and debate was the basic format. Ample time was available for in-depth consideration of varied points of view and perhaps most importantly, for familiarity with the specialized languages of each of the disciplines represented.

The meeting was held at the offices of METEO-FRANCE in Fort de France, Martinique, and was attended by 30 participants from 15 countries. Although the meeting was in the Caribbean, experts from the Atlantic, Pacific, and Indian Oceans were present. The Workshop was advertised in an international journal [*EOS*, 1993]. It is clear from the lack of participants from small island states at the Workshop that one of the most important issues is to develop training, education, and mutual assistance programs for SIDS scientists and administrators.

In order to best preserve the meeting results and to make them available to a wide audience, it was agreed to publish the lectures in the Coastal and Estuarine (book) Series of the American Geophysical Union (AGU). The AGU series was considered most likely to have a very wide audience, to provide high-quality printing, and since it is a non-profit organization, to keep the cost to a minimum. Consideration for translation into other languages is an option for the future. The title of the book was agreed to be *Small Islands: Marine Science and Sustainable Development*, and the effort was announced to the international scientific community at the annual ASLO/AGU Ocean Sciences Meeting held in San Diego, California, 21-25 February 1994 [Maul, 1994].

Main Results

Most SIDS undoubtedly will lack the infrastructure and personnel required to take advantage of modern coastal area management techniques. Indeed, some of the so-called "modern solution technology" may not be useful in real-life SIDS situations. The Workshop participants were made acutely aware of the level of problems and problem-solving by breaking into three groups and actually considering a typical coastal management issue recently encountered in the British Virgin Islands [Cambers, 1993]. The problem description and the solutions approached by the interdisciplinary teams form one part of this section of the chapter.

The conferees broke into five working groups to identify oceanographic science information essential to support coastal zone management in the following problem/issue areas:

1.   Water quality.
2.   Natural hazards prediction and response planning.
3.   Harvest of living marine resources.
4.   Shoreline coastal dynamics.
5.   Habitat conservation.

In addition to identifying that these five coastal zone management issues are of paramount importance to SIDS, important cross-cutting aspects were listed, including oceanic, atmospheric, and socioeconomic modeling, monitoring, education, training, data sharing, regional coordinating and networking, and appropriate methodology development.

The group discussed the World Coast Conference (The Netherlands, 2-5 November 1993), an issue not in Appendix II. Concern was expressed that the World Coast Conference lacked adequate technical background for many members of AOSIS (Association Of Small Island States), and that AOSIS may be politically driven rather than persuaded from a technical issues perspective. Workshop members strongly felt that such a direction may be unproductive for small island developing states, but which can be overcome by the three "c's" knitting together the participants in Martinique: communication, collaboration, and coordination.

A Practical Problem

The Workshop practicum (Appendix III) was designed by Dr. Gillian Cambers, based on a recent situation in the British Virgin Islands (BVI). The central issue involves using SCUBA to harvest conch and lobster, and the question is: *DOES SCUBA FISHING SIGNIFICANTLY DEPLETE CONCH AND LOBSTER STOCKS IN THE BRITISH VIRGIN ISLANDS?* In actual practice, not only is the question to be answered quickly (typically a month or less), but it must be done so with very limited personnel and equipment. The following paragraphs represent the summary response from the three groups.

Group 1: Achieving sustainable exploitation identifies a basic conflict between competing interests. In the short term, it is recommended: (a) to replace the ban on SCUBA with permits; (b) to create a seasonal ban; and (c) to create size limits and/or catch limits. For the longer term (six months) the suggestion was to conduct a resource study; to involve fishermen in the licensing procedure and in surveillance; to provide retraining and education of both tourists and fishermen; and to review a marine parks proposal.

Group 2: Create a mechanism for an annual permit by auction or lottery; allow permits to be resold. This would naturally result in enforcement by permit holders. The fishery needs reserves, therefore a temporary (seasonal) ban on SCUBA is warranted (up to three years); this can be achieved by regulating the buyers. Additionally, there is a need: (a) for basic fishing (fishermen independent) data; (b) to conduct needed research from license fees; and (c) to use fees to train fishermen for alternate employment. The BVI should not back away from existing legislation due to political pressure.

Group 3: There is insufficient information to establish the direction of the catch per unit effort, and a SCUBA ban may be unwarranted. In the short term in order not to create conflict: (a) involve the media to publicize an alarm; (b) make the public share-holders rather than regulators with draconian measures; and (c) strictly observe existing laws. In the long term: (a) a social cost must be put on the activity; (b) all SCUBA diving/fishing must be included in legislation; (c) work with the consumers not to buy illegal size catch; (d) set up a series of reserves; and (e) fund research and education with a possible shift towards mariculture.

Discussion following the group reports suggested that it was necessary to divide the regulation into three fisheries: conch, lobster, and fin-fish. The perception throughout the region is that fisheries are being depleted, and therefore certain international aspects need to be considered. A reliable and easily accessible database needs to be created and maintained. Overwhelmingly there is a need for scientists to work with the fishermen in order to prevent emergencies, and decisions need to be flexible as new research is reported.

Oceanographic Science Information Required

In this section, the oceanographic data required for problem definition and solution in the five SIDS primary problem areas identified during the Workshop are enumerated. Table 1 summarizes much of the discussion and is presented more as a point of departure for debate than a definitive statement of the attendees' opinions. In the tabulation, a blank (unchecked) element does not mean that the topic is unimportant for one of the five particular problem areas, but rather that it is not crucial.

Water Quality

Data needed to address water quality requirements include currents, bottom topography, water chemistry, temperature, turbidity, rainfall, waves, light, and salinity. The formats for data to be useful include paper copy, time series (trends), maps (spatial), and oceanic models of several types including conceptual box models, diagnostic data assimilation models, and prognostic (forecast) models. Education and training needs include oceanographic model operators, chemists, technicians, and graduate student support. In terms of data sharing and coordination, there needs to be better communication and collaboration between universities and public agencies, between university departments, and between regional agencies (in particular, it was suggested that the U.N. insure that the FAO, WMO, IMO, IOC, and UNEP meet at least annually to coordinate overlapping activities and to report to the member states on their progress).

TABLE 1:   Summary of information required for integrated coastal zone management decisions of small island developing states to achieve sustainable economic development.

Required Oceanographic Science Information	Water Quality	Hazards Prediction and Response Planning	Harvest of Living Marine Resources	Shoreline Coastal Dynamics	Habitat Conservation
Air-sea fluxes	✓	✓			
Bottom topography	✓	✓		✓	
Chlorophyll	✓		✓		
Currents	✓	✓	✓	✓	✓
Diastrophism		✓		✓	✓
Imagery	✓	✓	✓	✓	✓
Light	✓		✓		✓
Model, diagnostic	✓	✓	✓	✓	✓
Model, prognostic	✓	✓	✓		
Plankton	✓		✓		✓
Precipitation	✓	✓		✓	✓
Riverine input	✓	✓	✓	✓	✓
Salinity	✓		✓		✓
Sea level	✓	✓		✓	✓
Temperature	✓	✓	✓		✓
Tides		✓		✓	
Training	✓	✓	✓	✓	✓
Tsunamis		✓		✓	✓
Turbidity	✓				✓
Upwelling		✓	✓		
Water column chemistry	✓		✓		✓
Wind, waves, and swell	✓	✓		✓	
Winds	✓	✓	✓	✓	

Natural hazards prediction and response planning

Data needs include physical variables describing the ocean/atmosphere interface and the subsurface ocean in real time, climatological time series for analysis of past trends, and easy access to satellite imagery and image processing algorithms. There is a need for regional specialized marine centers with capability for on-line data acquisition, processing, and analysis, which can provide training with respect to specialized products that may be island specific. In general, there also needs to be improved observation networks, database management, and regional modeling, both oceanographic and meteorological. Storm-proof coastal sea level/weather observatories are required to adequately measure storm surge and wind waves.

Harvest of Living Marine Resources

The system is not deemed sustainable at the present, and much of the data needs have been made clear in the practicum discussed above (q.v. Appendix III). Of primary interest are ocean circulation prognostic models that are particularly tuned to forecast upwelling, and models that integrate biological variables in their results. Data to validate these models is continually required, and includes temperature, salinity, chlorophyll, winds, and fishing indices. Since some harvesting is pelagic, the 10 km resolution regional models are of the proper scale, but for the reef and coastal fisheries, there will have to be very fine resolution models nested into the regional models. Most importantly, there will have to be communication systems established to provide the model output to the fishermen and a feedback network to insure that the models are performing the needed tasks.

Shoreline Coastal Dynamics

In this problem area, information is required for physical planning including setbacks, natural habitat response to erosion, classification of shorelines, trends in shoreline changes (e.g., from mangrove to mudflat), reef ecology, riverine input to the coastal zone, bottom topography, sediment budgets, water column properties, waste disposal, wave directional spectra, sea level rise or fall, and mining activities. Long term tide gauge observations of sea level change are critically needed in SIDS in order to assess future shoreline change due to water level differences with time. Additionally, there is a need for information on coastal structures and reports from other agencies with good and/or bad experience in managing the shoreline on small islands.

Habitat Conservation

Oceanographic "information" rather than simply data is required, in particular circulation on the microscale (1-3 km), the mesoscale (10 km or so), and the ocean basin or microscale (order 100 km). The information needed includes temperature, salinity, turbidity, sea level, nutrients, and general planktonic statistics. Models need to be developed to give a physical foundation for interpretation of the biological data, in particular model packages that can be run on existing PC machines with limited data; a clearinghouse of models and modelers should be created; a case study repository should be developed; and a "standard" set of "circulation types" should be made available.

In general, there is a need to merge geophysical (*i.e.*, ocean and atmospheric) models, socioeconomic models, and observations. For example, if global climate change brings about a sea level rise, the effects need to be quantified in terms of the socioeconomic, natural habitat, shoreline, water quality, marine harvest, and hazards response impact. But global climate models and their predictions do not take into account very local effects, such as land uplift due to tectonic activity (for example). Therefore, for the issue of sea level rise, observations are also required since in the case of tectonic uplift, the local effect may be sea level *fall*. Hence, for SIDS, in particular, a balance between observations and models, and between regional networking and intra-country coordination, and between reason and tradition, are all necessary to progress towards sustainable economic development.

Schematicus

Figure 1 is a schematic describing the complex interactions of oceanic science in relation to sustainable economic development and coastal area management of small islands. It is always tempting to reduce three days of Workshop effort and months of post-Workshop reflection to a few simple notions. Such is the hazard of integrating scientific and socioeconomic issues into a common framework. Mindful of the traps one can set for oneself in such conceptualizing, it is hoped that the elements of Figure 1 do not lull the reader (or the writer for that matter) into a sense of intellectual security.

Conceptually, the five issue areas *water quality, natural hazards, living marine resources, shoreline coastal dynamics, and habitat conservation*, at the apex of Figure 1, are viewed as requiring management. The nature of computers leads to expecting quantitative results from which to make decisions, such as issuing storm warnings, beach closures, fishing restrictions, reforestation, *etc*. But these decision-making procedures are an evolving art based on improved *databases* and better understanding of the physical processes being forecast. Hence, it is central to modeling that *monitoring and surveys*, both socioeconomic and natural, be given long-term commitment by SIDS governments. Numerical models, whether ocean/atmosphere or socioeconomic, are no better than the data used to initialize them.

Numerical models, such as island-scale diagnostic and prognostic ocean and/or atmospheric circulation models, have boundary conditions that are determined from regional-scale models; similarly, the regional-scale models require information from the global-scale models. Thus, there is a "nesting" of the finer-scale model in the coarser-scale one. In addition, all such models parameterize physical processes, and again, the model is no better than the parameterization. Hence, *process studies* are essential to modeling and to *model development*. Moreover, it is also recognized that much of the information input to a model and output from it is very island specific, so while process

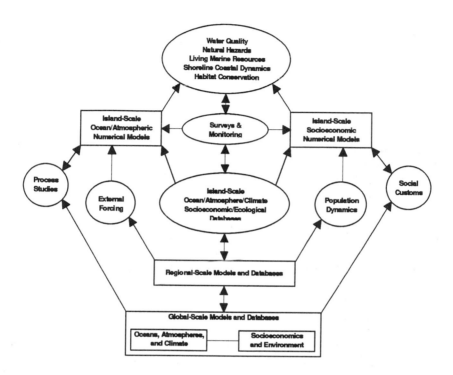

Fig. 1. Schematic of the interactions in modeling the effect of the ocean/atmosphere and socio-economic systems on five issues critical to sustainable economic development and coastal area management of small island developing states.

studies may have a generic nature, their implementation oftentimes is dependent on the locale.

Socioeconomic numerical models require input from ocean/atmosphere models and databases as well as other variables such as *social customs, external economics,* and *population dynamics.* In Figure 1 these inputs are shown to be quasi-independent, but clearly external issues such as human migration will affect population dynamics in a manner that may be beyond the control of local government. It is, however, the island-scale socioeconomic numerical model that allows the manager or politician to ask "what if" questions. The answer is a forecast! Confidence needs to be created in such prognostication by constant reference to the database where the results of earlier decisions are archived.

Perhaps most importantly, one needs to be aware that there may be more than one correct decision when faced with a question. Alternately, there may be no answer if chaotic forces are driving the system. Many systems contain nonlinear feedbacks that can make forecasting unreliable. One only has to look at the last weather or economic forecast to be cognizant of the inherent danger in putting too much confidence in a calculation; fancy computerized results can never be a substitute for human common sense.

Conclusions

A clear definition of "just what is a small island?" will probably be the focus of endless debate. Apparently the complete definition will require three geographical parameters. For example: total population of less than say 100,000 persons; total area of say less than 10,000 square kilometers; population density of say more than 25 persons per km^2 of "usable" land. Unfortunately, geography alone is inadequate to express the socioeconomic dimensions of such a definition, which may have to include income, opportunity, health, and a host of other intangibles. Ultimately "small island developing state" will be a political definition, and the challenge for scientists and managers will be to create a coordinated process to achieve sustainable development.

All coastal zone management decisions in small island developing states are ultimately political because the body of law is not well developed. SIDS coastal managers can usually only make recommendations. Therefore, the managers and scientists need to learn to work within the system; they need to become involved in the political process. Lack of coastal resources information is a major problem, and even the most egalitarian politician finds weakened conviction where there is no database. For this reason, the Environmental Impact Assessment is a little used tool, although it is in the early stages of development in many SIDS.

The time scale of change is typically one generation, *i.e.,* 30 years or so. Whether sustainable economic development is compatible with such time scales in the future is a matter for serious debate. Through the following steps the coasts can provide sustainability:

1. Create the political will.
2. Perceive the need for coastal zone management.
3. Embark on infrastructure building.
4. Create an integrated plan.
5. Implement the plan openly and fairly.

The challenge will be to continue the interdisciplinary dialog started during this Workshop on Small Island Oceanography in relation to Sustainable Economic Development and related Coastal Area Management of Small Island Developing States.

Acknowledgments.   Support for this chapter and indeed for the entire book was provided by the NOAA Atlantic Oceanographic and Meteorological Laboratory (AOML), and we express our appreciation to H. F. Bezdek and R. L. Molinari for their backing.   Word processing and editorial assistance was the contribution of G. Derr, (AOML), S. Heyse, and D. Hite of the Florida Institute of Technology; C. N. K. Mooers was the AGU Coastal and Estuarine Series Editor for this volume.   Finally, we wish to express our appreciation to the more than 50 persons who acted as peer referees; their unheralded contributions were essential to the success of the entire effort.

References

Cambers, G., Does SCUBA fishing significantly deplete conch and lobster stocks in the British Virgin Islands? Appendix III, this chapter, 1993.

EOS, Transactions, American Geophysical Union, 74(37), 427, 1993.

IOC, Intergovernmental Oceanographic Commission of UNESCO, *Workshop on Small Island Oceanography in relation to Sustainable Economic Development and Coastal Area Management of Small Island Developing States,* Fort-de-France, Martinique, French Antilles, 8-10 November 1993, Workshop Report No. 97, Paris, 5 pp. and VI Annexes, 1993.

Maul, G. A., Small island oceanography, invited abstract, *EOS, 75*(3), 208, 1993.

APPENDIX I: Invitational Workshop Description

Workshop on Small Island Oceanography in Relation to Sustainable
Economic Development and Related Coastal Area Management
Martinique, French Antilles, November 8-10, 1993

Organized by the
Intergovernmental Oceanographic Commission of UNESCO and its
Subcommission for the Caribbean and Adjacent Regions and the
Caribbean Environmental Programme of UNEP

Description

Small tropical and subtropical islands are particularly attuned to the effects of the surrounding ocean and the atmosphere. Frequently these islands are densely populated and rely on the sea for sustenance and economic viability. Oftentimes too, individual island nations do not have the resources to address the issues in oceanography and marine meteorology that are central to their future well-being. UNCED (Agenda 21, Chapter 17 G) and the U.N. General Assembly have recognized some of these issues by calling a "Global Conference on the Sustainable Economic Development of Small Island States" for April 1994 in Barbados.

Regional bodies of United Nations agencies are aware of their role in creating the infrastructure to collaborate with natural scientists and resource managers. This Workshop addresses this need with particular attention to the effects of the ocean and atmosphere on small tropical and subtropical islands and their economic sustainability. The biology, chemistry, geology, and physics of the ocean and the lower atmosphere are intertwined with the economics and health of small islands, thus demonstrating that an integrated approach to management is required for sustainable development. An interdisciplinary scientific workshop is envisioned that includes equally marine and atmospheric scientists and government and business decision-makers with environmental responsibilities.

The Workshop will last for three days and will primarily include invited papers in each of the disciplines mentioned above, covering time-scales ranging from those associated with storm damage and hazardous spills, to decade and longer climate changes. The concentration will be on quantitative effects and the use of oceanic and atmospheric events on small islands and their marine environs. A series of case studies are planned to be presented and discussed. Scientific sessions will emphasize transfer of information and application of the models to the decision-making sectors. The Workshop proceedings will be edited into a publication that will particularly be made available to the Global Conference on Sustainable Economic Development of Small Island States mentioned above.

The Workshop will emphasize quantitative models where the ocean's physical variability is coupled to biogeochemical economic effects. Envisioned are a series of nested numerical models each providing the boundary conditions from a coarse-resolution regional circulation model to a fine-resolution island-scale circulation model to a cellular automata socioeconomic decision support model. Some of these models are PC based, and thus are amenable to the available resources on most developing island states.

A typical day during the Workshop will start with one or more invited lectures on a topic followed by contributed papers on the same topic. After the technical session, working groups will be created to discuss in depth the presentations, and to formulate needed capacity building programs with research topics and technology transfer. Plenary sessions will be used to share the working group deliberations and to debate their conclusions. Ongoing report writing will insure a draft final report for plenary discussion.

The potential role of GOOS, the Global Ocean Observing System, will constitute one Workshop session. Reports on the status of monitoring networks will be reviewed and the implementation of GOOS will be discussed on a case by case basis. Specific recommendations will be made, and a funding and implementation plan will be developed.

Participation in the Workshop is primarily by invitation, but interested scientists are encouraged to contact the appropriate member of the organizing committee. In order to accomplish the Workshop's goals, participation of approximately 25 persons is expected. Those requiring financial assistance and lodging information should contact the IOC Senior Assistant Secretary for IOCARIBE, A.A. 1108, Cartagena, Colombia (Tel: 57 53 646399; FAX: 57 53 600407).

Scientific Organizing Committee:

George A. Maul (USA), Chairman; FAX 305-361-4582
Avril Suing-Chang (Trinidad and Tobago), Biology and Fisheries; FAX 809-634-4433
Artemio Gallegos (Mexico), Physics and Chemistry; FAX 52 5 548-2582
Calvin R. Gray (Jamaica), Climate and Meteorology; FAX 809-924-8670
Georges Vernette (France), Coastal Management and Geology; FAX 57 53-600407

APPENDIX II: Workshop Program

Workshop on Small Island Oceanography in Relation to Sustainable Economic Development and Coastal Area Management of Small Island Developing States

Martinique, 8-10 November 1993

Sunday, 7 November

1900   *Informal Gathering:* Discussion of draft agenda; time requirements for presentations; logistics of meeting rooms; plans for publication of meeting chapters in an AGU "Coastal and Estuarine Studies" volume.

Monday, 8 November

0830-0900   *Opening:* Statements by IOC (F. Robles); US-NOAA (G. Maul, Chairman of the Workshop); METEO-FRANCE (A. Soulan, host).

0900-1800   *Sessions 1 and 2:* Oceanography and Meteorology
              Papers by: Bowman, Gallegos, Lazure, Watlington, Michel, Maul and Hendry, Gray, Mandar, Daniel, and Bleuse.

Tuesday, 9 November

0830-1230   *Session 3:* Islands
              Papers by: Vernette, Klingebiel, Hendry, Granger, Holthus, and Durand.

1430-1800   *Session 4:* Chemistry, Biology, Fisheries

Papers by: Vicente, Mahon, Gourmelon, Proni, and Daniel.

Wednesday, 10 November

0830-1400   *Session 5:* Management
              Papers by: Robles, Engelen, Wolf, Cant, Templet, Garcia, Busby, and
              Cambers.

1430-1700   *Session 6:* Problems, solutions, program formulation, planning and
              input towards the Global Small Island Developing States Conference.

1700-1900   *Session 7:* Concluding discussion on follow-up, possible small islands
              oceanography program preparation, and contribution to (or input to) the
              Global Small Island States Conference on sustainable development.

2000   Fiesta!

Appendix III: Practicum

Does SCUBA Fishing Significantly Deplete Conch and Lobster
Stocks in the British Virgin Islands?

Background

The British Virgin Islands (BVI) are an archipelago of 50 islands, 160 km east of Puerto
Rico. The population is 20,000 people. The economy is based on tourism and offshore
financing, with fisheries third in order of importance. There is a fledgling Coastal Zone
Management Agency in the BVI, the Conservation and Fisheries Department (CFD).

The Fisheries Industry

The fisheries industry is based on a small scale artisanal (reef) fishery. Fisheries
contributes 3% to the gross domestic product. There are a total of 280 fishermen with
150 fishing boats. Fish traps are the most important gear. Sports fishing is an
important economic activity, but this fishery has not been included in the above
statistics. As yet, the BVI has not moved towards longlining or the exploitation of deep
water species, although this is being given consideration. A comprehensive data
collection system has only been in place since 1992, so few data are available.

Against a background of sparse data, there are signs that the reef fishery has been
depleted. Similarly, estimations are that the conch fishery has declined from the

maximum sustainable yield (MSY) of 188 tonnes to 59 from shallow habitats, and can only be found in +60 ft water depths. The lobster fishery is producing amounts of lobster equal to or slightly less than the MSY.

SCUBA Fishing

Diving with SCUBA gear has been conducted in the BVI since the late 1970's. In 1986, two Haitian boats were intercepted: there were two mother ships and 16 dinghies, and there were compressors on board with hoses going down to the divers. This operation was moving like a wall, taking everything on the shelf.

In addition, local SCUBA divers have begun operations. Following incidents such as that described above, the commercial fishermen and the Fisherman's Association made a very strong lobby to the government to ban SCUBA fishing. Their concerns were threefold:

1. Declining catch in the reef fishery which they attributed to SCUBA fishing, at least in part.
2. SCUBA fishermen interfere with their traps.
3. SCUBA fishing methods basically clean out everything on the seafloor.

Due to the concerns of the commercial fishermen, the CFD was asked to make a recommendation fast (within weeks). They attempted to discuss the matter with the local SCUBA fishermen, but these fishermen were not willing to discuss the matter. In one case guns were pulled on the fisheries officers.

A recommendation was made that SCUBA fishing be banned completely and this law was passed in 1990. It must be borne in mind that an election was imminent and commercial fishermen represented a significant number of votes.

Between 1990 and 1992 the law was publicized by the CFD, and warnings were given on several occasions to SCUBA fishermen. Yet, the fishing continued unabated. During this time many complaints were received about SCUBA fishing activities, and particularly the fact that foreigners were coming in to fish with SCUBA. In addition, there was a complete ban on conch fishing in the U.S. Virgin Islands between 1987 and 1992, yet this fishery continued in the BVI. Both territories share the same bank.

In 1993 on a surveillance patrol, a group of SCUBA fishermen were caught. In the following court case, which received considerable publicity, fines of $100 were levied. The magistrate was lenient since it was a first offense and since it was also the first time such a case had come to court.

Following this case, local SCUBA fishermen came to the CFD requesting the law be changed since their livelihood was being threatened. Meetings were held and their points were as follows:

- Divers do not interfere with traps.
- Decline in catch in fish traps may be a function of the number of traps, the prevalence of ghost traps, the mesh size, and environmental degradation (*e.g.*, pollution and loss of mangroves) so why blame the SCUBA fishermen for this?
- Divers exercise selectivity in their method of fishing.
- There is a lack of communication between divers and other fishermen.
- The SCUBA fishery is a local industry and provides lobster and conch for the visiting tourists.
- They were willing to work with the CFD.

Present Situation

The SCUBA divers, of which there are about ten operations, have lobbied the Minister responsible for fisheries and the Chief Minister, requesting that the law be changed such that permits can be given for local SCUBA fishermen. The Ministers are sympathetic to this request, and while they have asked the CFD for an opinion, they are leaning towards changing the law as a matter of priority.

Decision to be Made

The CFD has to make a decision within a matter of weeks: what should that decision/recommendation be? There is, of course, the opportunity to make a decision now and make provisions for a review perhaps in one or two years time after certain data have been collected.

You are asked to advise on what the immediate decision should be and, if a longer term data collection program is to be started, what should this involve. You are asked to bear in mind the following constraints:

- The CFD has a staff of 10 people, five of whom are graduates. There is one 35-foot boat and a small inflatable. They are responsible for the management of all coastal resources; their workload is heavy. There are, in addition, four fisheries inspectors with their own boats; however, these individuals do little more than collect data.

- The area of the BVI is 153 km^2, the territorial sea (out to three nautical miles) covers an area of 1,469 km^2. There is, in addition, a 200 mile exclusive fishing zone.

• There is a marine division of the police, with one boat, who assist with fisheries surveillance, but their main priority is drugs.

• A total ban on any activity is far easier to enforce than managing a particular fishery through selective licensing.

• Fisheries surveillance in the BVI is poor, bear in mind three years passed before a conviction for SCUBA fishing was made.

• One of the key SCUBA divers is a close friend/relative of the Chief Minister.

• Assistance could be sought from an AID agency for assistance in finding a long-term solution to this problem, but bear in mind such assistance takes at least one year, and usually more, to mobilize.

• There is internal inconsistency within the government system: there is a law banning SCUBA fishing yet the Immigration Department is allowing locals to sponsor foreign divers to come in and fish.

2

Observations and Modeling of Mesoscale Ocean Circulation Near a Small Island

Malcolm J. Bowman, David E. Dietrich, and Charles A. Lin

Abstract

Observations are presented of near surface circulations around the island of Barbados (13°10'N, 59°30'W) in the spring of 1990 and 1991, using a combination of hull-mounted ADCP and geostrophic calculations. The circulation showed highly variable characteristics between the two years. In 1990, the flow patterns suggested topographic steering in a clockwise sense around the Barbados Ridge located north of the island. In 1991, two island-scale eddies were observed in the offings of the west and east coasts (an anticyclone and cyclone, respectively), which were suggestive of Von Karman-type eddies. Preliminary three-dimensional numerical simulations of flow past an idealized island with Barbados-like characteristics (*viz.*, shape, depth, stratification, latitude) show that the disturbance to the ambient flow by the presence of the island is very extensive and exists for a distance of at least eight island diameters downstream. For realistic flow velocities and eddy viscosities, the island readily sheds Von Karman-type vortices, which have a shedding period of about 10 days (per pair), and which have a strong three-dimensional structure.

Introduction

This chapter reports on observations and modeling of the flow near Barbados (13°10'N; 59°30'W), a small isolated island located in the Lesser Antilles, lying in the path of the Guiana Current [Wust, 1964; Molinari *et al.*, 1980, 1981). Historical maps (Figure 1) show that the Guiana Current sweeps past the island from the southeast with a mean

Small Islands: Marine Science and Sustainable Development
Coastal and Estuarine Studies, Volume 51, Pages 18–35

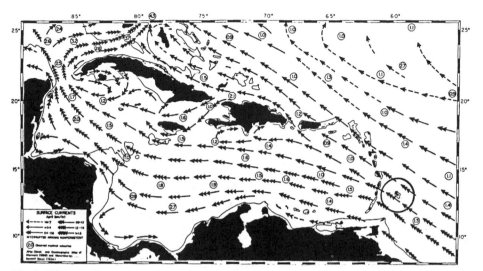

Fig. 1. Mean surface currents through the Caribbean Sea for April [from Wust, 1964]. Barbados is circled for identification.

speed in the range of 35-55 cm s^{-1}. The island, approximately 425 km^2 in area, lacks a significant continental shelf and is surrounded by water of depth 1000-2000 m (Figure 2).

Flows around isolated oceanic islands are affected by many factors including its size, the presence or absence of a continental shelf, the steepness of the bottom slope, the speed and vertical structure of the ambient flow impinging on the island, water column stratification, wind-induced effects [Cram and Hanson, 1974], and the latitude (*viz.*, the importance of Coriolis effects). Nearby features such as ridges and tall seamounts can steer the flow and lead to the formation of Taylor caps [see Hogg, 1980, for a review]. Island flows are often assumed to be quasi-geostrophic [*e.g.*, Gordon and Hughes, 1981; Hogg, 1972; Hogg *et al.*, 1978], unlike the frictionally-dominated wakes formed in shallow water where bottom drag generates relative vorticity, eddy spin up [*e.g.*, Pingree and Maddock, 1980; Wolanski, 1986, 1988; Wolanski *et al.*, 1984] and vortex streets in their lee [Maul, 1977, Fig. 7].

Little has been published to date on the regional oceanography of Barbados. Emery [1972] speculated that persistent attached eddies in the wake of Barbados might explain the distribution of tagged flying fish captured near the island. Murray *et al.* [1977] observed surface drifters released off the south coast in the summers of 1973 and 1974 and off North Point in the summer of 1973 for short periods (~1 day). A current meter was deployed 800 m northwest of Harrison Point (located near the northwest tip of the island) for the same period. They found zones of jet currents (with speeds of 50-80 cm s^{-1}), separated by stagnation zones of weak, disorganized flow (~2-10 cm s^{-1}) located around the island. They claimed that these zones were consistent with White's

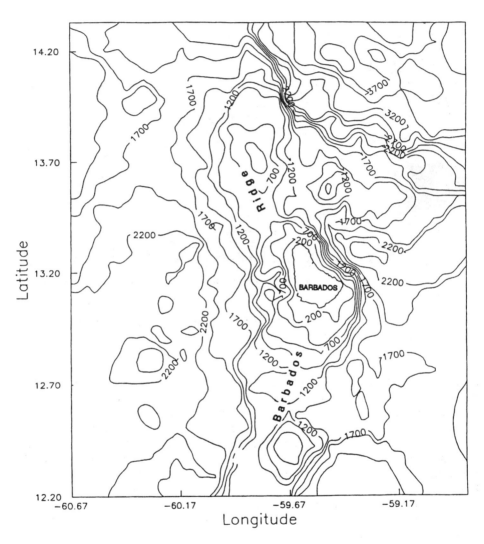

Fig. 2. Local bottom topography around Barbados determined from ship's fathometer and various navigation charts. Depths in meters.

[1971] theory of potential flow past a cylindrical island on an f-plane, if the coastline streamline was adjusted to allow for a net clockwise flow, and if an offshore ambient flow of 20 cm s^{-1} [reported by Warsh, 1971] was imposed.

During the Murray *et al.* [1977] survey, the current approached the island from the east-southeast, split on the southeastern corner of the island with one arm flowing northwards along the west coast, a weak flow up the eastern coast, a belt with very high speed currents (~60 cm s^{-1}) concentrated at the southwestern corner of the island, and a stagnation zone of weak currents off North Point. These features were essentially in agreement with the flow patterns depicted by the drifter trajectories. Tidal velocities were calculated to be ~10 cm s^{-1} in the near-vicinity of the coast (*i.e.*, within a few hundred meters of shore) but dropped off rapidly offshore.

In this chapter, results are presented from two cruises made in April-May 1990 (aboard R/V *Endeavor*) and April-May 1991 (aboard R/V *Columbus Iselin*) to measure the physical properties and near surface circulation and to relate these to the distribution of tropical reef fish larvae. The near surface circulation was mapped using a combination of hull-mounted acoustic Doppler current profiler (ADCP) measurements and geostrophic calculations based on numerous CTD stations taken in dense grids around the island. Further details of the cruises with a discussion of the physical properties measured are presented in Bowman *et al.* [1994].

Near Surface Currents in 1990

Surface geostrophic streamlines derived from CTD stations for the period May 5-12, 1990, calculated with respect to 250 dbar are shown in Figure 3. The broad scale patterns are quite similar to the surface (16 m) ADCP velocities (calculated with respect to 250 m) shown in Figure 4. The current approached the island from the south and east, split into two branches, with the western branch looping around the northern part of the Barbados Ridge, before rejoining up with the southern branch to the east of the island and flowing off in an eastward direction.

One of the most striking features of the survey was the presence of an intense shear line located north of the island, where the flow reversed itself as it swept around the north of the island. Figure 5 illustrates vertical geostrophic (upper panel) and ADCP (lower panel) velocity sections for the transect shown in Figure 4 (stations 61-65). The geostrophic section shows strong shears, with a subsurface northward jet current at 100 m lying between stations 63 and 64, and a southward jet located between stations 61 and 62.

Near Surface Currents in 1991

In general, the 1991 circulation was qualitatively different from that measured during the spring of 1990. Sea surface topography calculated with respect to 250 dbar is shown in Figure 6 (no CTD profiles were obtained south of the island due to faulty equipment). The surface flow appeared to be organized into two island-scale eddies; an anticyclone to the west and a cyclone off the east coast. These eddies can be observed in the ADCP plot of surface currents labeled H and L (Figure 7). The northwest directed undisturbed

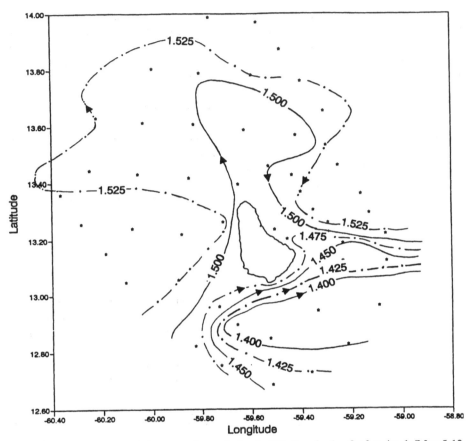

Fig. 3. Topography (m) of the sea surface relative to 250 dbar for leg 2 of cruise 1 (May 5-12, 1990). The stars represent CTD station locations.

flow further to the east was similar in structure to that revealed by the ADCP with a broad northwestward sweep past the island. In general, there appeared to be good qualitative agreement between the ADCP surface currents and the geostrophic streamlines.

ADCP/Geostrophic Sections

Several ADCP and geostrophic sections were prepared to investigate the vertical structure of currents in the upper 250 m of the water column. Not all transects contained sequential stations, introducing some lack of synopticity. The sections were chosen to slice through significant features revealed by the surface dynamic topography. For example, section 1 cuts through the northern edge of the west coast anticyclone (Figure 6).

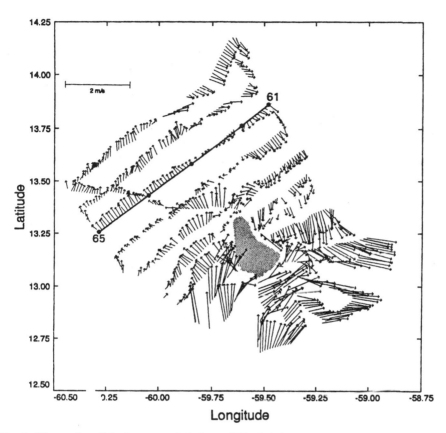

Fig. 4. Near surface (16 m) currents (relative to 250 m) derived from ADCP 5-minute ping data ensembles during leg 2 of cruise 1 (May 5-12, 1990). The line running across the diagram north of the island is the location of the geostrophic and ADCP sections illustrated in Fig. 5. The dots represent the locations of CTD stations 61-65.

The geostrophic velocity across the section (relative to 250 m) is shown in Figure 8a. At the southwestern end of the section the flow was concentrated in the upper 50 m as a jet with surface velocities in excess of 40 cm s^{-1}. Between stations 82 and 75 the current was weak, as a consequence of the flow around the northern side of the anticyclone being almost parallel to the section (Figure 4). Between station 75 and 64 there was a return jet which represented the continuation of the circulation around the anticyclone. Centered between stations 64 and 65, there was a subsurface jet current at ~75 m below msl which appeared to be associated with the cyclone off the east coast. This jet also showed up in the ADCP plot (Figure 8b), but with peak velocities ~35 cm s^{-1}. The ADCP section seems very similar to the geostrophic section and exhibits all three jet currents in approximately the same positions, although the center jet (near station 75) appears to be considerably weaker than that depicted in the geostrophic section.

Preliminary Modeling Simulations

The DieCAST z-coordinate three-dimensional model [Dietrich *et al.*, 1993] was used to simulate the flow in the lesser Antilles over a region 400 km by 400 km surrounding the island, oriented parallel to the historical mean flow direction (Figure 1). Barbados and the nearby Martindale Seamount located north-northeast of the island (Figure 2) were approximated as elliptical Gaussian bells, rising out of an abyssal plain depth of 2000 m (Figure 9). The DieCAST model uses a semi-collocated modified Arakawa "a" grid which has advantages in lateral boundary fitted curvilinear coordinates, is hydrostatic, incompressible, partially implicit, fully conservative, and uses a rigid lid.

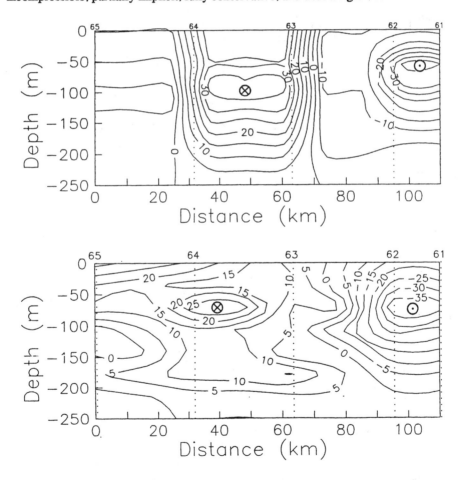

Fig. 5. Vertical section of geostrophic flow relative to 250 dbar (upper panel) and ADCP currents w.r.t. 250 m (lower panel) normal to the section (see Fig. 4 for location), showing the subsurface jets which define the cores of the intense shear zone north of the island.

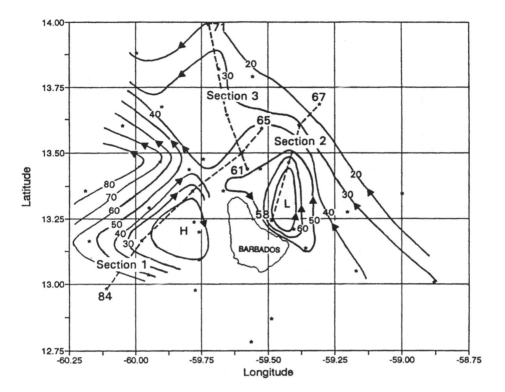

Fig. 6. Sea surface topography (cm) relative to 250 dbar for leg 2 of cruise 2 (April 25-May 2, 1991). The three dashed lines were selected to construct vertical sections, revealing the subsurface velocity structure, both from geostrophic calculations and ADCP measurements. The stars represent CTD station locations.

The mass conservation equation and pressure gradient terms, which are the weak points of collocated grids, receive careful treatment by use of fourth order approximations. The model was run with 5 km horizontal resolution with solid side walls, and with 10 levels in the vertical (increasing geometrically with depth). Details of the model and examples of its application to circulation in the Gulf of Mexico are given in Dietrich *et al.* [1993] and Dietrich and Lin [1993]. A series of numerical experiments were run to validate the models for flows around two dimensional cylinders and for islands with more realistic shapes. These recent studies [Dietrich *et al.*, 1994] have shown that the model can produce non-eddying motions very close to laboratory flows at moderate Reynolds numbers (e.g., ~ 40), as well as classical von Karman vortex streets for higher Reynolds numbers. Further, for sloping topographies, we have found that in model runs with sufficient resolution, the vertical component of vorticity is generated mainly by vortex stretching and bottom drag rather than lateral non-slip conditions, and that vorticity is not numerically induced on corner grid-points.

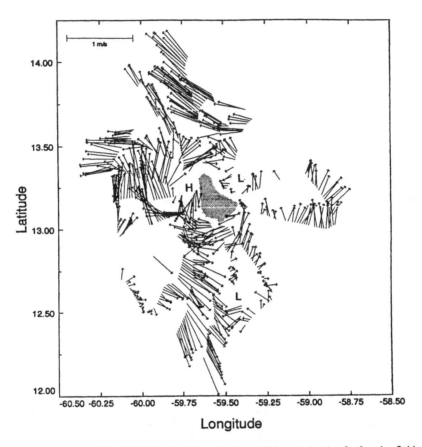

Fig. 7. Near surface (16 m) ADCP velocities relative to 250 m during leg 2 of cruise 2 (April 25-May 2, 1991). The H and L west and east of the island represent locations of the anticyclone and cyclone respectively, suggested in the sea surface topography map (Fig. 6). A second cyclone (also labeled L) was located south of the island.

All available NODC CTD data for the region were averaged and used to initialize the model. A steady surface inflow of 30 cm s^{-1} was imposed, which dropped off with depth in a near Gaussian fashion with an e-folding scale of 400 m. Horizontal density gradients were adjusted to a thermal wind balance across the upstream boundary. The downstream boundary condition used was a modified Neuman outflow condition with slow restoring over the last 10 zones to upstream conditions with a 30-day period. Vertical eddy viscosity and diffusivity were set at 1 to 10 cm^2 s^{-1}, and lateral viscosity and diffusivity at 10^5 cm^2 s^{-1}. The model was run with a 30 minute time step for 180 days.

Figure 10 is a plot of surface (10 m) currents at day 140. The flow diverges around the island and encloses an elongated wake in which is embedded a Von Karman vortex street of alternating cyclones and anticyclones, with swirl velocities ~ 10-15 cm s⁻¹, which form behind the island and swim away with the downstream flow. The spacing of the eddies is of the order of one island length, slowly increasing downstream from the island. Figures 11 and 12 display surface currents at days 145 and 150, respectively. It can be seen that the shedding is a regular process which goes through one cycle of shedding eddy pairs in about 10 days. No net circulation around the island is observed.

The flow at day 145 just north of the island (Figure 11) shows a similarity to the observed currents in 1990 (Figures 3 and 4) in that the currents sweep northward to the west of the island and rotate clockwise in a large loop north of the island. It is possible that what was observed in 1990 was an anticyclonic eddy in the process of being generated, rather than topographic flow around the Barbados Ridge and the nearby Martindale Seamount.

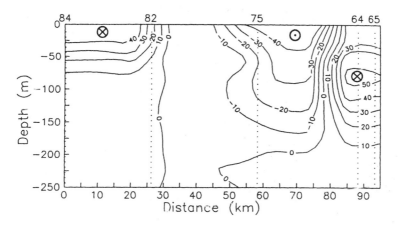

Fig. 8a. Geostrophic velocity (cm s⁻¹) relative to 250 dbar for section 1 (see Fig. 6 for location).

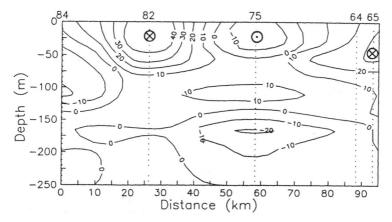

Fig. 8b. ADCP velocity (cm s⁻¹) relative to 250 dbar for section 1 (see Fig. 6 for location).

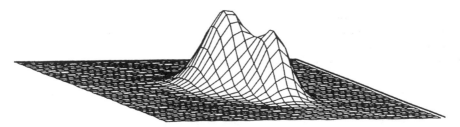

Fig. 9. Simplified topography of Barbados and the adjacent Martindale Seamount used in the modeling simulations, as viewed from the northeast. The island rises out of an abyssal plain of depth 2000 m. Only the top 200 m of the island is exposed. The incident flow impinges on the island from the south.

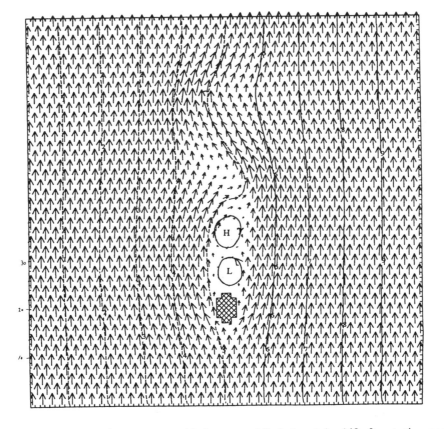

Fig. 10. Simulated surface currents and isobars around Barbados at day 140 after starting model from rest. The incident flow has a speed of 30 cm s^{-1}. The resolution is 5 km, and the modeled region is 400 km by 400 km. A pair of Von Karman-type eddies with closed circulation cells can be seen just north of the island, as well as a long sinuous wake extending to the downstream boundary of the model domain. The maximum current is 39 cm s^{-1}.

From Figures 10, 11, and 12 it is possible to estimate the translation speed of the eddies as they are swept away with the ambient flow. Initially the translation velocity is ~11 cm s^{-1}, increasing slowly away from the island to ~18 cm s^{-1} near the outflow boundary. Thus, the eddies are propagating with a downstream velocity of about 35% to 60% of the ambient surface velocity.

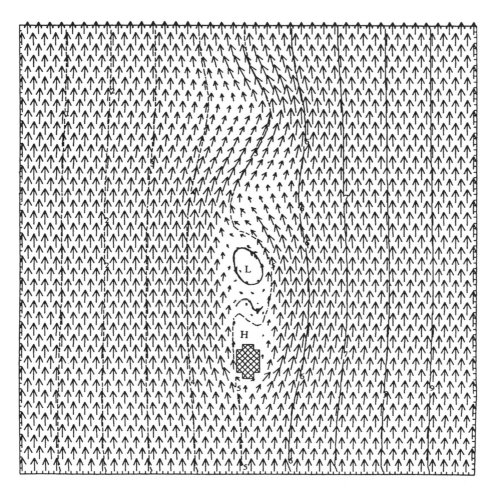

Fig. 11. Simulated surface currents and isobars around Barbados at day 145 after starting model from rest. The two eddies shown in Fig. 11 have migrated away from the island and a third eddy (anticyclone) is developing near the island. This plot shows some similarities to the observations of surface currents during leg 2 of the 1990 cruise (Figs. 3 and 4). The maximum current is 40 cm s^{-1}.

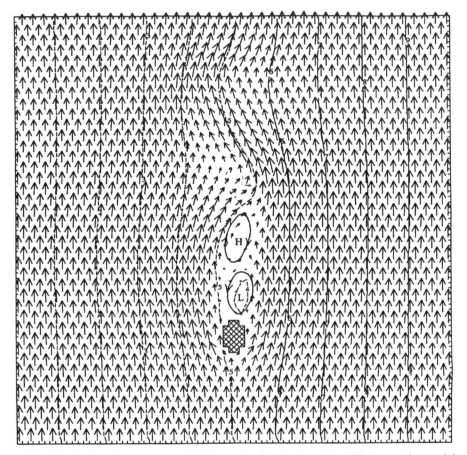

Fig. 12. Simulated surface currents and isobars around Barbados at day 150 after starting model from rest. The eddy shedding has gone through a complete cycle, and a new cyclone has just shed from the island to replace the anticyclone shown in Fig. 10. The maximum current is 39 cm s^{-1}.

Figure 13 is an example of the flow pattern in the thermocline at a depth of 332 m below the surface for day 140. The eddies are quite elongated in the direction of the mean flow to the extent that adjacent cyclones and anticyclones at times merge into one another to form even longer eddies. The eddies clearly have strong three-dimensionality, and tend to lean forward in the ambient flow due to the vertical shear in the mean current.

The extent of the island wake is clearly seen in Figures 14 and 15, which are isotachs of along-axis and cross-axis currents. Figure 14 shows a zone of weak reverse flow behind the island which extends at least six island lengths downstream. The dashed line labeled "1" encompasses this wake region, which is a persistent feature in time, and might provide an effective return route for reef fish larvae that have been swept away from the island in surface eddies, but which are capable of swimming from one eddy to another to take advantage of the return flow. The cross-axis current fluctuations are also extensive in extent (Figure 15), filling much of the domain downstream of the island. This plot illustrates the sinuous nature of the wake as well as the lobes of accelerated flow on both flanks of the island.

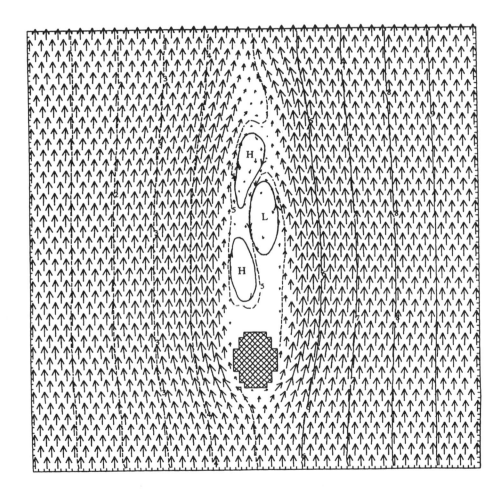

Fig 13. Currents and isobars at 332 m below mean sea level around Barbados at day 140 after starting model from rest. At least three eddies can be seen, elongated in the direction of the mean flow, and confined in lateral extent to the width of the island at that depth. There are large areas behind the island where the currents are quite weak and could serve to trap reef fish larvae for long periods of time. The maximum current is 25 cm s^{-1}.

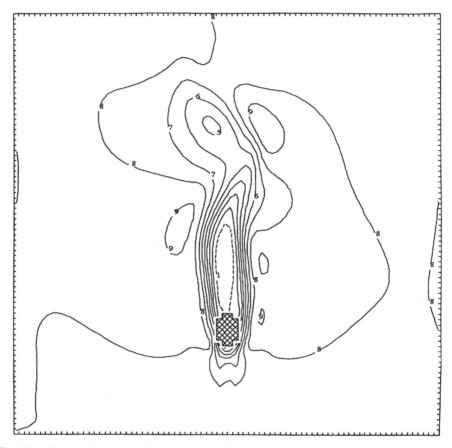

Fig. 14. Along-axis surface isotachs at day 180 after starting model from rest. The contour interval is 5.5 cm s^{-1}, ranging from a minimum of -4.5 cm s^{-1} (line 1), to a maximum of 39.5 cm s^{-1} (line 9).

Discussion

Based on the 1990 and 1991 cruises [Bowman *et al.*, 1994], the mesoscale circulation near Barbados shows highly variable characteristics. Off the east coast the currents set in opposite directions; south and eastward in 1990 and north and northwestward in 1991. In 1991, two island scale eddies were observed in the offings of the west coast (anticyclonic) and east coast (cyclonic), neither of which were observed in 1990. The flow patterns in 1990 suggested topographic steering around the Barbados Ridge with a complete reversal of the west coast current extension around the ridge perimeter [Bowman *et al.*, 1994], although results presented in this chapter of preliminary

Fig. 15. Across-axis surface isotachs at day 180 after starting model from rest. The contour interval is 4.2 cm s^{-1}, ranging from a minimum of -14.8 cm s^{-1} (line 1) to a maximum of 18.8 cm s^{-1} (line 9).

modeling simulations suggest that what was observed might have been a developing anticyclonic wake eddy. Unfortunately, we have no way of confirming this hypothesis with the present data set. It would require a multi-ship ADCP survey with a large number of CTD stations, fixed current meter moorings and satellite tracked drifters to map the region rapidly enough to obtain an accurate picture of the circulation. Satellite derived sea surface temperature images show very little contrast in the tropics and are of limited use. Future SeaWifs ocean color images may prove to be a powerful tool in identifying and classifying mesoscale features near the island.

To what extent the observed wake structure can be attributed to vortex generation by local topography, versus the advection into the region of mesoscale features of remote origin [*e.g.*, Richardson *et al.*, 1993; Limeburner *et al.*, 1993] will continue to be a subject of investigation. Simple wake theory [*e.g.*, Hogg, 1972; Hogg *et al.*, 1978; Gordon and Hughes, 1981; Pingree and Maddock, 1980; White, 1971; Wolanski, 1988] provides a dynamical framework and a basis for the numerical simulations, but cannot at this stage be expected to account for all of the observed structure. A more complete understanding of small, isolated island wakes must await the integration of Large-Scale circulation studies (models and observations) with more localized surveys and modeling experiments such as we have undertaken. Such experiments will include sensitivity studies of model parameters (e.g., flow velocity, stratification, latitude, bottom slope, vertical shear).

The simulations presented in this chapter show that the disturbance to the ambient flow field around an isolated oceanic island is very extensive, and at this latitude exists for a distance of at least eight island diameters downstream. The structure of the island's wake, including the formation of eddies and associated recirculating currents is likely to be very significant for the local larval reef fish populations including their initial dispersal, pelagic stages, and eventual recruitment phase back to their natal reef habitats. Further expeditions are presently being planned to investigate these physical and biological phenomena in an integrated field and modeling study.

Acknowledgments. The contributions of the Captains and crew of the R/V *Endeavor* and R/V *Columbus Iselin* are gratefully acknowledged. K. M. M. Lwiza is thanked for advice in interpreting the ADCP records, and T. C. Wilson is thanked for technical assistance at sea. This project was supported by NSF grant OCE 8911120 to R. K. Cowen and M. J. Bowman.

References

Bowman, M. J., K. L. Stansfield, S. J. Fauria, and T. E. Wilson, Coastal ocean circulation near Barbados, West Indies: Spring 1990 and 1991, *J. Geophys. Res.*, *99*, 16131-16142, 1994.

Cram, K., and K. Hanson, The detection by ERTS-1 of wind-induced ocean surface features in the lee of the Antilles islands, *J. Phys. Oceanogr.*, *4*(4), 594-600, 1974.

Dietrich, D. E., and C. A. Lin, Numerical studies of eddy shedding in the Gulf of Mexico, *J. Geophys. Res.*, *99*, 7500-7615, 1994.

Dietrich, D. E., D.-S. Ko, and L. A. Yeske, On the application and evaluation of the relocatable diecast ocean circulation model in coastal and semi-enclosed seas, Tech. Rep. 93-1, Mississippi State University Center for Air Sea Technology, 1993.

Dietrich, D. E., M. J. Bowman, and C. A. Lin. Numerical studies of small island wakes. *J. Geophys. Astrophys. Fluid Dynamics*, 1994, submitted.

Emery, A. R., Eddy formation from an oceanic island: Ecological effects, *Carib. J. Sci.*, *12*(3/4), 121-128, 1972.

Gordon, H. B., and R. L. Hughes, A study of rotating baroclinic non-linear flow around an island, *J. Phys. Oceanogr.*, *11*, 1011-1014, 1981.

Hogg, N. G., Steady flow past an island with applications to Bermuda, *Geophys. Fluid Dyn.*, *4*, 55-81, 1972.

Hogg, N. G., Effects of bottom topography on ocean currents, in *Orographic Effects in Planetary Flows*, GARP Publ. Series No. 23, 167-205, 1980.

Hogg, N. G., E. J. Katz, and T. B. Sanford, Eddies, islands, and mixing, *J. Geophys. Res., 83*, 2921-2938, 1978.

Limeburner, R., R. C. Beardsley, I. D. Soares, S. J. Lentz, and J. Candela, Lagrangian flow observations of the Amazon River discharge into the North Atlantic, *J. Geophys. Res.*, 1993, submitted.

Maul, G. A., The annual cycle of the Gulf Loop Current, Part I: Observations during a one-year time series, *J. Mar. Res., 35*(1), 29-47, 1977.

Molinari, R. L., D. K. Atwood, C. Duckett, M. Spillane, and I. Brooks, Surface currents in the Caribbean Sea as deduced from satellite-tracked drifting buoys, *Proc. Ann. Gulf Carib. Fish. Inst., 32*, 106-113, 1980.

Molinari, R. L., M. Spillane, I. Brooks, D. K. Atwood, and C. Duckett, Surface currents in the Caribbean Sea as deduced from Lagrangian observations, *J. Geophys. Res., 86*(C7), 6537-6542, 1981.

Murray, S. P., H. H. Roberts, D. M. Conlon, and G. M. Rudder, Nearshore current fields around Coral Islands: Control on sediment accumulation and reef growth, in Proceedings of Third International Coral Reef Symposium, Rosenstiel School of Marine and Atmospheric Science, University of Miami, Florida, 53-59, 1977.

Pingree, R. D., and L. Maddock, The effects of bottom friction and the earth's rotation on an island's wake, *J. Mar. Biol. Assn. U.K., 60*, 499-510, 1980.

Richardson, P. L., G. Hufford, and R. Limeburner, North Brazil Current retroflection eddies, *J. Geophys. Res.*, 1993, submitted.

Warsh, K. L., K. L. Echternacht, and M. Garstang, Structures of near surface currents east of Barbados, *J. Phys. Oceanogr., 1*, 123-129, 1971.

White, W. B., A Rossby wave due to an island in an eastward current, *J. Phys. Oceanogr., 1*, 161-168, 1971.

Wolanski, E., Island wakes in shallow seas. *J. Geophys, Res., 93*, 1335-1336, 1988.

Wolanski, E., Water circulation in a topographically complex environment, in *Physics of Shallow Estuaries and Bays*, edited by J. Van de Kreeke, 280 pp., Springer-Verlag, New York, 1986.

Wolanski, E., J. Imherger, and M. L. Heron, Island wakes in shallow coastal waters, *J. Geophys. Res., 86*, 10553-10569, 1984.

Wust, G., *Stratification and Circulation in the Antillean-Caribbean Basins*, 201 pp., Columbia Univ. Press, New York, 1964.

3

Descriptive Physical Oceanography of the Caribbean Sea

Artemio Gallegos

Abstract

The Caribbean Sea, characterized by its fast transition from a cooling to a warming phase during winter and early spring and a relatively abrupt period of intense thermal energy release to the atmosphere from mid-summer to late fall, is the largest marginal sea of the Atlantic Ocean. Its density structure may be modeled by a surface layer (0-50 m) of homogeneous density that swiftly responds to atmospheric forcing, a transition layer (50-250 m) of strong vertical density gradient (≈ 0.020 σ_t-units m^{-1}), and a thick layer (250-2000 m) where density increases very slowly and almost linearly with depth (≈ 0.001 σ_t-units m^{-1}). Below the deepest sill depth (Yucatan Channel; 2040 m) and down to the maximum depth (Cayman Trench; 7100 m), the stratification of the Caribbean Sea deep water is virtually null. The upper layer circulation is dominated by the Caribbean Current, which contributes to the heat flux from the tropic to the mid-latitudes of the North Atlantic Ocean. Present research is aimed at identifying and modeling the regional modes of motion and the physical processes that control both present and future oceanographic conditions of the Caribbean Sea.

Physiography

The Caribbean Sea (CS) is the largest marginal sea of the Atlantic Ocean. It has a surface extension of 2.52×10^6 km^2; almost twice as large as that of the Gulf of Mexico, and its volume (6.48×10^6 km^3) is twice that of the Mediterranean Sea.

Small Islands: Marine Science and Sustainable Development
Coastal and Estuarine Studies, Volume 51, Pages 36–55
Copyright 1996 by the American Geophysical Union

The north and eastern boundaries of the CS are the Greater and Lesser Antilles, respectively. It is limited to the south by the irregular coasts of Venezuela, Colombia, and Panama. The western boundary of this major sea is the Central American eastern zig-zag littoral.

The CS is located between 8°N and 22°N latitude and 60°W and 89°W longitude, which implies north-south extensions close to 1500 km and east-west breadths of the order of 3000 km. In fact, connecting distances between pairs of selected diametrically-opposed ports within the region are: Bridgetown (Barbados) to Cancún (Mexico), 3030 km; Colón (Panama) to Cienfuegos (Cuba), 1440 km; Puerto Estrella (Colombia) to Oviedo (Dominican Republic), 600 km.

The CS has an average depth of 4400 m. It consists of five principal basins. From east to west, the first is the Grenada Basin, with an average depth of 3000 m, immediately to the west of the Lesser Antilles. The second (and largest) is the Venezuela Basin, which has an average depth of 5000 m and connects with the Grenada Basin trough wide channels across the Aves Ridge at depths not greater than 1800 m. Waters from the Atlantic Ocean have direct access to the Venezuela Basin through the Jungfern-Anegada (sill depth: 1815 m) and Mona (sill depth: 475 m) passages.

The Beata Ridge, a north-south submarine escarpment between the island of Hispaniola and Colombia, separates the Colombia Basin from that of Venezuela, but there is a wide submerged pass that allows water exchange at depths in excess of 3600 m, close to the average depth (4000 m) of the Colombia Basin.

The fourth basin is the Cayman Trench with a record depth of 7100 m. The average depth of this elongated basin is 6000 m and has direct seawater exchange with the Atlantic Ocean through the Windward Passage (sill depth: 1690 m). It also connects with the Colombia Basin across the Jamaica-Haiti Passage (sill depth: 1475 m) and the various submarine channels that cut across the ridge between Jamaica and the Honduras-Nicaragua continental shelf, with maximum sill depths of 1600 m.

The Yucatan Basin has an average depth of 5000 m and links the CS with the Gulf of Mexico via the Yucatan Channel. Since the sill depth of this passage is 2040 m, only under unusual dynamic conditions, such as large amplitude internal waves or extreme compensatory flows, can seawater of depths greater than the sill be exchanged between these two large bodies of water. The Cayman Ridge, with its axis marked by the Cayman Islands, the Rosario Reefs and the Misteriosa Bank, separates the Yucatan Basin from the Cayman Trench but allows seawater exchange from the surface down to depths close to 4000 m.

There are other smaller marginal basins, gulfs, and bays within the CS which are very important on their own, have particular characteristics, pose controversial problems--

real Pandoras' boxes that demand detailed attention. For example, the Cariaco Basin is a unique feature in the world ocean. It has a shallow sill depth (200 m) and is ordinarily ventilated by water from the Venezuela Basin. In recent years, geochemical evidence suggests that the ventilation has ceased, and this can be interpreted as being caused by a decrease in the strength of the Caribbean Current [Richards, 1975]. This has profound global climatic implications because the Caribbean Current contributes significantly to the oceanic heat flux from the tropics to mid-latitudes where air-sea interaction is the primary mechanism of moderating North Atlantic climate.

Unfortunately, the scope of the present regional description is far too general to include local aspects with further detail. Just to mention some of the most important, apart from the Cariaco Trench, there are the basins of Bonaire and Tobago, and the major bays and Gulfs of Paria, Venezuela-Maracaibo, Mosquitos, Darien, Honduras, Batabanó, Guacanayabo and Gonave.

Probably the most important physiographic feature of the CS is the barrier formed by the Antillean Arc that does not permit much exchange with the deep waters of the North Atlantic Ocean. In fact, excluding flows through narrow submarine channels and along stringent lateral boundaries, horizontal motion below the average sill depth of the Antillean Arc (1200 m) is close to stagnant. This suggests that a simple two-layer model with the lower layer at rest may be used to model the surface circulation of the CS.

Climate

The climate of the CS is the integrated result of basic ocean-atmosphere momentum, energy and mass transfer processes, such as wind stress, absorption and emission of radiation, evaporation/precipitation, and diffusion plus advection of heat, acting and interacting over a wide range of space and time scales. To a large extent, the rates of these transfers regulate the lag times and the feedback loops of this thermodynamic system and ultimately determine the nature, magnitude, tendency and variability of the atmospheric and oceanic climate of the CS.

The wind system over the North Atlantic Ocean is dominated by a high sea-level pressure cell, known as the Azores High (AH), whose magnitude, geographic location, and distribution are basically controlled by the regular inter-tropical excursion of the sun across the Earth's equator. Statistical studies [e.g., Tucker and Barry, 1984] reveal that the monthly-averaged location of the center of the AH traces a simple closed and elongated trajectory, with its longitudinal axis along a west-southwest to east-northeast direction. Starting from the northeast extreme (35°N, 25°W) in January (Figure 1), the center of the AH takes only three months to move to the opposite end (30°N, 45°W), about 2200 km away. In spring the center is displaced due north and slowly veers to the

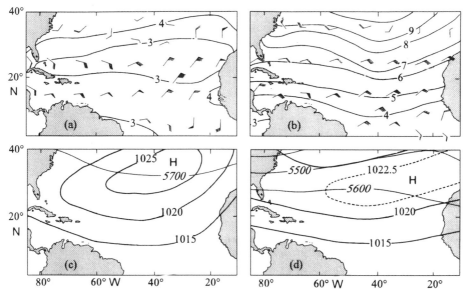

Fig. 1. Surface wind (1 barb = 1 m s^{-1}; solid triangle = 5 m s^{-1}) and isopleths of standard deviation (m s^{-1}); mean sea level pressure (mb) and average thickness of the 100-500 mb layer (in geopotential meters, gpm) for the months of July (a,c) and January (b,d). Reproduced from Tucker and Barry [1984].

east to close the path traversed annually. The averaged sea-level pressure values at the center of the AH reach a high maximum in July (up to 1030 mb) and a low maximum in January (1025 mb).

The North Atlantic Trades are the geostrophically-balanced winds in the southern half of the sea-level pressure gradient distribution established by the AH at sea level. The CS is located in the southwestern quadrant of this wind system so that average winds over the Caribbean region are persistent both in direction (from the east and northeast in winter; from the east and southeast in summer) and in magnitude (3.5-10.5 m/s or 3-5 Beaufort) throughout the year.

Nonetheless, there are interruptions to the average wind conditions over the CS. In fact, daily weather maps resemble monthly-averaged charts of atmospheric conditions over the tropical Atlantic Ocean about 60% of the time in summer, less than 30% of the time in winter, near 50% of the time in fall and close to 40% of the time in spring [Stommel, 1965]. The rest of the daily weather maps are complex arrangements of high and low sea-level pressure cells of different sizes typified as fronts, stagnant lows, linkages of the AH to the continental highs, tropical depressions, easterly waves (tropical storms) and hurricanes.

Frequent interruptions to the Trades, two or three per month in the winter months, are due to eastward-moving, week-long fronts, which generate violent gales over the Gulf of Mexico and the northern half of the CS, and are known as 'northers'. Other shifts in wind are ascribed to the coalescence of high pressure continental cells with the AH. These events usually occur also during the winter months, half as often in spring and fall and are rare in summer.

In contrast, tropical storms and hurricanes occur in summer and fall. September stands out as the "hurricane month" because the three or four tropical storms that normally happen in this period have a probability greater than 0.70 to reach hurricane intensity (Figures 2 and 3).

The oceanic climate of the CS builds upon the heat gained by the surface layer due to the net absorbed radiation (NR), the heat lost to the atmosphere by evaporation (E) and turbulent conduction (C), and the convergence or divergence of heat due to currents and heat exchange with deeper layers of the ocean (A). The sum of these quantities is the "heat storage" term (S). The sum will be positive if the heat gains are greater than the heat losses. In this case the surface layer warms up. If heat losses exceed heat gains, the storage term is negative and the temperature of the surface layer decreases. The heat-budget equation may be simply written as

$$S = NR - (E + C + A).$$

The implicit variables in the heat-budget equation above are many and have complex interrelations among them, yet to be determined. The ones thought to be more important to the climate of the ocean are: wind, currents, sea surface temperature, air temperature, solar radiation, back radiation, vapor pressure, vapor content of the lower atmosphere, cloudiness, atmospheric sea-level pressure, depth of the mixed layer, and entrainment at the base of the mixed layer. Each of these variables has its own temporal and spatial structure within the Caribbean region. Frequent attempts are made to depict and understand such distributions using whatever technology is available, from simple and direct "in situ" measurements to modern satellite radiometry and remote sensing [Gruber and Arkin, 1992].

The climate of the CS is appropriately described using Figure 4, which shows the integrated effect of the implicit variables in the heat-budget equation viewed through the explicit integrated processes of net radiation (NR), evaporation (E), turbulent conduction (C) and diffusion plus advection of heat (A). The "residual" term is the heat storage term (S). According to this figure, the net radiation is the major source of heat of the surface layer. The annual average rate of heat transfer is about 300 Ly/day (cal/cm^2-day), equivalent to a gain of 2.76×10^{18} cal/year for the entire CS. The net radiation absorbed varies throughout the year. A maximum of 20% of the annual average above the average value occurs in May and a minimum with the same amplitude below the annual average

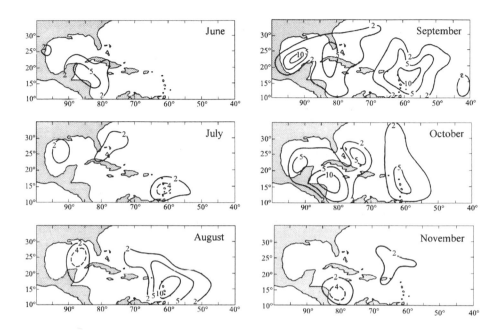

Fig. 2. Total frequency of tropical cyclones with a track starting from each 5° square during the period 1887-1950 for the months of June through November. Reproduced from Colon [1953].

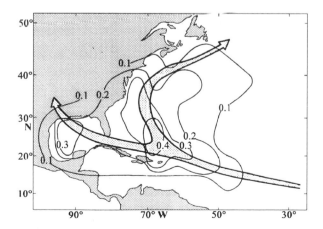

Fig. 3. Preferred tracks and normalized mean frequency of mid-September tropical storms and hurricanes per 5° squares during the period 1899-1971. Reproduced from Crutcher and Quayle [1974].

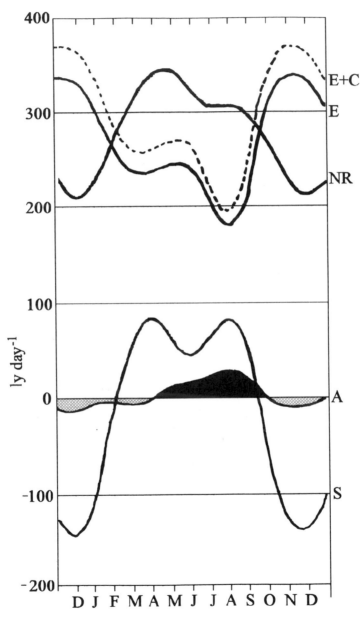

Fig. 4. Annual variations of the components in the heat balance for the Caribbean Sea: NR = net radiation; E = evaporation; C = turbulent conduction; A = advection plus diffusion of heat; and S = heat storage component. Units are Ly day^{-1}. Adapted from Colon [1963; Figure 3].

value happens in December. The presence of clouds might explain why the NR decreases during the summer months and the fact that it takes seven months to decay from a maximum to a minimum, but only five months to recuperate the next year's maximum.

Opposingly, evaporation (E) plus turbulent conduction (C) reach their maximum rates of heat transfer (370 Ly/day) to the atmosphere (the main loss of heat of the upper layer of the ocean) in mid-fall and gradually decay from November to March, maintain values close to the annual average (280 Ly/day) during the spring months and decay to a minimum (190 Ly/day) in August. Cloudiness is again responsible for the marked difference between the time it takes the term "E + C" to go from a maximum to a minimum, nine months (fall to summer), and the time it takes to go from the minimum to the consecutive maximum, only three months (mid-summer to mid-fall).

The heat-budget equation is basically the balance between net radiation (heat gain) and evaporation plus turbulent conduction (heat loss). The advection plus diffusion term is nearly one order of magnitude smaller, and may have positive or negative sign, depending upon the convergence or divergence of the warm waters of the CS. Colon [1963] found in his study that (Figure 5), typically, from April to October (summer track) the Caribbean Current is divergent, exporting heat to the Gulf of Mexico. From October to next year's April (winter track), this same current imports heat due to convergence. The heat source is necessarily in the equatorial Atlantic Ocean.

The heat storage term, considered as a residual term, as shown in Figure 6, reveals that the upper layer of the CS warms from early February to mid-September. It has maximum values in April and August (80 Ly/day) and a relative minimum in June (40

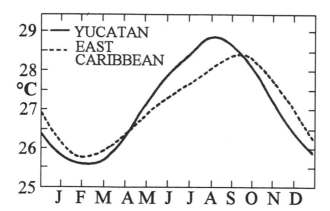

Fig. 5. Annual variation of surface water temperature (°C) at the Yucatan Channel (continuous curve) and at the Lesser Antilles area (dotted curve). Reproduced from Colon [1963; Figure 2].

Ly/day) during this seven month-long warming phase. This same upper layer cools to its annual average temperature in only five months (October to February). It reaches a maximum cooling rate of 140 Ly/day in December, probably as a result of the largest imbalance between net radiation and evaporation plus conduction.

The spatial and temporal structure of the sea surface temperature (SST) and the temperature of the surface layer of the CS follow closely the annual cycle of the terms in the heat-budget equation. This is demonstrated with Figures 6 and 5, where the annual variations of water temperature at several depths averaged over the CS, and the annual variation of SST at the east and west extremes of the CS, respectively, can be related to the energy transfer processes that appear in the heat-budget equation.

Surface salinity is the result of by precipitation, evaporation, river discharge, upwelling, and currents. The winter distribution of surface salinity shows relative high salinities (36.0 to 36.4 psu) along a well-defined narrow north-south band in the central CS (from Colombia to the Windward Passage, over the continental shelf of Nicaragua, the waters south of Cuba, along the north-central coast of Venezuela and south of Puerto Rico). In summer most of the high salinity winter patches are replaced by lower salinity continuous tongues. There is a gradual increase in surface salinity values from the Lesser Antilles (33.0-34.0 psu) to the Yucatan Strait (36.0 psu). Typical sea surface salinities for winter and summer are shown in Figures 7 and 8. Upwelling areas of the CS are easily identified with the salinity maximum along the northern coasts of Venezuela and Colombia.

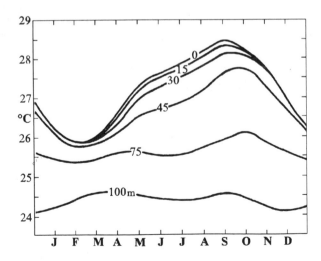

Fig. 6. Annual variations of sea water temperature (°C) at the indicated depths (m), averaged over the Caribbean Sea. Reproduced from Colon [1963; Figure 1].

Precipitation within the CS has a strong annual signal in amplitude and a relatively weak signal in spatial distribution. The central Caribbean is relatively dry throughout the year. To the east and west of this north-south dry band, precipitation increases. The rainy season peaks in August and is minimum in February (Figures 9 and 10). Rainfall concentrates along the coastal waters of Panama and Nicaragua and also along the Leeward Islands of the Lesser Antilles [Sukhovey, 1980], as shown in Figure 11.

Hydrography

Knowledge of the spatial distribution of the water masses in any region of the ocean provides helpful cues to figure out the three-dimensional circulation in that part of the ocean. There are a large number of water masses, generally accepted to originate in the mixed surface layer of the ocean, in a particular source region, under particular air-sea exchange conditions that persist for a significant period. In this way, water masses acquire their distinct values of temperature, salinity and surplus density, and then sink to the level of their neutral buoyancy.

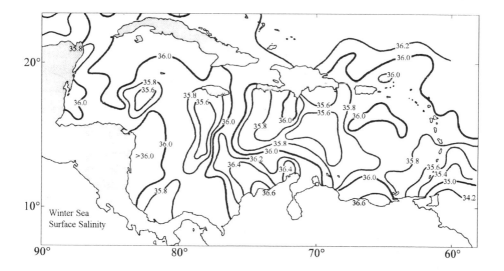

Fig. 7. Sea surface salinity (psu) of the Caribbean Sea in winter. Reproduced from Sukhovey [1980].

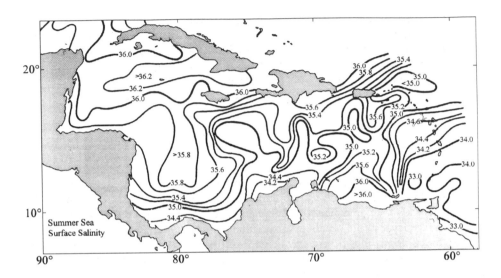

Fig. 8. Sea surface salinity (psu) of the Caribbean Sea in summer. Reproduced from Sukhovey [1980].

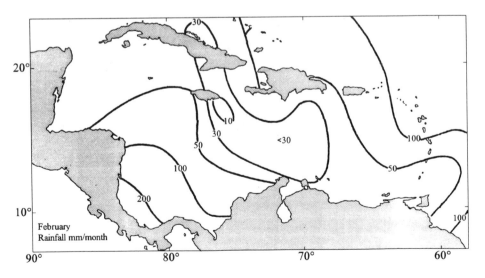

Fig. 9. Rainfall over the Caribbean Sea in a "typical" February (mm month⁻¹). Reproduced from Sukhovey [1980].

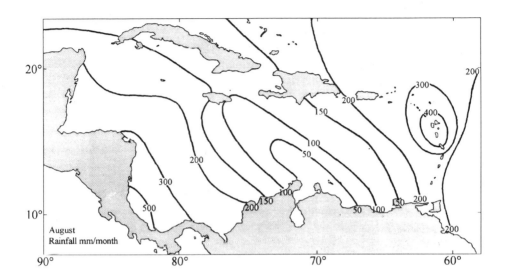

Fig. 10. Rainfall over the Caribbean Sea in a "typical" August (mm month^{-1}). Reproduced from Sukhovey [1980].

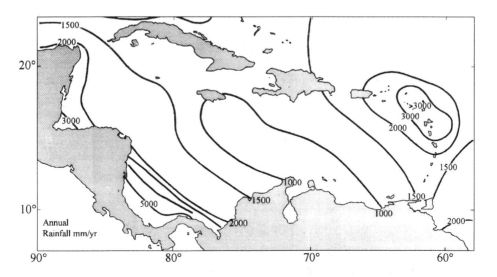

Fig. 11. Annual rainfall over the Caribbean Sea (mm year^{-1}). Reproduced from Sukhovey [1980].

Below the surface layer (0-50 m), four distinct water masses fill the basins of the CS. Each one flows throughout this region in a different manner and proceeds at its own pace. In order of increasing density (with their depths), they are: the Subtropical Underwater (SUW; 50-250 m), the Western North Atlantic Central Water (WNACW; 250-750 m), the Antarctic Intermediate Water (AAIW; 750-950 m) and the North Atlantic Deep Water (NADW; 950 m to bottom). Except for the NADW, the average depths of the cores of these water masses within the region are at a depth which is the average of the corresponding depth range (see Figure 12 and Table 2).

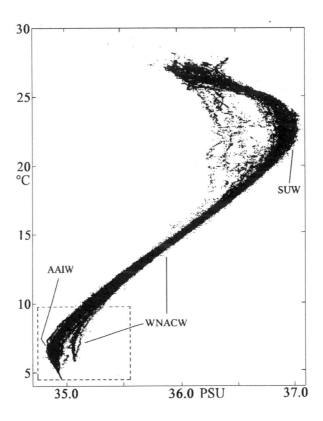

Fig. 12. Temperature-salinity diagram using all CTD data from an oceanographic survey to the northern Caribbean Sea (oceanographic cruise YUCA III, R/V *Justo Sierra*, UNAM, Mexico, April 1991).

TABLE 2. The Water Masses of the CS

Name	Temperature (°C)	Salinity (psu)	Depth Range (m)	Volume of CS (%)	Source Region
Subtropical Underwater (SUW)	21-23	36.6- 37.0	50- 250	5	Subtropical North Atlantic Ocean
Western North Atlantic Central Water (WNACW)	7-20	35.0- 36.7	250- 750	12	Labrador Sea and North Atlantic Ocean
Antarctic Intermediate Water (AAIW)	2-6	33.8- 34.8	750- 950	6	South Atlantic Ocean
North Atlantic Deep Water (NADW)	1.5-4	34.8- 35.0	950- bottom	73	Greenland and Norwegian seas

Within the CS, the waters below 1200 m are practically homogeneous in temperature and salinity. Therefore, the vertical stability is almost neutral, a condition which favors large vertical excursions of water if unstable conditions are produced at any depth in this range. The temperature and salinity values of this layer are 3.8-4.2°C and 34.90-35.00 psu, respectively. This water mass accounts for 73% of the total volume of seawater in the CS.

The AAIW forms in the Antarctic Polar Frontal Zone in a region of convergence in the southern extreme of the South Atlantic Ocean (55°S). It is the most widespread inter-mediate water mass in the world ocean, characterized by low temperatures (3.0°C) and low salinities (34.20 psu). In the Atlantic Ocean this water-mass sinks and spreads northward as far as the Tropic of Cancer. Remnants of AAIW enter the CS through those passages of the Antillean Island Arc whose sill depths exceed 900 m (see Table 1). The AAIW is found everywhere in this region in a layer 200 m thick and is easily identified as the salinity minimum at an average depth of 850 m. This minimum is 34.70 psu at the Lesser Antilles and 34.85 psu at the Yucatan Strait. This salinity erosion is due to diffusion of salt along its path across the CS. The AAIW represents 6% of the total volume of the seawater in the CS (see Table 2).

Recent hydrographic data indicate that the present northern boundary of the AAIW cuts through the Windward Passage (detail at left bottom of Figure 12). This circumstance may be used to study seawater exchange across this passage [Gunn and Watts, 1982].

The WNACW amounts to 12% of the total volume of the seawater in the CS. Its temperature-salinity values are 7.0-20.0°C and 35.0-36.7 psu, respectively (Table 2). An oxygen minimum at the sigma-t surface of 27.1 σ_t-units helps to identify this water mass. The most salient feature of the WNACW within the CS is that it separates the layer of minimum salinity (AAIW) below, from the layer of maximum salinity above, which corresponds to the Subtropical Underwater.

TABLE 1. Major Passages and Basic Dimensions of the CS

Name	Sill Depth (m)	Adjacent Basins
Yucatan	2040	Yucatan Basin-Gulf of Mexico
Anegada-Jungfern	1815	Venezuela Basin-North Atlantic
Windward	1560	Cayman Basin-North Atlantic
Jamaica-Haiti	1475	Cayman Basin-Colombia Basin
Dominica	1370	Grenada Basin-North Atlantic
St. Lucia	980	Grenada Basin-North Atlantic
Martinique	950	Grenada Basin-North Atlantic
St. Vincent	890	Grenada Basin-North Atlantic
Grenada	740	Grenada Basin-North Atlantic
Guadeloupe	650	Grenada Basin-North Atlantic
Mona	475	Venezuela Basin-North Atlantic
Florida*	760	Gulf of Mexico-North Atlantic

*For comparison only.
Surface area = 2.52×10^6 km^2; volume of water = 6.48×10^6 km^3; average depth = 4.40 $\times 10^3$ m.

In fact, the Subtropical Underwater is marked by a salinity maximum at an average depth of 150 m throughout the CS. The source region of this water mass is located under the Azores High, where evaporation exceeds precipitation. The relatively high salinity (and therefore denser) water formed in this area becomes a part of the North Equatorial Current and, as it sinks, flows westward and enters the CS across most of the passages between the Leeward Islands of the Lesser Antilles. Both the SUW and the AAIW, with their salinity extrema, are used to depict the principal paths that these water masses follow along their excursion throughout the CS [Sukhovey, 1980], as shown in Figures 13 and 14.

Currents

Charts of surface current of the tropical Atlantic Ocean indicate that the flow which results from the confluence of the Guiana Current and the North Equatorial Current splits downstream, at the verge of the Lesser Antilles, in two branches. One is a northerly flow, known as the Antilles Current, that streams along the eastward side of the Antillean Island Arc to merge into the Florida Current. The other branch is a west-northwesterly inflow, known as the Caribbean Current, coming through the various passages between the Windward Islands of the Lesser Antilles.

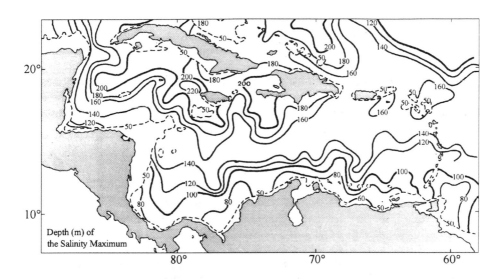

Fig. 13. Depth of the salinity maximum (psu) in the Caribbean Sea. Reproduced from Sukhovey [1980].

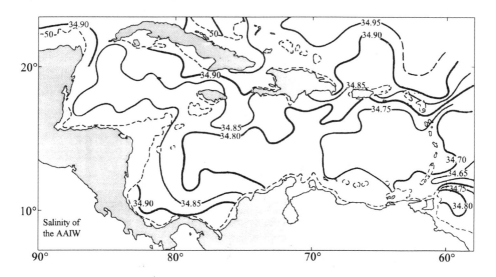

Fig. 14. Salinity (psu) at the core of the Antarctic Intermediate Water (salinity minimum) in the Caribbean Sea. Reproduced from Sukhovey [1980].

All students of the circulation of the CS agree that there is indeed a Caribbean Current, but their descriptions rapidly diverge as soon as detailed features of it are revealed. This expected situation is explained by Figure 15 [Molinari *et al.*, 1981], which shows a composite of the trajectories of satellite-tracked drifters within the CS. It is clear in this figure that all drifters (released inside and near the Lesser Antilles) migrate to the west, some of them along smoothed and uniform tracks, while others describe cyclonic and/or anticyclonic loops of different diameters and show cusps and abrupt reversals and meanders all along or in some parts of their trajectories.

One is, therefore, impelled to recognize in the westward drift shown in Figure 15 the overall effect of the Caribbean Current, and in the loops, cusps, meanders and reversals, the presence of eddies, filaments of currents and countercurrents and other typical fluid motions, including turbulence, within a wide range of time and space scales. Hence, only in a statistical sense may one refer to the trajectory of the Caribbean Current, its width, depth and intensity.

Traditional descriptions of the circulation of the upper CS proceed in steps, the first of which separates the CS waters in two strata. One is the upper layer, that extends from the surface to a depth ascribed to the sill-depths of the main passages of the Antillean Island Arc. The average sill depth of the 10 deepest passages that link the CS with the Gulf of Mexico and the Atlantic Ocean is close to 1200 m (see Table 1). The upper circulation, therefore, refers to the currents and mesoscale motions of the waters above this depth. Surprisingly enough, Gordon's [1967] classical paper on the geostrophic circulation of the CS refers the surface dynamical topography to the 1200 db isobaric

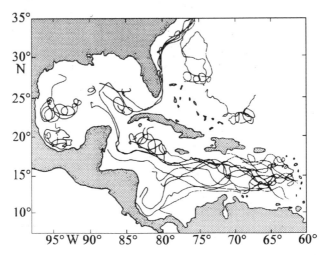

Fig. 15. Composite of satellite-tracked drifter trajectories in the Intra-Americas Sea. This chart delineates the mean surface circulation and suggests a sense of its variability. Reproduced from Gallegos *et al.* [1993].

surface. The other stratum is the deep layer, extending from the average sill depth to the bottom. This layer is weakly stratified and practically motionless, except for an almost imperceptible upward water displacement forced by renewal flows across the deepest passages, in very thin layers, just over the respective sills [Kinder, 1985].

The second step distinguishes the density-driven circulation from the wind-driven circulation. The third and last step is to discuss the upper circulation of those areas not under the direct influence of the Caribbean Current, such as the eastern Cayman Sea and the southwestern corner of the CS.

The density-driven circulation focuses on those kinematic features that are consistent with the observed average density field, under the strict condition of geostrophic balance. Accordingly, the Caribbean Current is the central kinematic feature of the CS. It is a warm, persistent, and powerful current setting west-northwestward throughout this region, with a gentle increase in velocity as it flows from the Windward Islands to the Yucatan Channel. The axis of the main flow, which is about 20 km wide, extends from the surface to a few tens of meters below and streams about 200-300 km off the coast of Venezuela, then veers northwest across, over and beyond the various submarine channels of the Jamaica-Honduras Ridge and finally exits through the Yucatan Channel. The axis of this current maintains an annual average velocity of 0.50 m/s, being faster in spring-summer (0.80 m/s) than in autumn-winter (0.40 m/s). Maximum velocities in excess of 2.0 m/s have been measured. The magnitude of this current decreases with depth to speeds not greater than 0.05 m/s at a depth of 1000 m. The annual average volume transport of the Caribbean Current is estimated at 30×10^6 m^3/s. This is the gross description of the Caribbean Current on the basis of the observed average density field within the CS.

The wind-driven circulation completes the description of the upper circulation of the CS, particularly the kinematics of the mixed layer (0-200 m). It explains the observed ageostrophic motions that result from the effect of the wind stress and wind stress curl over the entire region. It is shown elsewhere [Gordon, 1967] that if the wind blows parallel to and in the same sense as the geostrophic surface current, the net flow veers a few degrees towards higher pressure, defining a small component along the subsurface pressure gradient ("uphill" motion). This process has two important effects. First, it accumulates available potential energy in the density field, sufficient, in the conversion to kinetic energy, to maintain a surface geostrophic flow, even under relaxed Trade winds. Second, it increases the pressure gradient in the layers below, accelerating deeper waters of the upper layer (0-1200 m) and forcing them to flush across relatively narrow submarine passes and channels, as those in the Jamaica-Honduras Ridge. This process stimulates the filamenting of broader flows, eddy shedding, the generation of internal waves, flow instabilities, lateral and vertical mixing and mesoscale turbulence, all in the upper layer of the CS. Furthermore, the wind-driven circulation also explains the upwelling processes along the northern coasts of Venezuela and Colombia and also in those south of Cuba.

Conclusions

Fluid motion in any region of the world ocean may be described as an endless sequence of "flow snapshots," each different from the previous one. If in a sufficiently long sequence of snapshots of a region, peculiarities such as persistent patterns or coherence among groups of successive snapshots are observed, we say that the regional circulation has a set of preferred "flow configurations" or "modes" of motion. This fact raises interesting questions, for example: what are the most persistent modes, what characterizes a better given mode, how many different modes occur in that region, and what is the sequence of such modes. There are harder questions such as: are modes predictable, do they correspond to solutions of specific mathematical models, and what causes the transition from one mode to another.

Answers to these and many other similar questions cannot be worked out until additional knowledge of the elements of ocean circulation is available. This knowledge will come through the analysis of historic and recent hydrographic data, measurements of ocean currents, drifter trajectories, tides and sea level, satellite imagery, and other complementary ocean and atmosphere observations. With this information one may eventually identify and model regional modes of motion and use them to estimate time and space scales, magnitude and timing of those physical processes that control the actual oceanographic conditions and the thermodynamical evolution of the ocean system. In the Caribbean region such an effort progresses thanks to the collaboration of researchers working for academic institutions and governmental agencies of the region concerned with the expected impacts of global climate change in the Intra-Americas Sea [Maul, 1993].

Acknowledgments. I am grateful to Dr. George A. Maul for his kind invitation to collaborate in this book. I would also like to thank Augustín Fernández and Olivia Salmerón for their technical assistance in the preparation of the figures that appear in this modest contribution.

References

Colon, J. A., A study of hurricane tracks for forecasting purposes, *Mon. Wea. Rev., 81,* 53-66, 1953.

Colon, J. A., Seasonal variations in the heat flux from the sea surface to the atmosphere over the Caribbean Sea, *J. Geophys. Res., 68*(5), 1421-1430, 1963.

Crutcher, H. L., and R. G. Quayle, Mariners worldwide climatic guide to tropical storms at sea, *Naval Weather Service Command*, NAVAIR 50-IC-61, Washington, D.C., 114 pp. and 312 charts, 1974.

Gallegos, A., S. Czitrom, J. Zavala, and A. Fernández, Scenario modeling of climate change on the ocean circulation of the Intra-Americas Sea, in *Climate Change in the Intra-Americas Sea*, edited by G. A. Maul, pp. 55-74, Edward Arnold Publishers, London, 1993.

Gordon, A. L., Circulation of the Caribbean Sea, *J. Geophys. Res.*, 72(24), 6207-6223, 1967.

Gruber, A., and P. A. Arkin, Reviews of modern climate diagnostic techniques, WMO/TD-No. 519, WRCP-76, 54 pp., 1992.

Gunn, J. T., and D. R. Watts, On the currents and water masses north of the Antilles/Bahamas arc, *J. Mar. Res.*, 40(1), 1-18, 1982.

Kinder, T. H., G. W. Heburn, and A. W. Green, Some aspects of the Caribbean circulation, *Mar. Geol.*, 68, 25-52, 1985.

Maul, G. A. (Ed.), *Climatic Change in the Intra-Americas Sea*, 389 pp., Edward Arnold Publishers, London, 1993.

Molinari, R. L., M. Spillane, I. Brooks, D. Atwood, and C. Duckett, Surface currents in the Caribbean Sea as deduced from Lagrangian observations, *J. Geophys. Res.*, 86(C7), 6537-6542, 1981.

Richards, F.A., The Cariaco Basin (Trench), *Oceanogr. Mar. Biolo. Rev.*, 13: 11-67, 1975.

Stommel, H., *The Gulf Stream*, 248 pp., Cambridge University Press, London, 1965.

Sukhovey, V. F., Hydrology of the Caribbean Sea and the Gulf of Mexico, *Ed. Gidrometeoizdat*, Leningrad, 182 pp. (in Russian), 1980.

Tucker, G. B., and R. G. Barry, Climate of the North Atlantic Ocean, in *Climates of the Ocean, World Survey of Climatology*, edited by H. Van Loon, pp. 193-262, Vol. 15, Chapter 2, 1984.

4

Oceanic Features Influencing Small Island Circulation Patterns: Case Studies

Roy A. Watlington and Maria C. Donoso

Abstract

Several observed oceanic features may have particular impact on the ability of small island developing states in the Intra-Americas Sea to pursue sustainable development and conduct effective coastal management. Among these are the variability of the circulation of the Panama-Colombia Counter Current, the highly variable, local scale circulation patterns that are reflected in unpredictable movement of pollutants among the islands of the Antillean chain, and the existence throughout the Caribbean of ocean thermal gradients. Recent events have demonstrated the need for an improved understanding of these features. Cooperative studies in different fields of oceanography and the sharing of resources by the institutions of the Intra-Americas Sea are recommended as means of meeting the increasing regional demand for better knowledge of the sea. This multilateral approach will benefit both island and coastal states.

Introduction

The Intra-Americas Sea (IAS) has been the subject of investigation for over a century. Interest in the Caribbean Sea, as part of the IAS, is sometimes attributed to its role as the source of the Gulf Stream, but its oceanographic characteristics have significance to the economic and strategic development of the states of the region. Consequently, the precise knowledge of features such as water mass distribution, currents, and upwelling systems is of interest not only from a scientific perspective, but also as a necessary element in generating and implementing management and development plans and/or programs for the region. This is especially true for small islands, particularly in the case of small island states.

Small Islands: Marine Science and Sustainable Development
Coastal and Estuarine Studies, Volume 51, Pages 56–70

Focused studies of small-island oceanography are needed to:

1. Provide the environmental information necessary to maximize the productivity and health of fishery stocks and nurseries, including sea grass beds and coral reefs, and to respond to crises affecting fisheries such as massive, wide-range marine epidemics.
2. Predict the paths taken by pollutants such as oil and non-point pollution, such as sediment-laden run-off.
3. Identify and assess available but untapped renewable resources, such as the thermal energy stored in the sea in close proximity to IAS shorelines.

We will illustrate these concepts with three case studies. The first concerns the Panama-Colombia Counter Current (PCCC) and its effect on the national park "Corales Islas del Rosario" in Colombia. The second briefly covers the implications of the *Vesta Bella* oil spill of 1991. Finally, we will consider the regional potential for ocean thermal energy conversion (OTEC). Figure 1 identifies the locations of the regions under consideration. The discussion of these examples stresses the necessity for continuing and increasing cooperative studies among island and coastal nations of the IAS.

Fig. 1. The Intra-Americas Sea: (1) national park "Islas Corales del Rosario," Colombia; (2) *Bella Vista* sinking site; and (3) OTEC site near St. Croix [after Maul, 1989].

Background

The information available today on Caribbean oceanography is the result of studies dating back to the last century. Relatively recent and lasting contributions to the body of knowledge have been made in investigations associated with the oceanographic cruises of research vessels such as *Atlantis, Meteor, Lomonosov, Crawford, Knorr, Endeavor, Providencia, Malcolm Baldrige, Trident, Justo Sierra* and, most recently, the *Columbus Iselin* and the *Malpelo* in experiments relevant to this presentation. In the last two decades, important contributions to the general understanding of global and site specific aspects of the oceanography of the IAS have been accomplished by the regional scientific community.

An important aspect of the oceanography of the Caribbean Sea is the exchange of water with the Atlantic Ocean. Water enters and leaves the Caribbean via a number of passages between the islands and shallow plateaus. The major surface and near-surface exchange with the Atlantic takes place through the eastern passages. Schmitz and Richardson [1991] estimate that 22×10^6 m^3/s of transport into the Caribbean takes place through the Grenada, St. Vincent, St. Lucia, and Dominica Passages where sill depths are in the 740 to 980 m range. Surface flow is fed by the Guiana and North Equatorial Currents and exhibits variability associated with the annual migration of the Intertropical Convergence Zone (ITCZ). In addition, as stated in Chapter 3 by Gallegos, the Antarctic Intermediate Water (AAIW) is largely admitted into the Caribbean through these passages.

Exchange over a range of depths also takes place through the Windward, Mona and Anegada/Jungfern Passages in the northern boundary of the Caribbean. Sill depths are 1600 m, 475 m, and 1825 m, respectively. Substantial inflow of the Subtropical Underwater takes place through the Windward and Anegada/Jungfern Passages. The latter is also the primary pathway for the densest water entering the Caribbean--the Upper North Atlantic Deep Water (UNADW). Originating in the Labrador Sea and traveling to the Caribbean as a component of the Deep Western Boundary Current, UNADW enters the central basins of the Caribbean through the Virgin Islands Basin (4685 m) passing through the Anegada Passage (1975 m) and the Jungfern Passage (1825 m). Some deep inflow also occurs over the Windward Passage sill, but is prevented by topography from contributing to the replenishment of the bottom waters of the Venezuela and Colombia Basins. Although the sill at the Yucatan Passage is deep (1900 m), the 800 m sill at the Straits of Florida precludes back flow of deep Atlantic water into the Caribbean through the Gulf of Mexico.

Whereas deep inflow to the Caribbean is small (10^6 m^3/s) with none of this water exiting the Caribbean before being diluted to a less-dense state, the surface currents through the Caribbean are relatively robust. The familiar image of a broad and persistent stream crossing the Venezuela and Colombian Basins from southeast to northwest remains accurate. A total of approximately 30×10^6 m^3/s exit the western end of the Caribbean through the Yucatan Passage [Schmitz and Richardson, 1991].

Case Studies

Case Study I: The National Park "Corales Islas del Rosario" (Colombia)

The national park "Corales Islas del Rosario" (Colombia), founded in 1988, is a natural complex formed by the archipelago "Islas de Nuestra Senora del Rosario" and the western coastline of Baru (Figure 2). This study site is centered at 75.75°W and 10°N.

The El Rosario Islands were formed by the progressive colonization of the continental shelf of diapiric origin by coral species that took place during the maritime transgression of the last glaciation period. From this process originated the Pleistocenic coral fossils that form the islands that are today bordered by live coral reefs [Vernette, 1986]. These islands fall under the category of cay reefs [Quintero, 1993]. As indicated by Vernette in Chapter 11, reef development is enhanced on the windward side of the islands. Here wave action and water qualities favor the growth of corals.

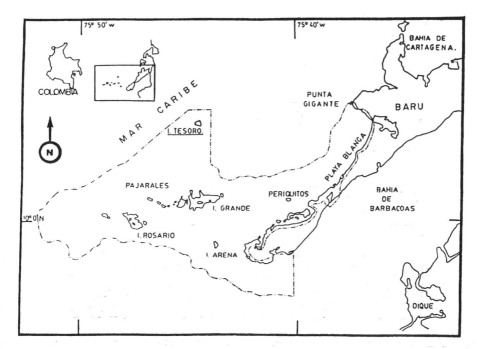

Fig. 2. Map of the national park "Islas Corales del Rosario," Colombia [after Sanchez, 1989].

The waters in the vicinity of the island complex are characterized by a low surface salinity, slightly higher than 35 psu, especially in the southwestern part where continental waters nourish the study area through the Barbacoas Bay. The sea surface temperature varies between 27°C-29.5°C. The surface circulation in the region is dominated by the seasonal variations of the PCCC. The PCCC is a regional secondary current, very susceptible to seasonal fluctuations, that derives from the main Caribbean Current at the southern part of the Colombia sub-basin and flows along the Panama and Colombia coastline from the southwest to the northeast [Donoso, 1990]. A fuller description of the general circulation of the Caribbean Sea is presented by Gallegos in Chapter 3. Figure 3 shows schematics of the surface currents of the Caribbean Sea.

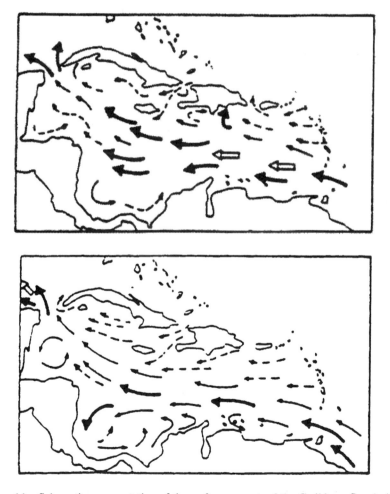

Fig. 3 a and b. Schematic representation of the surface currents of the Caribbean Sea during (a) the dry season and (b) the rainy season [modified from Donoso, 1990].

As is the case in the entire Caribbean Sea, the circulation in the Colombia sub-basin is influenced by the variability of the trade winds which, in turn, respond to the north-south migrations of the ITCZ.   During the winter, the PCCC is weak and circulation in the vicinity of the El Rosario archipelago is mostly from the northeast, while in the summer or rainy season the PCCC is strong and extends eastward beyond the study area, reaching as far as the Guajira Peninsula [Donoso, 1990]. Figure 4 presents the seasonal variability of the alongshore range of influence of the PCCC. The need for a better understanding of the hydrodynamics of the PCCC arises from the need to predict the impact of anthropogenic (*e.g.*, oil spills) and non-anthropogenic (*e.g.*, sediment transport) agents over the natural environment of the national park.

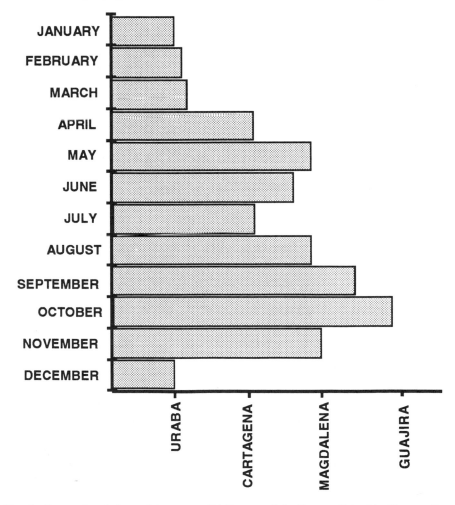

Fig. 4. Seasonal variations of the range of influence of the Panama-Colombia Counter Current [after Donoso, 1990].

A joint research effort to address the circulation dynamics of the southwestern Caribbean was initiated in 1992 between the University of Miami and the Colombian Navy. As part of the scientific activities of this bilateral venture, two scientific cruises were carried out in March-April (CaribVent II on board the University of Miami's *Columbus Iselin*) and April-May (Caribe I on board the ARC *Malpelo* of the Colombian Navy) of 1992 [Cabrera and Donoso, 1992]. The results from the CaribVent-Caribe cruises show that, during the transition dry-to-rainy period (March-May), the PCCC reaches the study site and is considered to be partly responsible for the intermittent (northeast-southwest) circulation pattern observed in this area [Cabrera *et al.*, 1994]. Geostrophic calculations yield a mean surface velocity of 15-35 cm/s for the PCCC.

Cabrera and Donoso [1993] report that during the transition from its minimum to maximum eastward extent, the contrast between the warm PCCC water and the colder water arising from wind-driven upwelling in the proximity of the Guajira Peninsula creates an east-west front. The variability of the PCCC and the dynamics of this front may have impact on the movements of fish and larvae in the region.

Within the range of the PCCC, Garzon and Acero [1986] found that species of fish indigenous near the El Rosario Islands were surprisingly absent from the Santa Marta (11.5°N, 74.5°W) region to the east. The "relative environmental instability" to which these researchers attribute the absence of these species has to be associated with the PCCC and upwelling dynamics. The front between these two oceanographic features constitutes a natural barrier that divides biological regimes. This suggests that the migration of the PCCC exerts fundamental control on fish and larval movement. Consequently, increased knowledge of the PCCC-upwelling system is essential to protect and optimally exploit the fisheries resources of the region. In addition, the spread of pollutants (*e.g.,* chemicals, oil, debris/trash, sediments from rivers) along the Colombian coast and the degree and extent of their impact over this sector of the Caribbean also depend on the PCCC dynamics. To improve our understanding of the biological regimes associated with the PCCC-upwelling system, joint oceanographic and fisheries experiments need to be designed in which accurate velocity measurements should be taken (*e.g.,* using acoustic Doppler current profilers).

Case Study II: The Vesta Bella Oil Spill

Several studies have suggested other circulation patterns in addition to the broad, diagonal stream of the Caribbean Current. In Chapter 3 Gallegos provides a strong sense of the variability when he describes "...loops, cusps, meanders and reversals, the presence of eddies, filaments of currents and countercurrents..." Analysis by Duncan *et al.* [1982] revealed a narrow counter current centered around 16°N flowing eastward between 69°W and 62°W. Atwood *et al.* [1976] observed eastward flow south of Puerto Rico. The acoustic Doppler current profiler (ADCP) survey conducted as a part of the Subtropical Atlantic Climate Studies (STACS) project revealed lateral variability of the surface flow through the Caribbean as well. As shown in Figures 5a and 5b, eastward flow is evident in STACS/ADCP transects south of Puerto Rico and east of the Lesser Antilles in the vicinity of Martinique [Wilson and Routt, 1992]. Eddies have been revealed in satellite altimetry [Nystuen and Andrade, 1993], satellite-tracked drifters [Molinari *et al.*, 1981], drift bottle studies [Brucks, 1971; Duncan *et al.*, 1977], and numerical models [Heburn *et al.*, 1982]. These mesoscale and other variable features increase the uncertainty in forecasting the direction and rate of pollutant spread or the dispersal of passive biological materials such as zooplankton or pathogens affecting marine life.

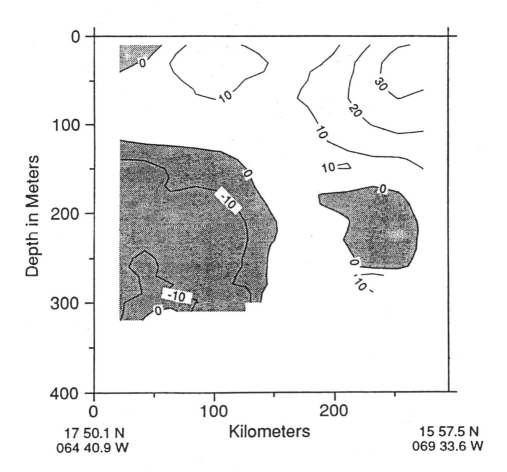

Fig. 5. Vertical velocity sections showing the east-west flow through the eastern Caribbean: (a) south of Puerto Rico. Flow to the west is shaded [after Wilson and Routt, 1992].

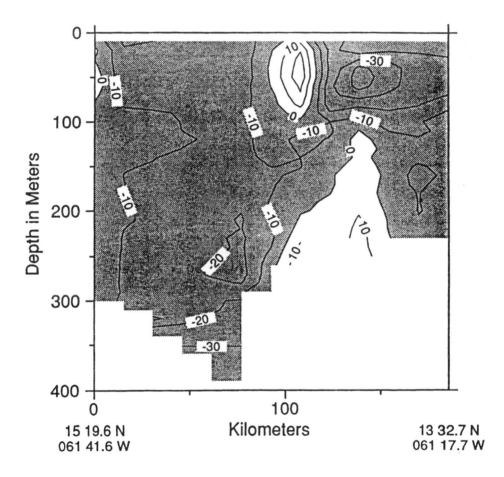

Fig. 5. Vertical velocity sections showing the east-west flow through the eastern Caribbean: (b) west of the Antillean arc in the vicinity of Martinique. Flow to the west is shaded [after Wilson and Routt, 1992].

The *Vesta Bella* oil spill of 1991 focused our attention on unanswered questions about eastward flow and mesoscale variability in surface flow of the northeast Caribbean. The accident occurred on March 6, 1991, when the Trinidadian tanker sank with 13,500 barrels of crude oil in 600 m of water near the eastern Caribbean island of St. Kitts. Oil and tar advanced through the northeastern Antilles, reaching the beaches of St. Martin and St. Bartholomew in less than a week and the eastern beaches of St. John in the Virgin Islands (75 km away) by March 24th. Twenty-one days after the spill, the eastern shores of Puerto Rico were affected. En route, the oil seemed to follow unpredictable paths. For example, reports indicate that after starting 25 km from the St. Kitts shore, a portion of the spill that had been moving away from the island reversed direction and was making impact on St. Kitts by March 14th. Uncharted mesoscale variable features may have been the cause of this unpredictable occurrence.

Unconvinced by the appearance of tar and oil 75 km away from this relatively small spill, French Navy officials argued that the oil appearing on Virgin Islands beaches might not have come from the *Vesta Bella* but may have instead been released from tankers whose crews were using the oil spill to cover prohibited flushing of their tankers' ballasts at sea [Roberts, 1991]. A study by Duncan *et al.* [1977] showed that surprisingly large speeds can be attained by flotsam in parts of the Caribbean at certain times.

Information can be obtained on pollutant dispersal from drift bottle studies of the 1970's [Duncan *et al.*, 1977; Brucks, 1971; Metcalf *et al.*, 1977]. These studies represented realistic means of inexpensively acquiring synoptic scale information. However, many questions cannot be answered with these inexpensive devices because their actual trajectories between release and recovery cannot be known.

Continuously tracked drifters, though more expensive, provide much more detailed information. In a state-of-the-art drifter experiment, 19 satellite-tracked drifting buoys were released in the eastern Caribbean and tracked for several months [Molinari *et al.*, 1981]. Their trajectories verified the trajectories assumed for many of the drift bottles and revealed both cyclonic and anticyclonic features. Buoy speeds up to 80 cm/s were observed near the South American coast. Eddies and meanders were found to occur more frequently over topographical features such as the Aves Rise, Beata Ridge, or the Jamaica-Nicaragua Rise. Despite these studies, however, many essential details about the flow among the islands in the northeast corner of the Caribbean remain incomplete.

As long as tankers travel through the IAS, bringing much needed fuel, there will be a need to predict the possible directions in which pollutants from a particular spill may be moved by the wind and the currents. The modeling of flow around Barbados presented in Chapter 2 by Bowman *et al.* demonstrates the formation of wakes, cyclones, and anticyclones as the mean flow encounters this oceanic island. Although the oceanic topography around St. Kitts may differ substantially from that around Barbados, a similar modeling effort focused on the flow around St. Kitts may reveal similar mesoscale features that could clarify the observed movement of oil and tar from the *Vesta Bella*.   Comprehensive computer models that can consider pollutant characteristics, as well as the variable winds and currents, would be extremely useful in cases such as these.

Case Study III: OTEC Potential in the Caribbean Region

The third case study involves energy, which must be regarded as very important if development is to be sustainable. The core of intermediate water that lies within 1000 m of the surface throughout the Caribbean and close to the shorelines of many islands presents a thermal resource that has been considered for exploitation in the region.

OTEC utilizes the temperature difference between surface and intermediate depth waters, about which much is already known. Although the technology for OTEC has been advanced considerably since D'Arsonval first experimented with the concept in 1881, few OTEC plants have been made operational worldwide. In the Caribbean, where many island states are economic hostages to the high price of imported fuel, utilizing OTEC can be realistically considered because the criteria for optimal utilization of OTEC are met at a number of locations. One criterion, for example, is a mean surface water temperature of 25°C [Thomas and Hillis, 1991]. Figure 6 shows the temperature profile for three regions: (a) the Colombia Basin near Cartagena; (b) the Virgin Islands Basin near St. Croix; and (c) the tropical Atlantic just north of Anegada Passage in the Virgin Islands. In each case, the 6°C isotherm is found above a depth of 1000 m. This ocean thermal resource is reliable at many locations throughout the region. Given the economic incentive and the availability of the resource in the Caribbean, OTEC technology is expected to be significantly utilized eventually. The U.S. Virgin Islands, for example, have been considering such a project. Competing companies have proposed building an OTEC plant that would tap the thermal resource at 800 m depth, which can be found very close offshore of the island of St. Croix [Munier et al., 1978; Thompson, 1992; Johnson, 1993].

In spite of the attractiveness of OTEC technology, its exploitation entails some problems relating to the great cost of building an on-line OTEC power plant and its likely vulnerability to severe storms in a region prone to hurricanes. In addition, there exists a small but non-negligible pollution threat from the chemicals used as efficient working fluids, most of which are corrosive or otherwise hazardous. The necessity of using anti-fouling procedures on submerged OTEC machinery poses yet another problem for the environs of a plant. A National Oceanic and Atmospheric Administration report identifies, among others, biocide and other toxic chemical release as one of the most significant threats to the environment by OTEC technology [Flanagan, 1992]. Further understanding of the resource and of its potential interaction with the environment may be crucial to the success of this method of energy conversion.

Operating an efficient and safe OTEC facility will require the monitoring and integration of a great deal of information about the atmosphere and ocean, as well as about equipment. In order to allow mitigating measures to be taken quickly in the event of a sudden catastrophic storm or the release of hazardous fluid, the weather, currents, and other variables in the region surrounding the OTEC plant have to be charted at

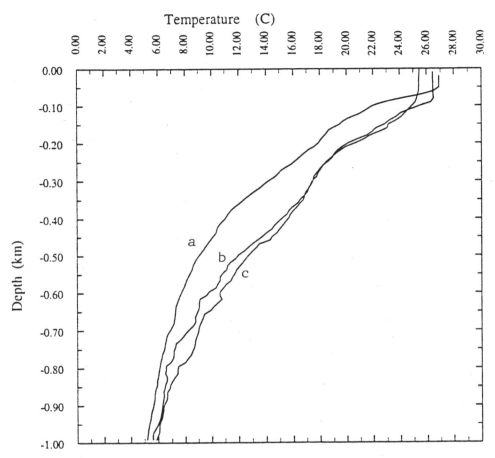

Fig. 6. Depth profile of temperature down to 1000 m in three locations: (a) the southern Colombian Basin; (b) the Virgin Islands Basin near St. Croix; and (c) the tropical Atlantic north of Anegada Passage.

every moment. Problems have to be anticipated on a continual basis. Once again, the development and use of prognostic models is strongly suggested.

Conclusions

These examples of recent and current investigations into the circulation around the islands of the Caribbean Sea are significant to coastal resource management and marine-oriented sustainable development in the region. Yet, either intensive or extensive oceanographic studies are beyond the means of most island states. In many instances, the advancement of knowledge can be greatly facilitated only by the pooling of resources in cooperative ventures. The Caribe I/CaribVent II exercise demonstrates the scale and scope that marine research often requires in order to be productive. This multinational collaboration brought together the Centro de Investigacciones Oceanograficas e Hidrograficas (CIOH) of the Colombian Navy and the University of Miami. The University of Miami is also instrumental in the Windward Island Passages Monitoring Program, which is measuring the current structure and transport through the major passages of the southern Antillean arc. This study is jointly conducted with the University of the West Indies in Barbados and the U.S./NOAA Atlantic Oceanographic and Meteorological Laboratory in Miami [Wilson *et al.*, 1993, 1994].

There are other successful examples of sharing and cooperation among insular states and territories. Laboratories, field stations and vessels, such as those run by the University of Puerto Rico, the University of the West Indies, and the University of the Virgin Islands, have been made available for multinational marine research for some time. The availability of the Barbadian patrol vessel, HMBS *Trident,* for the Windward Island Passages study mentioned above is an outstanding example of a cost-saving, innovative approach to oceanographic research.

In the future much can be done to share or pool computer facilities, communication lines, and the laboratory facilities necessary for the analysis of hydrographic data. In addition to minimizing the great cost of ship time, we may be able to reduce the waste inherent in duplicating standard chemical laboratories throughout the region or in assembling multiple similar computer capabilities. Perhaps one insular university or agency can specialize--and achieve superiority--in the analysis of particular tracers, such as oxygen or silica, while another may specialize in the expertise necessary to interpret unique data sets, such as ADCP data. The savings earned by avoiding duplication can be directed towards developing virtuosity in particular fields. The capability of one insular state group may rest in its facilities or its fleet, while in another community the resource may consist of a collection of specialists. Conducting successful modern oceanographic research at this time of relatively inexpensive but powerful personal computers and excellent global satellite communication may increasingly depend on identifying the capable research specialists in each of our countries and establishing a framework for us to divide research efforts and share results.

Only through bilateral and multilateral ventures will it be possible to accelerate the accumulation of knowledge of the oceanography of the entire Intra-Americas Sea and to pursue our development and management goals.

Acknowledgements. The authors wish to acknowledge Dr. Claes Rooth (University of Miami) for his comments and suggestions. Our thanks to Dr. William Johns (University of Miami) and LCDR Edgard Cabrera (Colombian Navy) for providing informative input related to the Caribe and CaribVent cruises.

References

Atwood, D. K., J. R. Prolifka, and C. P. Duncan, Temporal variations in current transport in the eastern Caribbean Sea, *Proceedings, 11th Meeting of the Association of Island Marine Laboratories of the Caribbean, 11*, 37, 1976.

Brucks, J. T., Currents of the Caribbean and adjacent regions as deduced from drift-bottle studies, *Bull. Mar. Sci., 21*(2), 455-465, 1971.

Cabrera, E., and M. C. Donoso, Estudio de las caracteristicas oceanograficas del Caribe Colombiano, Region III, Zona 1, PDCTM, *Memorias, VIII Seminario en Ciencias y Tecnologias del Mar*, Colombia, 150-165, 1992.

Cabrera, E., and M. C. Donoso, Caracteristicas oceanograficas dek Caribe Colombiano. *Boletin Cientifico CIOH*, Colombia, 19-32, 1993.

Cabrera, E., C. Rooth, W. Johns, and M. C. Donoso, Mesoscale oceanographic characteristics of the Colombian Caribbean sub-basin, *EOS, Trans. Amer. Geophys. Un., 75*(3), 208, 1994.

Donoso, M. C., Circulacion de las aguas en el Mar Caribe, *Memorias, VII Seminario en Ciencias y Tecnologias del Mar*, Colombia, 245-251, 1990.

Duncan, C. P., D. K. Atwood, J. R. Duncan, and P. N. Froelich, Drift bottle returns from the Caribbean, *Bull. Mar. Sci., 27*(3), 580-586, 1977.

Duncan, C. P., S. G. Schladow, and W. G. Williams, Surface currents near the Greater and Lesser Antilles, *Internat. Hydro. Rev., 59*(2), 67-78, 1982.

Flanagan, J. P., A summary of environmental conditions in the vicinity of a potential ocean thermal energy conversion (OTEC) site on the north coast of St. Croix, NOAA Tech. Memo., NOS-OCRM-1, 43 pp., 1992.

Garzon, J., and A. Acero, Fishes from the Rosario and San Bernado Islands, *An. Inst. Invest. Mar. Punta-de-Betin, 15-16*, 67-77, 1986.

Heburn, G. W., T. H. Kinder, and J. H. Allender, A numerical model of eddy generation in the southeastern Caribbean Sea, in *The Hydrodynamics Enclosed Seas of Semi-Enclosed Seas*, edited by J. C. J. Nihoul, pp. 299-327, Elsevier Scientific Publishing Company, Amsterdam, 1982.

Johnson, K., Two companies seeking WAPA contract, *Virgin Islands Daily News*, April 5, 1993, p. 5, 1993.

Maul, G. A., Implications of climatic changes in the Wider Caribbean region: Preliminary conclusions of the task team of experts, Caribbean Environmental Programme Tech. Rep. No. 3, 25 pp., 1989.

Metcalf, W. G., M. C. Stalcup, and D. K. Atwood, Mona Passage drift bottle study, *Bull. Mar. Sci., 27*(3), 586-590, 1977.

Molinari, R. L., M. Spillane, I. Brooks, D. Atwood, and C. Duckett, Surface currents in the Caribbean Sea as deduced from Lagrangian observations, *J. Geophys. Res., 86*, 6537-6542, 1981.

Munier, R., S. C. Hess, T. Lee, H. B. Michel, St. Croix: Supplement to a source book of oceanographic properties affecting biofouling and corrosion of OTEC plants at selected sites, Univ. Miami Tech. Rep. TR 78-2, 56 pp., 1978.

Nystuen, J. A., and C. A. Andrade, Tracking mesoscale ocean features in the Caribbean Sea using GEOSAT altimetry, *J. Geophys. Res. 98*, C5, 8389-8394, 1993.

Quintero, R., Cartografia bioecologica de Isla Tesoro, Caribe Colombiano, *Boletin Cientifico CIOH*, Colombia, 45-64, 1993.

Roberts, A., Oil despoils St. John, *Virgin Islands Daily News*, March 25, 1991, p. 1., 1991.

Sanchez, F., El Parque Nacional Natural Corales islas del Rosario, *Bull. Inst. Bassin d'Aquitaine*, Bordeux, 205-213, 1989.

Schmitz, W. J., Jr., and P. L. Richardson, On the sources of the Florida Current, *Deep-Sea Res.*, *38*, suppl. 1, 379-409, 1991.

Thomas, A., and D. L. Hillis, Ocean thermal energy conversion--an update, *Sea Tech.*, *32*(10), 19-24, 1991.

Thompson, B., Genotec, WAPA set to sign power-plant pact, *Virgin Islands Daily News*, September 18, 1992, p. 5, 1992.

Vernette, G., La plataforme Continentale Caraibe de Colombie. Importance du diapirism aegileux sur la Morphologie et la Sedimentation, Doctoral thesis, Mem. del Inst. Geologie de la Bassin d'Aquitaine, Bordeux, 387 pp., 1986.

Wilson, W. D., and A. Routt, Shipboard acoustic Doppler current profiler data collected during the Subtropical Atlantic Climate Studies (STACS) Project (1989-1990), NOAA Tech. Memo., ERL AOML-71, 141 pp., 1992.

Wilson, W. D., W. E. Johns, M. D. Hendry, and J. A. Routt, Windward Island Passages Monitoring Program: Physical oceanographic data collected on cruise WI-91-01, HMBS *Trident*, 15-21 December 1991, NOAA Tech. Memo., ERL AOML-76, 55 pp., 1993.

Wilson, W. D., W. E. Johns, J. A. Routt, and M. D. Hendry, Windward Island Passages Monitoring Program: Physical oceanographic data collected on cruises WI-02, HMBS *Trident*, 6-10 May 1992 and WI-03, HMBS *Trident*, 19-23 September 1992, NOAA Tech. Memo., ERL AOML-79, 99 pp., 1994.

5

Subtidal Circulation in Fort-De-France Bay

Pascal Lazure, Jean Claude Salomon, and Maruerite Breton

Abstract

Two numerical models have been used to study coastal circulation around Martinique and Fort-de-France Bay. The circulation over the continental shelf of the island has been roughly simulated with a bidimensional model and a three-dimensional model has been used for the bay. The last 3-D model is more advanced than the global study because the bay is partly isolated from external influences. After a review of past studies, the first results of the global model and the three-dimensional model used for the study of the bay are briefly presented. The main aspects of the circulation are then outlined and an application to the bacteria transport is presented.

Introduction

Industrial and tourist activities in Martinique need more and more local studies in order to expand with limited impacts on the environment. To answer questions that may arise in coastal management, IFREMER has begun a program to model the main aspects of coastal circulation around Martinique. Due to the high population concentration in Fort-de-France, special attention to its bay has been planned.

The bay (Figure 1) is the largest along the Caribbean coast of Martinique. The northern side is greatly urbanized, and most industries stand near Cohe du Lamentin. Conversely, the southern part of the bay is sparsely populated, with the exception of Pointe du Bout, which has numerous touristic activities.

Small Islands: Marine Science and Sustainable Development
Coastal and Estuarine Studies, Volume 51, Pages 71–82
Copyright 1996 by the American Geophysical Union

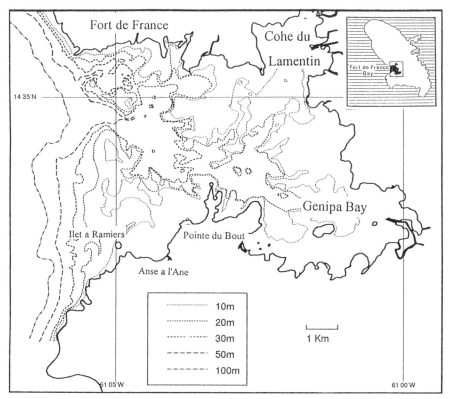

Fig. 1. Location and bathymetry of Fort-de-France Bay.

Physical Background

Three main dynamical processes have been considered for the study of coastal circulation. The first is the large-scale circulation which seemed able to act near the coast because the continental shelf may be very narrow, especially along the Caribbean. The island of Martinique is a part of the Lesser Antilles. These islands interrupt the North Equatorial Current; some of this flow enters the Caribbean Sea through passages and continues westward as the Caribbean Current. Stalcup and Metcalf [1972] have estimated the flow through the St. Lucie passage (south of Martinique) at 6 Sv (1 Sv = 10^6 m^3/s), whereas Brooks [1978] has measured one-third of this. Both noted that this flow is highly variable. The flow through St. Dominique passage (north of Martinique) has been less investigated. Stalcup and Metcalf [1972] estimated it at 1 Sv but highly variable with the tide.

The tide is the other dynamical process. The Caribbean Sea has a micro-tidal range, and the tidal type shows great variation. In Martinique it is of mixed mainly diurnal type [Kjerfve, 1981], and the mean tidal range varies from 10 to 30 cm. Tidal currents in the

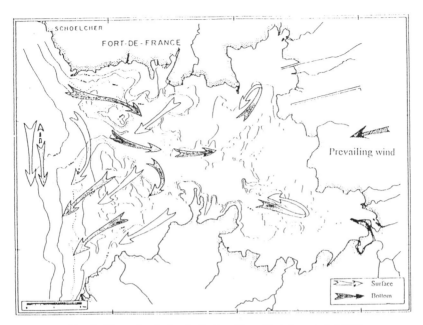

Fig. 2. Mean circulation in the bay [from Castaing *et al.*, 1986].

St. Lucie passage are able to reverse the flow [Stalcup and Metcalf, 1972]. Near the coast, tidal currents of around 20 cm/s have been measured at the mouth of Fort-de-France Bay in the north [Castaing *et al.*, 1986] and in the south [Breton *et al.*, 1993]. However, inside the bay tidal currents are very weak [Castaing *et al.*, 1986].

The last dynamical process is the circulation induced by local winds. According to Castaing *et al.* [1986], it is the main circulation process in Fort-de-France Bay. The winds in Martinique are fairly constant in direction: 86% of the time they blow from northeast to east, their speed varies during the year, and the maximum occurs in June.

The main aspects of the circulation (Figure 2) issued from drifter trajectories and local current measurements have been summarized by Castaing *et al.* [1986]. Due to the prevailing wind, the surface currents are directed toward the west. This outflow is compensated by an inflow near the bottom where the currents are topographically guided and follow channels. Near the coast, in the eastern part of the bay, this circulation is likely to give rise to vertical movement. At the entrance, a southward general flow prevails, and the currents are uniform from the surface to the bottom.

Other processes which act occasionally like cyclone-induced circulation or massive freshwater input may be temporarily of greater importance.

Global Model

The global model of the whole of Martinique has been built by Salomon [personal communication]. This model is vertically integrated. The large-scale westward current has been reproduced by applying a sea level slope at the boundaries, which have been adjusted to provide the correct mean transport. In addition to this pressure gradient, a uniform wind stress and tidal oscillation extracted from tidal charts [Schwiderski, 1983] has been imposed. Figure 3 shows an extracted current field. The main results may be summarized as follows:

1. Under wind and pressure gradient forcings only, the model shows that the island acts as an obstacle and that currents diverge near the southeast extremity of the island. A first vein goes north along the coast accelerating over shallows. A second vein creates a gyre along the southern coast. Along the Caribbean coast, the velocities decrease considerably and Fort-de-France Bay becomes mainly sheltered from oceanic influences. Over the eastern shelf, these features are not validated due to the lack of reliable current measurements.
2. The tide being added into the computation gives correct elevation, but current intensities are generally three to five times lower than measurements. It seems to show that tidal currents are not a uniquely barotropic mechanism.

For this reason and because previous studies have shown that the tide is very weak inside Fort-de-France Bay, this forcing has been neglected in the following computations.

Model of the Bay

The numerical model employed is three-dimensional and assumes hydrostaticity [Lazure and Salomon, 1991].

$$\frac{\partial u}{\partial t} + u\frac{\partial u}{\partial x} + v\frac{\partial u}{\partial y} + w\frac{\partial u}{\partial z} - fv = -\frac{\partial P}{\partial x} + \varepsilon\left(\frac{\partial^2 u}{\partial x^2} + \frac{\partial^2 u}{\partial y^2}\right) + \frac{\partial}{\partial z}\left(Nz\frac{\partial u}{\partial z}\right)$$

$$\frac{\partial v}{\partial t} + u\frac{\partial v}{\partial x} + v\frac{\partial v}{\partial y} + w\frac{\partial v}{\partial z} + fu = -\frac{\partial P}{\partial y} + \varepsilon\left(\frac{\partial^2 v}{\partial x^2} + \frac{\partial^2 v}{\partial y^2}\right) + \frac{\partial}{\partial z}\left(Nz\frac{\partial v}{\partial z}\right)$$

$$\frac{\partial P}{\partial z} = -\rho g$$

$$\frac{\partial u}{\partial x} + \frac{\partial v}{\partial y} + \frac{\partial w}{\partial z} = 0$$

where x,y, and z form a right-handed system, with x and y in the horizontal directions and z pointing vertically upward. Continuing, u, v, and w are velocities, f is the Coriolis parameter, ε is the horizontal viscosity, g is gravity, P is pressure, ρ is density, and Nz is the vertical eddy viscosity. It has free surface and includes a sigma coordinate system. The equations are solved on a traditional staggered grid in the horizontal by using a mode-splitting technique [Blumberg and Mellor, 1981a]. The vertical turbulent viscosity

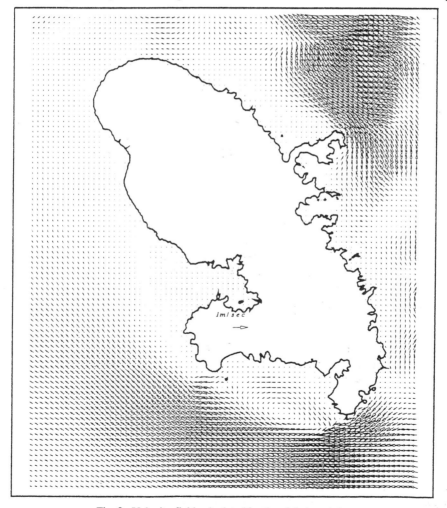

Fig. 3. Velocity field calculated by the global model.

is calculated according to the Prandtl [1925] mixing length model. This model is simple and economical, and many successful calculations have been obtained for boundary layer type flows [Rodi, 1987]. The mixing length formulation is attributed to Sauvaget [1985].

$$ Nz = \ell^2 \left| \frac{\partial u}{\partial z} \right| \quad 1/\ell = \frac{1}{0.4(z)} + \frac{1}{0.7(H-z)} $$

where H is the total depth and e is the mixing length.

The open boundary condition is a simple radiation condition for the velocity, and the levels have been calculated by the global model. Because of the previously mentioned problems, the tide has been removed. The wind stress at the surface have been taken as uniform over the entire bay due to the lack of data about the possible effects of the local topography.

A grid size of 125 m has been chosen because a good spatial resolution may be reached, whereas the computing time remains reasonable. The vertical discretization consists of 10 levels equally spaced from the surface to the bottom. The simulations have been performed on a medium-sized workstation (SUN SPARCstation 10).

Results

The bay is supposed initially at rest and an east-northeast 7 m/s wind is suddenly imposed at the surface. A seiche is generated and oscillates for a few hours. However, its influence is very weak on the sea level (the amplitude of the oscillation is less than 1 cm), as well as on the currents (less than 1 cm/s). After 20 hours of simulation, the permanent state is almost achieved, and the computed surface and bottom currents are presented in Figure 4.

Comparison of the results with Figure 2 shows that the model is able to reproduce the main features of the circulation. It also provides interesting information that could not be obtained by measurements.

The surface currents may reach 20 cm/s north of Ilet à Ramiers. In general, it is directed toward the west or the south. Most of the variability may be attributed to topographical effects. The strength of the current increases as the depth decreases.

Near the bottom, the circulation is very different and often inverted. A general flow enters the bay from the northwest by the central channel. North of Pointe du Bout, it separates into two branches: one directed toward Cohe du Lamentin and the other toward Genipa Bay.

In the shallow parts of the bay, the vertical shear prevent the flow from reversing near the bottom. This can be observed south of Genipa Bay, for example. However, over most of the bay, the current depends on the depth. Below the surface, the influence of the wind stress decreases with the depth.

Between the surface and the bottom, currents vary differently. Figure 5 presents a vertical profile from Pointe du Bout to Fort-de-France. One may observe the diversity of the vertical structures of the currents. In the north, the current reverses near the bottom. Over the shallow depths, the speed slightly decreases from the surface to the bottom. In the channel, the surface current is very weak because the pressure gradient and wind stress act in opposite directions. Below, the countercurrent increases until a maximum is reached at 3 m above the bottom.

When the wind direction is slightly shifted to the north or to the east, the main features of the circulation remain the same.

Application

The transport and mixing of bacteria in the bay have been calculated by adding to the numerical model the three-dimensional, advection-dispersion model:

$$\frac{\partial \, HC}{\partial \, t} + \frac{\partial H \left(U \, C \text{-} Kx \, \frac{\partial C}{\partial x} \right)}{\partial x} + \frac{\partial H \left(V \, C \text{-} Ky \, \frac{\partial C}{\partial y} \right)}{\partial y} + \frac{\partial H \left(W \, C \text{-} Kz \, \frac{\partial C}{\partial z} \right)}{\partial z} = I \text{-} M$$

where C is the concentration, Kx and Ky are horizontal diffusivities, Kz is the vertical diffusivity, I represents the input fluxes, and M is the mortality rate. The horizontal advection terms are calculated with the Takacs [1985] scheme, which is very efficient and produces small numerical dispersion. All vertical derivatives are treated implicitly. In the absence of stratification, vertical diffusivity has been taken to equal the vertical viscosity.

Most of the bacteria inputs are located in the northern part of the bay, in Fort-de-France and its suburb. Tourist activities, including many ships mooring near Pointe du Bout, also produce a significant input. In the absence of measurements, each flux has been roughly estimated at 10^9 fc/s (fc: fecal coliform). Finally, north of Anse à l'Ane, a new sewage treatment plant will be built; the predicted flux is $1.7 \; 10^9$ fc/s. As long as the spreading plume doesn't interact with other plumes, it has been simulated.

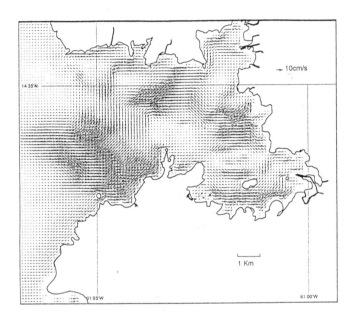

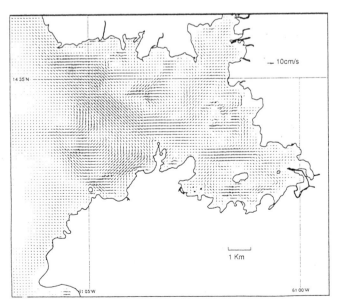

Fig. 4. Velocity field near the surface (top) and near the bottom.

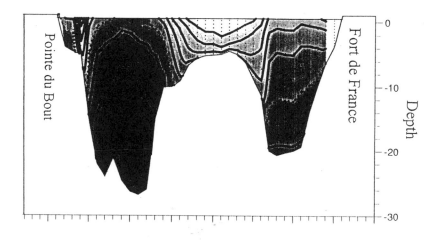

Fig. 5. Vertical current profile. Contour lines are at 0.02 cm/s. The dotted line is the zero velocity. The pattern lightens when the current is westward. At the opposite, it darkens when it is eastward.

Bacteria cannot be considered as a conservative dissolved matter since they die when they enter the sea. This mortality may be partly explained by the light effect; mortality increases with the light. The T90 quantifies the mortality rate and represents the time needed for a decrease of one log (90%) of a population of bacteria. According to recent measurements in the bay, this time has been estimated at three hours during daylight (10 hours) and at 20 hours otherwise [Pommepuy *et al.*, 1993].

The concentrations are shown (Figure 6) for the end of the night and at midday. The depth at which these results are presented is not important, since the water is well-mixed and the concentrations are the same from the surface to the bottom.

At the end of the night, before the sun becomes efficient, the concentrations are the greatest over the entire bay. However, one may observe that most of the bay remains clean from a bacteriological point of view. As expected, in regard to the input locations, the most important concentrations are found near Fort-de-France and do not extend for

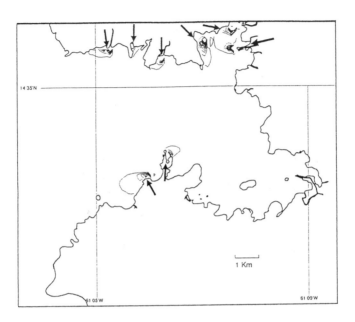

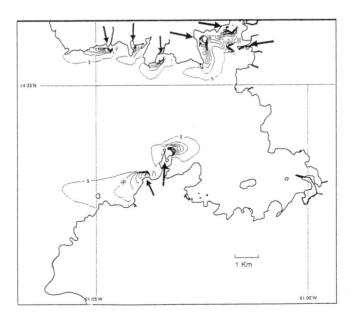

Fig. 6. Concentration of fecal coliform/100 ml at the end of the night (bottom) and at midday. The arrows show inputs. Contour lines are at 50 fc/100 ml intervals.

more than 1 km to the south, the mean transport being directed to the west along the northern coast. The eastern part of Cohe du Lamentin is not affected by the three inputs.

In the south, according to the mean currents, the plume which originates at Pointe du Bout extends eastward, whereas the plume which will originate from the new sewage treatment plant will spread to the west. Because of the strong currents in this part of the bay, its extension will be the greatest and probably affect the Anse à l'Ane beach during east-northeast winds.

As the sun begins to shine, bacteria die and, after a few hours, concentrations are constant and reduced by more than a factor of 10. All the bacteria are located within a radius of about 1 km from the input.

Some recent bacteriological measurements (Pommepuy personal communication) have shown that the higher concentrations are found very close to the northern coast as in the model. However, the input fluxes and the death rate have been too crudely estimated to expect a quantitative comparison.

Conclusions and Perspectives

The three-dimensional model presented shows that it may be a useful and economical tool in coastal management. The main aspects of the circulation have been reproduced. However, more comparisons with measurements are needed to ensure that it is correctly calibrated. An acoustic Doppler current profile (ADCP) survey has been planned in 1995 in order to provide a large set of data at different depths.

In order to improve the global model, the density structure which has been taken as constant over the whole ocean will be schematically taken into account and a three-dimensional calculation will be tested. It is expected that baroclinic effects will notably increase the speed of tidal currents near the coast.

However, many uncertainties will remain in modeling both mechanisms and many measurements near the coast and farther seem indispensable. At the moment, reasonable confidence can be put into a local study of a semi-closed area which is apparently protected from large-scale circulation influence. The study of the Fort-de-France Bay is an example, which can probably be extended to the other islands under the same conditions.

References

Blumberg, A. F., and G. L. Mellor, A numerical calculation of the circulation in the Gulf Mexico, Rep. 66, pp. 153, Princeton N.J., 1981a.

Breton, M., P. Douillet, P. Lazure, and J. C. Salomon, Compilation de mesures de courantologie sur le littoral occidental de la Martinique, Rapport IFREMER, 1993 (in press).

Brooks, I. H., Transport and velocity measurements in St. Lucia Passage in the Lesser Antilles (abstract), *Eos Trans. AGU*, *59*, 1109, 1978.

Castaing, P., A. de Resseguier, C. Julius, M. Parra, J. C. Pons, M. Pujos, and O. Weber, Qualité des eaux et des sédiments dans la baie de Fort-de-France (Martinique): Compte-rendu de fin d'étude, Action concertée, CORDET, 1986.

Kjerfve, B., Tides of the Caribbean Sea, *J. Geophys. Res.*, *86*(C5), 4243-4247, 1981.

Lazure, P., and J. C. Salomon, Coupled two-dimensional and three-dimensional modeling of coastal hydrodynamics, *Oceanologica Acta*, *14*(2), 173-180, 1991.

Pommepuy, M., A. Derrien, F. Le Guyader, D. Menard, M. P. Caprais, E. Dubois, E. Dupray, and M. Gourmelon, Microbial water quality on a Caribbean island (Martinique) (submitted).

Prandtl, L., Über die ausgebildete Turbulenz, Zeitschrift für angewandte Mathematik und Mechanik, 5, 136, 1925.

Rodi, W., Examples of calculation methods for flow and mixing in stratified fluids, *J. Geophys. Res.*, *92*(C5), 5305-5328, 1987.

Sauvaget, P., A numerical model for stratified flows in estuaries and reservoir, M.A. thesis, University of Iowa, 1985.

Schwiderski, E., Atlas of ocean tidal charts and maps, Part I: The semi-diurnal principal lunar tide M2, *Mar. Geod.*, *6*(3-3), 219-265, 1983.

Stalcup, M. C., and W. G. Metcalf, Current measurements in the passages of the Lesser Antilles, *J. Geophys. Res.*, *77*, 1032-1049, 1972.

Takacs, L.L., A two-step scheme for the advection equation with minimized dissipation and dispersion errors, *Mon. Wea. Rev.*, 1050-1065, 1985.

6

Sea Level, Tides, and Tsunamis

George A. Maul, Malcolm D. Hendry, and Paolo A. Pirazzoli

Abstract

Small islands have intimate contact with oceanic phenomena, and in many cases their geography is a totally marine environment. Accordingly, catastrophic events such as tsunamis can affect their entire land area, and rising global sea level is feared to flood whole nations. In a survey of sea level at numerous small islands, it is shown that in many cases sea level is falling and has been for centuries, and that any effect of global change is very site specific. Tides, in general, have a smaller range at islands than at continental sites, but even though tidal observations are essential for determining vertical datums and for predictions, many small island developing states do not operate tide gauges. Tsunami prediction, for example, requires improved bottom topography information at most sites, but again the observational network for issuing warnings and improving such forecasts is oftentimes absent. A commitment by small island developing states to initiate observations and to participate in regional research and monitoring programs is considered essential for effective decision-making to assist sustained economic development.

Introduction

Historical Context

Ancient stories, such as the Biblical Great Flood, the Gilgamesh Epic, and the Atlantis fables, echo humankind's concern with water level from time immemorial. Modern scientific evidence surrounding the eruption of Mt. Santorini *ca.* 1470 BC and the subsequent destruction of the Minoan civilization by a tsunami (seismic sea wave) probably gave rise to the story of Atlantis [Emery and Aubrey, 1991]. Events such as these are deeply imbedded in our psyche and suggest that we have been concerned with related environmental events far longer than recorded history.

Small Islands: Marine Science and Sustainable Development
Coastal and Estuarine Studies, Volume 51, Pages 83–119
Copyright 1996 by the American Geophysical Union

Approximately 18,000 years ago, Earth was at the height of the last ice age. Humans were traveling across the land bridge between Asia and North America because the ocean's water was locked up in the great continental ice sheets; sea level stood approximately 120 m lower than today [Pirazzoli, 1991]. As Earth warmed, the ice melted and sea level rose, slowly at first then most rapidly to a maximum rate of over 10 mm/yr *ca.* 15,000-7,000 years ago, and then slowed down to a rate of less than 0.5 mm/yr throughout the course of recorded history. Humankind created great coastal civilizations during this latter period of very slow sea level change, and acquired the mistaken notion that the sea was a fixed datum against which vertical land measurements could be referenced.

During the last several centuries, and in particular since the mid-19th century, scientists and engineers have invented and installed instruments to observe the vertical motion of the sea. Their interest was, of course, the tides, knowledge of which had great commercial value for marine transportation. An illustration of the attention paid to tides during the last century is given in Figure 1, which shows the world-wide range of spring tides as understood *ca.* 1900. The original plate was republished in color from a physical atlas by Berghaus reported by Harris [1901], and it shows that the tidal range over most of the small islands considered herein is less than 1 m. Figure 1 also shows the paucity of information concerning pelagic tides, and how very few tide gauges were deployed at central ocean island sites.

These tide gauges are now used extensively to determine more than just the semidaily and daily ebb and flow of Earth's gravitational tides; today they are used for the vertical geodetic datum certainly, but equally important for purposes herein, tide gauges are used to measure the relative motion of the sea with respect to the land. This relative change is called relative sea level (RSL) because at a single tide gauge, it is never certain whether the sea is rising or falling or the land is subsiding or being uplifted, or both. Water level measuring instruments have many other applications in today's society besides determining RSL and the tides. Two other purposes which are of concern *vis-a-vis* water level are storm surges (*cf.* Chapters 8 and 9) and tsunamis.

In the tropical regions, where many small island developing states are situated, the environmental event of greatest concern is the tropical cyclone. These energetic storms, when fully developed, are called *hurricanes* in the Atlantic, *typhoons* in the Pacific, and *cyclones* in the Indian Oceans, and off Australia they have the colorful name of *willy-willy*. They are anything but colorful to the people affected by their devastating winds and accompanying storm surge. Storm surges are caused by a combination of factors: low atmospheric pressure in the center of the storm; wind driven currents; and coastal stretches that do not allow the water to advect away. Tide gauges that survive storm surges have measured water levels in excess of 5 m, on top of which, are wind waves of nearly equal height. Tsunamis, on the other hand, can grow to heights in excess of 15 m and cause great loss of life and property. Warning systems such as exist in the Pacific, based on seismographs and tide stations, are an essential component of tsunami mitigation.

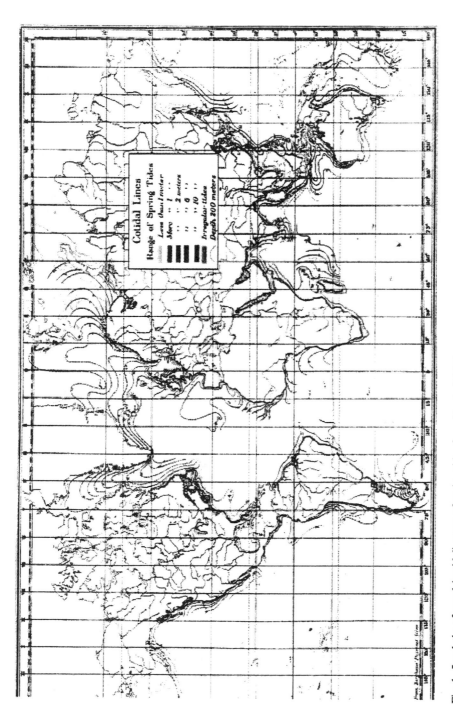

Fig. 1. Isopleths of equal (cotidal) range of spring tides from Harris [1901].

Spectrum of Sea Level

Each of the features of the ocean's surface level, RSL, tides, tsunamis, storm surges, *etc.*, have different time and space scales that define their energy. One way to represent this continuum of amplitudes and frequencies is through a spectrum; an example that generalizes the basic information is given in Figure 2. Based on hourly and monthly RSL observations from Key West, Florida (24.5°N, 81.8°W; *q.v.* Figure 1), the vertical axis shows the energy (cm^2/cpd) and the horizontal axis the frequency. The inset [from Pond and Pickard, 1983] schematically illustrates spectral energy at the highest frequencies and is included to identify wind waves, swell, tsunamis, and long gravity waves. Events such as the rise in water level associated with hurricanes (storm surge) do not appear in a spectrum because they are episodic but not periodic (regularly recurring) events. Similarly, tsunamis appear to have little relative energy but that is again because they are episodic compared with the tides which are periodic.

There are two most important things to notice about Key West spectral variability: First, the overall shape rises from low values at high frequencies (short periods) to higher values at the lowest frequencies (longest periods); this makes the spectrum "red," which is a term used in analogy to the spectrum of light which would appear reddish in color if light waves were arrayed similarly. Second, this spectrum is punctuated with very narrow and sharp lines that represent specific geophysical processes. Each of these discrete lines are identified as (right to left): quarter-diurnal, terdiurnal, and semidiurnal (quarter-day, thrice, and twice daily, respectively) tides, diurnal tides, inertial motions, fortnightly and monthly tides, semiannual and annual gravitational tides, steric (heat and salt) and geostrophic current variations, interannual variability associated with ocean circulation and climatic events such as El Niños (five-six years), the North Atlantic Oscillation (13-14 years), very low frequency tides such as the lunar nodal tide of 18.61 years, and finally long-term change in RSL associated with climate change and land motion.

It should be reiterated that some of the different types of ocean waves shown in Figure 2 are very regular and predictable (*e.g.*, the tides), and others are quite irregular and episodic (*e.g.*, storm surge, tsunamis, and El Niños). Perhaps least predictable at this stage of oceanographic knowledge is the redness of the spectrum itself, because this is the aspect of long term change associated with Earth's climate. One of the forcing terms of climate is the earth-sun orbital relationship, which would lead to a simple expectation that Earth is entering a 5,000 year cooling period. Climate change is more complicated than just caused by orbital variations, in particular, the view that human activity is inadvertently modifying Earth's atmosphere and that humankind is causing a global warming.

Comparison with Continental Sea Level

Sea level spectral variability at islands is, in general, less energetic than that at continental sites. The famous tides of Canada's Bay of Fundy, where the mean spring

KEY WEST, FLORIDA SEA LEVEL

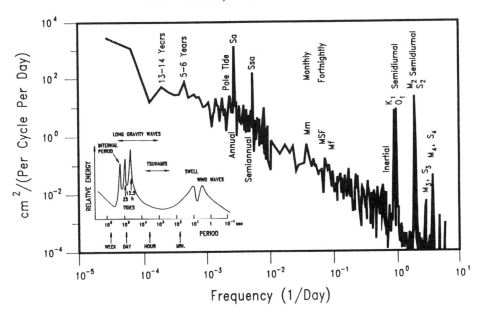

Fig. 2. Spectrum of sea level from Key West, Florida (USA), using monthly means (1913-1992) for the seasonal and lower frequencies and hourly heights (1981-1983) for the daily-to-monthly frequencies.

(fortnightly) range is 12.9 m, are more than twice an order of magnitude larger than most islands (*q.v.* Key West, Figure 2, where the mean spring range is 0.5 m). The tidal range is more related to coastal geometry and bottom topography than the fundamental (equilibrium) forces causing the tides, and is a complex problem in geophysics that is still an area of research. For purposes herein, it is important to recognize that most islands have small RSL ranges because the gravitational and steric waves are least perturbed by land in the open sea and near the tropics.

RSL variations of many frequencies cause water levels to change, but the non-tidal changes perhaps are of greater interest, particularly those changes associated with decade to century scale time-frames. These long-term changes in the instrumental record are prevalent at all measuring sites, whether island or continental. Key West, for example, has experienced a sea level rise of about 2 mm/yr since the mid-19th century but other islands have different rates. Many continental sites have larger century-scale changes, mostly due to vertical land motion caused by removal of the great continental ice sheets. In many such places sea level is falling at rates five times greater than it is rising at Key West. To better understand this phenomenon, a geological perspective is warranted.

Long-Term Sea Level

The term "sea level" has many uses in oceanography. In this section, however, aspects of RSL associated with decades, centuries, and millennia, are discussed. In terms of Figure 2, these time scales represent the very lowest frequencies and with few exceptions are lower than 10^{-5} cycles per day (cpd) at the left-hand of the abscissa. It is from the geological rather than the instrumental record that information about RSL at these low frequencies is available.

Geological Control of Sea Level

During the last 20,000 years, important changes in relative sea-level have been produced on all the coasts of the world, though at different rates, by the melting of land-based ice sheets, especially in North America and Scandinavia. This melting caused a global (eustatic) sea-level rise of the order of 10^2 m and a series of isostatic adjustments. The latter include uplift of the areas previously depressed by ice load, subsidence in a wide belt situated around former ice caps, again subsidence of the ocean floor due to the melt-water load (which in coastal areas varies with water depth), and finally less significant but variable vertical displacements, depending on the rheology of Earth, in regions remote from former ice caps. Superimposed on the effects of the above processes are vertical movements resulting from other causes (tectonics, sediment compaction, etc.), which often became significant in the Late Holocene, after the deceleration of eustatic and isostatic processes. Therefore, trends in relative sea-level often vary from place to place [Pirazzoli, 1991] and, consequently, isolated small islands can be affected by very different sea-level histories. Figure 3 depicts the variability in local RSL that can occur in circumstances of differential vertical tectonic displacement.

Holocene Sea-Level Changes in Small Islands

An example of this variability can be obtained by comparing Holocene sea-level trends reported from Bermuda (Atlantic Ocean) with those obtained in the Mascarene Islands (Indian Ocean) (Figure 4). In Bermuda, a relative sea-level rise of about 23 m occurred since 9200 yr B.P., at rates gradually decreasing with time (curves A, B, C); such a trend may relate to the position of Bermuda within the peripheral subsidence belt of the former Canadian ice sheet. In Reunion and Mauritius, on the other hand, where no peripheral bulge effect is possible, the relative sea-level rise came to an end about 5000 yr B.P. and no emergence occurred since that time (curves D, E).

In the central Pacific area, which is also remote from former ice caps, the present sea level seems to have been reached between 5500 and 6500 yr B.P., in Polynesia as well as Melanesia. In Polynesia (Figure 5) a slight emergence, though of variable amount, predominated in several islands during the Late Holocene: between 2.5 and 3.8 m in the

Tonga Islands, according to Nunn [1991] (band C); less than 1 m in French Polynesia (curve G); but no emergence is reported from Rarotonga and Aitutaki (Cook Islands), where the present sea level seems to have been reached only 3000 yr B.P. (curve D); in

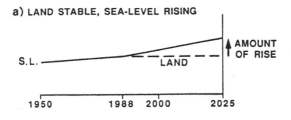

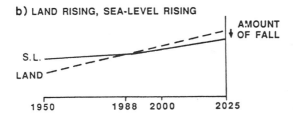

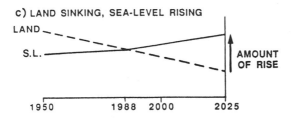

Fig. 3. Examples of variability in local sea-level change. These scenarios assume a greenhouse-induced increase in the rate of sea level rise from about 1990-on. The effect of differential vertical tectonic displacement on the direction and amount of sea-level change are demonstrated for three hypothetical situations [from Hendry, 1993].

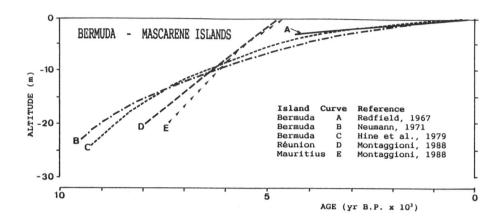

Fig. 4. Holocene relative sea-level changes in Bermuda (Atlantic Ocean) and in the Mascarene Islands (Indian Ocean).

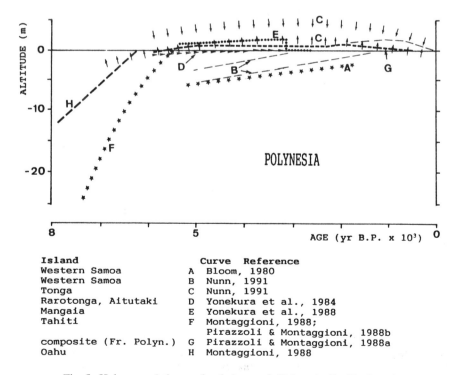

Fig. 5. Holocene relative sea-level changes in Polynesia (Pacific Ocean).

Mangaia (Cook Islands), a sudden tectonic uplift of about 2 m, producing an emergence of the same order, is claimed to have occurred about 3150 yr B.P. by Yonekura *et al.* [1988] (curve E). In the Western Samoa Islands, on the other hand, curve A by Bloom [1980] suggests the occurrence of rapid submergence since 5500 yr B.P., with sea level still 2 m below the situation 1500 yr B.P.; however, the presence of a sea level 2.1 m above the present one about 1300 yr B.P. seems unequivocal to Nunn [1991] (band B) and this supports the indications of 1-2 m Late-Holocene emergence previously reported by Sugimura *et al.* [1988b] and Rodda [1988] in the same area.

Sequences of sudden or of more or less gradual uplift movements, in some cases probably related to seismic events, interspersed with quiescent intervals, has been reported from the tectonically active Vanuatu Islands in Melanesia (Figure 6, curves A and B). These movements have raised a reef with a probable age of 6,000 years to almost 33 m above sea level in Santo Island, while a reef crest 6,700 years old is now at +19 m in Tangoa Island [Bloom and Yonekura, 1985]. In other Melanesian islands, sea-level data suggest trends comparable to those reported from Polynesia: a 2 m emergence at Vanua Levu (curve D), ascribed by Miyata *et al.* [1988] to a sudden (co-seismic) relative sea-level drop which occurred some time later than 3400 yr B.P.; a 1 m emergence earlier than 3000 yr B.P. at Viti Levu (curve E); or an emergence between 0.7 and 2.3 m (band F) summarizing the situation in the Fiji Islands area.

In the West Indies, relative sea-level histories show trends similar to those from Bermuda, the present sea level representing the maximum reached in the Holocene. The various curves of Figure 7 appear roughly parallel to each other; for the Late Holocene, altitudinal differences between curves are not easily accounted for--most authors suggest tectonic stability during the Holocene, though this claim may need to be revisited.

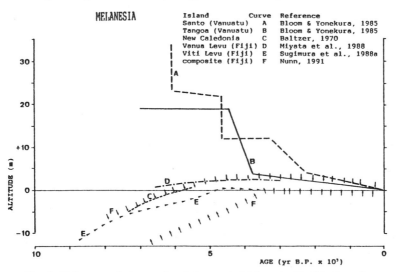

Fig. 6. Holocene relative sea-level changes in Melanesia (Pacific Ocean).

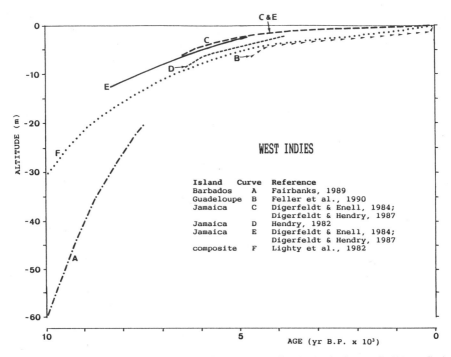

Fig. 7. Holocene relative sea-level changes in the West Indies (Atlantic Ocean-Caribbean Sea).

Digerfeldt and Hendry [1987] pointed out the risks associated with extrapolation of RSL data between sites, particularly in chronological re-construction of coastal history, as vertical differences of meters are apparent in the curves. Digerfeldt and Hendry [1987] also suggested that geoidal variation be considered to explain the differences: this mechanism has been invoked for the northeast coast of Brazil to account for three higher-than-present late-Holocene sea-levels [Suguio *et al.*, 1988].

Though Baffin Island cannot be considered a "small island," relative sea-level changes reported from some of its coasts have been summarized in Figure 8 to show that diverging trends may affect nearby localities (and, therefore, many small islands of the Arctic Sea), even several thousand years after complete melting of the Canadian and Scandinavian ice sheets. In Baffin Island, emergence since 6000 yr B.P. may vary from about 100 m on the central south coast (which was nearer to the Canadian ice sheet center) (curve A), to about zero near the southeastern coasts of the island (curve E); the same deglacial history may, therefore, produce gradual emergence trends during the last few thousand years (curves A, F, G) and, at relatively short distances, also gradual submergence trends (curves B, E); finally, glacial loading and unloading, which may have accommodated along pre-existing fault systems, can explain, according to Andrews and Miller [1985], important relative sea-level fluctuations such as those shown by curve B around 8200 yr B.P.

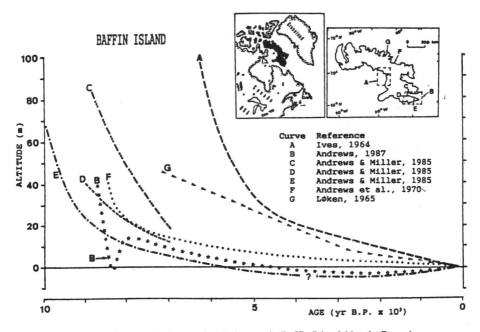

Fig. 8. Holocene relative sea-level changes in Baffin Island (Arctic Ocean).

To summarize, relative sea-level variations correspond to a complex of local, regional, and global processes which interact at various temporal, areal, and vertical scales. In order to predict the impacts of a possible near-future sea-level rise on an island, it is essential to clarify the past sea-level history in this island and to interpret correctly the causes of observed past changes. In the following section, some results from the instrumental (tide gauge) record of the current century are discussed.

Instrumental Trends

Direct measurement of sea level from tide gauges was originally for commercial purposes associated with determining the relationship between upper meridional passage of the moon and high tide. The *establishment of the port*, related to this luni-tidal information, was a well-kept secret by mariners, as was the tide range itself. With the instrumental record, mostly since the 19th century, came the analysis and prediction of tidal height. The notion of determining long-term trends in sea level is much more related to studies of climate change which have been given contemporary urgency.

Tide gauges only indirectly measure the decadal-to-centennial scale change. It is the tide staff, to which both the tide gauge record (the marigram) and the local land elevation benchmarks are referenced, that give information about sea level. The key elements involved are (1) the observer who regularly notes on the marigram the water level from the tide staff, (2) the surveyor who determines the stability of the staff with respect to the fixed tidal benchmarks in the vicinity, and (3) the oceanographer who combines the observations to create a summary datum and then analyzes the result for geophysical purposes. The Permanent Service for Mean Sea Level (PSMSL) [Woodworth, 1991] is one such agency that performs this work, and it is PSMSL records that will be discussed next.

The Revised Local Reference (RLR) PSMSL data-set represents one of the most carefully constructed records of monthly and annual mean RSL. From these RLR records the linear least-squares trend at selected small islands that are representative of each ocean area has been computed. Table 1 is a summary calculated from this subset of the PSMSL RLR station file (chosen to give representation to a wide variety of small islands), giving the station name, location, the years of operation, the number of complete years (n), the trend (mm/yr), the standard error (ε) of the trend line ($\pm$ mm/yr), and the linear correlation coefficient (r). A cursory study of Table 1 shows that sea level at most islands selected is rising (positive trend), but that at a substantial fraction of them, it is falling (negative trend). Where $\pm 2\varepsilon$ is large enough to change the sign of the trend, the null hypothesis of no trend cannot be rejected at the 95% confidence level. This uncertainty is illustrated in the next drawing.

In Figure 9, the thin line connects the annual mean sea level at each of 11 of the stations in Table 1, and the heavy line shows the linear least squares trend for each record respectively. Interannual variability, which is a naturally occurring oscillation of sea level, makes determining the trend uncertain; short records also contribute to this uncertainty. Each curve is arbitrarily offset by 10 cm in order that individual records can be seen. Stations plotted in Figure 9 are selected to represent small islands from the Antarctic and Arctic Oceans (Argentine Islands and Barentsburg, respectively), the eastern and western North Atlantic Ocean (Santa Cruz and St. Georges, respectively), the Indian Ocean (Port Louis), and a wide range of islands from the South and North Pacific (Guam, Pago Pago, Canton, Midway, Hilo, and Unalaska).

As with the Holocene RSL curves (q.v. Figures 4-8), one is struck with the wide variety of trends and uncertainties in the instrumental record vis a vis small island sea level during this century. Even islands near to each other have opposite trends (cf. Guam and Truk in Table 1) suggesting that intra-country variations are to be expected for small island developing states covering substantial areas. One is also struck with the paucity of records from small island developing states and with the shortness of many records at critical places. Hendry [1993] found that tectonic motion was in many locations an order of magnitude lower than historical sea level changes, but definitive interpretation was impossible because of the time-averaged, extrapolated nature of the tectonic data, much of it based on movement of 120,000 year old elevated reefs, and the lack of coincidence between locations where tectonic and tide gauge measurements were made. Clearly there are serious deficiencies in oceanographic and geological knowledge of past sea level and, as will be considered below, in scientific ability to predict changes at individual sites based on global projections to the year 2050 and beyond.

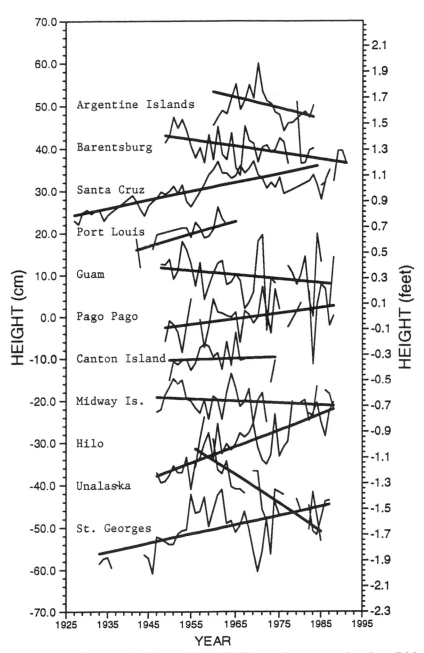

Fig. 9. Summary of linear sea level change from PSMSL annual means at selected small islands where the record length was at least 18.61 years long and where there were at least 10 full years in each record.

TABLE 1.

Summary of linear least squares trend statistics for selected small islands from the Revised Local Reference (RLR) file of annual mean sea level archived by the Permanent Service for Mean Sea Level (PSMSL), Bidston Observatory, Birkenhead L43 7RA, United Kingdom.

PSMSL RLR Station	Latitude	Longitude	Years	n	Trend (mm/yr)	ε ±	r
Adak	51°52'N	176°38'W	1943-1988	35	0.1	0.6	0.02
Argentine Islands	65°15'S	64°16'W	1958-1988	20	-2.6	1.3	-0.42
Barentsburg	78°04'N	14°15'E	1948-1991	40	-1.6	0.5	-0.45
Cagliari	39°12'N	09°10'E	1896-1934	26	1.6	0.4	0.65
Canton Island	02°48'S	171°43'W	1949-1974	20	0.3	1.0	0.07
Eniwetok	11°22'N	162°21'E	1951-1972	20	0.8	1.8	0.11
Friday Harbor	48°33'N	123°00'W	1934-1988	51	1.1	0.3	0.46
Guam	13°26'N	144°39'E	1948-1988	37	-1.0	0.8	-0.22
Guantanamo Bay	19°54'N	75°09'W	1937-1968	26	1.9	0.5	0.63
Hilo	19°44'N	155°04'W	1946-1988	41	3.9	0.5	0.80
Hong Kong	22°18'N	114°12'E	1929-1985	30	-0.8	1.0	-0.16
Honolulu	21°19'N	157°52'W	1905-1988	84	1.6	0.2	0.73
Jolo	06°04'N	121°00'E	1947-1991	23	-1.0	0.8	-0.28
Johnston Island	16°45'N	169°31'W	1950-1988	35	0.6	0.6	0.19
Keelung	25°08'N	121°44'E	1956-1989	34	-0.9	0.5	-0.31
Key West	24°33'N	81°48'W	1913-1988	74	2.2	0.1	0.89
Kwajalein	08°44'N	167°44'E	1946-1988	42	1.1	0.5	0.32
Magueyes Island	17°58'N	67°03'W	1955-1988	26	1.7	0.4	0.65
Midway Island	28°13'N	177°22'W	1947-1988	33	-0.5	0.5	-0.17
Montauk	41°03'N	71°58'W	1947-1988	36	2.1	0.4	0.71
Naha	26°13'N	127°40'E	1966-1988	21	2.4	1.1	0.45
Naos Island	08°55'N	79°32'W	1949-1968	19	1.3	1.2	0.27
Pago Pago	14°17'S	170°41'W	1948-1988	35	1.3	0.6	0.35
Palermo	38°08'N	13°20'E	1896-1922	15	1.0	0.6	0.40
Port Louis	20°09'S	57°30'E	1942-1965	14	3.0	0.9	0.67
Puerto Plata	19°49'N	70°42'W	1949-1969	14	4.4	1.1	0.74
Queen Charlotte	53°15'N	132°04'W	1957-1984	13	0.6	1.6	0.11
Resolute	74°41'N	94°53'W	1957-1977	11	-3.3	1.6	-0.56
Reykjavik	64°09'N	21°56'W	1956-1989	27	4.2	0.8	0.70
Russkaya Gavan	76°12'N	62°35'E	1953-1990	35	-1.4	0.6	-0.34
Santa Cruz	28°29'N	16°14'W	1927-1987	46	2.0	0.2	0.80
St. Georges	32°22'N	64°42'W	1932-1988	45	2.2	0.5	0.56
Torshavn	62°00'N	06°46'W	1957-1990	14	1.8	0.8	0.56
Truk	07°27'N	151°51'E	1947-1988	31	1.1	0.9	0.22
Unalaska	53°53'N	166°32'W	1955-1988	21	-6.7	-1.1	-0.82
Victoria	48°25'N	123°22'W	1909-1984	71	0.8	0.2	0.46
Wake Island	19°17'N	166°37'E	1950-1988	32	1.9	0.6	0.50
Wellington	41°17'S	174°47'E	1918-1991	15	-2.2	0.4	-0.82

Tides

Planet Earth experiences tides in all three of its fluid elements: atmosphere, ocean, and "solid" earth. It is the gravitational tides of the earth-sun-moon system with which this section is concerned. In Figure 2, these primary gravitational tides are identified: semidiurnal, M_2 and S_2; diurnal, O_1 and K_1; fortnightly, MS_f and M_f; monthly, M_n; semiannual and annual, S_{sa} and S_a respectively. In addition, other periodic motions such as the inertial period (12 hours ÷ sine latitude) and the pole tide associated with the (Chandler) wobble of Earth on its axis of rotation (436 days) are identified. A complete discussion of all these and many other periodic motions is beyond the purposes of this chapter [cf. Pugh, 1987], but the spectrum in Figure 2 serves to illustrate the principal tidal constituents and some other of Earth's periodic motions.

Semidiurnal to Annual

The most important gravitational tides are the M_2, the principal lunar semidiurnal tide with a period of 12.42 hours and the S_2 or the principal solar semidiurnal whose period is 12.00 hours. Both the M_2 and the S_2 can be clearly recognized in time series of ocean, atmospheric, and earth tides and are associated with the upper and lower transit of the moon and sun respectively past the meridian of the geographical point under consideration. However, due to complexities in the configuration of the continents and the ocean's bottom topography, the actual amplitude and phase of each oceanic constituent is complex. At amphidromes, the amplitude is zero and the phase may rotate in either a clockwise or an anticlockwise direction. Figure 10 [Parke, 1983] illustrates the complexities of these two of dozens of significant tidal constituents, each of which have different maps showing their respective amplitudes and phases.

Some tidal constituents are complicated by being forced by non-gravitational factors. Two simple illustrations are the daily (S_1) and annual (S_a) periods of 24.00 hours and 365.25 days respectively. At these two periods, regular effects of heating, non-tidal currents, and winds can complicate the picture. When two or more cycles with the same period are superimposed upon each other, the result is a harmonic of the same period but with different amplitude and phase. The annual cycle of RSL at Key West, for example (cf. Figure 2), is the sum of the annual gravitational tide plus the annual steric cycle plus the annual Gulf Stream speed cycle, plus the annual wind cycle plus, etc. Without independent measurements, it is impossible to identify each contribution to the mean S_a observed with the Key West tide gauge.

For most practical (engineering) purposes, the mean range (or mean spring range) of the tide is of interest. The range of the tide being twice the tidal amplitude, the sum $2(M_2+S_2)$ is often used to characterize this most important variable. This sum is used because near an M_2 amphidrome (for example) that constituent's range is zero (cf.

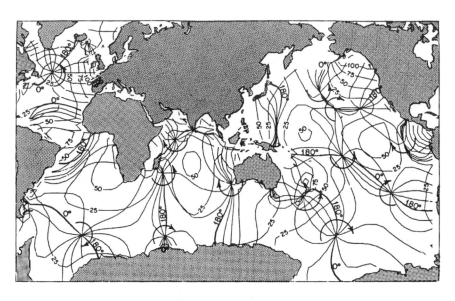

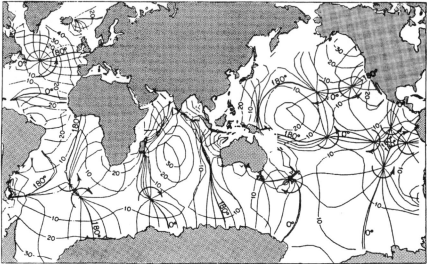

Fig. 10. The upper panel shows the amplitude in centimeters (fine lines) and phase in degrees (heavy lines) of the M_2 tide; the lower panel is for the S_2 tide [from Parke, 1983].

Figure 10), yet there usually is a semidiurnal tide due to the S_2 (again, for example) (cf. Figure 2); these 12.42 and 12.00 hour semidaily tides account for most of the water level variation. The mean range varies widely from place to place, but fortunately is well documented in the tidal tables of many agencies.

Regional Characterization

Typical of global maps of tidal constituents such as shown in Figure 10 is the lack of detail in a specific region of interest. The Caribbean Sea is one such illustration, but for which there are investigations that provide detailed information. Kjerfve [1981] studied the tides of the Caribbean Sea and computed average amplitudes for the six major constituents, summarized in Table 2. Given that semidiurnal tidal ranges are typically $2(M_2+S_2)$, one calculates a mean range of 27 cm for the region, which categorizes Caribbean tides as microtidal [Davies, 1964]. An inherent danger in the statistics of Table 2 is that the range varies markedly from station to station, even within the confines of a small area such as the Caribbean Sea.

Another aspect of regional tides is their type; that is, are they primarily diurnal, semidiurnal, or mixed? Kjerfve [1981] followed classical tidal analysis and computed the form number $F=(K_1+O_1)/(M_2+S_2)$, and contoured the result. For the Caribbean as a whole, $F=1.1$ (q.v. Table 2), characterizing the tide as mixed primarily semidiurnal. Again, care should be taken in using a single characterization, even for a small region as this can be quite misleading: Caribbean tides are mixed semidiurnal in the Cayman Sea ($0.25 \leq F < 1.5$), mixed diurnal in the Colombian Basin ($1.5 \leq F < 3.0$), diurnal in the Venezuelan Basin ($F \geq 3.0$), and mixed semidiurnal in the southern Lesser Antilles [Kjerfve, 1981]. Primarily semidiurnal tides ($F < 0.25$) are not prevalent in the Caribbean Sea.

TABLE 2.

Summary of major diurnal and semidiurnal tidal amplitudes (cm) for the Caribbean Sea [abstracted from Kjerfve, 1981].

Constituent	Symbol	Period (hours)	Theoretical Coefficient	Mean Amplitude
Principal lunar	M_2	12.42	100	10.4
Principal solar	S_2	12.00	47	3.1
Larger elliptical lunar	N_2	12.66	19	2.8
Luni-solar	K_1	23.93	58	8.0
Principal lunar	O_1	25.82	41	5.7
Principal solar	P_1	24.07	19	2.7

Finally, the tidal amplitudes of the Caribbean Sea can be compared with the theoretical coefficients for the global equilibrium tide; again see Table 2. It is fortuitous that the theoretical coefficient and the M_2 mean amplitude are almost both multiples of 10, but it allows visual inspection of the relative amplitudes and theoretical coefficients. For this region, there is a rather good correlation between the theoretical and observed amplitudes, but this is more a reflection of the Caribbean being oceanic in character than a goodness of fit.

Predictions

Tidal predictions are based on observations at each site. Tables are routinely available from government or inter-governmental agencies such as the International Hydrographic Office located in Monaco. Table 2 shows six terms used in most prediction schemes, but Figure 2 shows that many other terms would be necessary for a site such as Key West because of the terdiurnal and higher frequency terms, and because of the lower frequency terms such as the fortnightly (MS_f, M_f), monthly (M_n), and annual/semiannual (S_a, S_{sa}) terms. Many tide prediction algorithms use 37 such terms in order to accurately represent the tides. Without the observations, however, the tides cannot be adequately predicted.

Tidal prediction usually requires 19 year records in order to include the effects of the lunar nodal cycle. If a nearby reference station is operated during shorter observations at a site, records as short as 29 days can be useful in determining the ratio between the amplitude and phase of each constituent. Such observations provide useful predictions of the daily tides, but do not provide useful predictions of future sea level rise because RSL is very site specific (*q.v.* Figure 9). Tidal models, however (*q.v.* Figure 10), play a critical role in remote sensing of the ocean from satellite altimeters because tides constitute a large fraction of the sea level signal.

Tsunami

Characteristics

The cause and nature of tsunamis is still confused in the public literature through use of the term "tidal wave." The name derives from the Japanese for "great wave in harbor," which is a descriptive nomenclature. In terms of process, however, tsunamis have no connection with tidal forcing, though a form of solitary progressive wave can develop during flood tides in estuarine areas with high tidal range. This wave form is a "tidal bore," which typically will measure in the order of decimeters compared to meters for tsunami waves. Further confusion may occur because the shoaling tsunami wave is also often referred to as a "bore."

Tsunami formation is independent of the wind and gravity effects which generate sea waves and tides. They are impulse-generated, long-period waves initiated by submarine earthquakes, landslides, and volcanic eruption. Tsunamis develop when the still-water surface is displaced by seafloor flexuring, water-body displacement, or propagation of seismic shock from the seafloor into the water column. Hence, source areas tend to be associated with convergent boundaries of tectonic plates where subduction, vertical and strike-slip displacement and volcanism occur, and where slumping occurs on unstable island or continental margins and delta fronts. Tsunamis can also be formed by meteorite impact or nuclear explosion.

Historical analysis indicates that earthquakes are the most important tsunami-generating source, and have been the cause of over 90% of the great tsunamis of the world. Although only about 5% of tsunami events have been caused by volcanic eruption, many of these have been catastrophic [Soloviev, 1982].

Tsunami wavelength may be hundreds of kilometers, while amplitudes in deeper parts of the oceans may be only a meter. Tsunamis travel at several hundred kilometers per hour, at a speed proportional to the square root of the water depth. Consequently, they can travel across an ocean basin the size of the Pacific in less than 24 hours (Figure 11), and therefore can have a profound influence at locations great distances from the source areas.

In the open ocean tsunamis behave like shallow water waves due to their great wavelength. Refraction effects are caused by interaction with the ocean floor, and calculation of changes in wave crest travel direction are possible using the same methodology as for shorter wavelength swell waves in shallower water. In addition, shoaling, dispersion, diffraction, scattering, and reflection also occur in deep-ocean conditions.

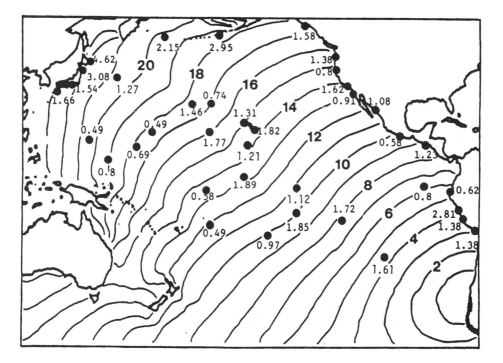

Fig. 11. Refraction diagram of the Chilean tsunami of May 22, 1960, showing positions of the tsunami front at 1 hour intervals. Solid circles are tide gauges where tsunami height is measured in meters [from Nagano *et al.*, 1991].

Shallow Water Effects at Islands

Tsunamis occur in groups, rarely singly, and may appear at the coastline in a sequence. The first to appear are the forerunner waves, perhaps preceded by withdrawal of water from the shore. These early waves are short-period, low-amplitude oscillations. The main tsunami waves follow, and secondary undulations that are caused by resonance effects may also occur. Not all of these events may occur together, however [Murty, 1977].

Tsunami run-up distance and elevation depend on island size and bottom topography. With regard to shoaling and refraction effects, Brandsma *et al.* [1974] suggested that tsunami behavior tends to uniformity for very long waves so that oceanic islands with a diameter smaller than the tsunami wavelength are not generally affected. However, tsunami behavior becomes extremely complex once the wavelength is comparable to the size of the island. Consequently, focusing and spreading of tsunami energy and amplitude occurs in response to variation in shelf depth and shoreline configuration [Murty, 1977].

Where narrow shelves flank an island, the run-up heights are likely to be greater than on broader shelves, where energy dissipation through shoaling over a wider approach area can occur.

Historical Tsunami Records in Islands

To illustrate the effects of tsunami on islands, examples are taken for three ocean areas - the Pacific, Mediterranean and Caribbean.

Pacific

Tsunami occurrence in the Pacific islands is extensive, with the Hawaiian experience in particular being well documented. The susceptibility of these islands to tsunami is illustrated in Figure 12, which shows run-up heights and approach directions for earthquake-generated tsunamis at Oahu for the years 1946, 1952, 1957, and 1960.

Amongst the biggest killers was the 1946 Aleutian earthquake, tsunami from which killed 173 persons in Hawaii, prompting the establishment of a tsunami warning system. Public failure to appreciate the meaning of a warning resulted in 61 deaths during the 1960 tsunami [Dudley and Lee, 1988] which was generated by an earthquake in Chile. Improvements in the warning system followed.

The 1960 Chilean earthquake was also observed in Tahiti and the Samoan and Cook Islands. Figure 11 shows the positions of the tsunami front at 1 hour intervals and the tsunami elevations recorded at tide gauges on several islands. Gauge records do not

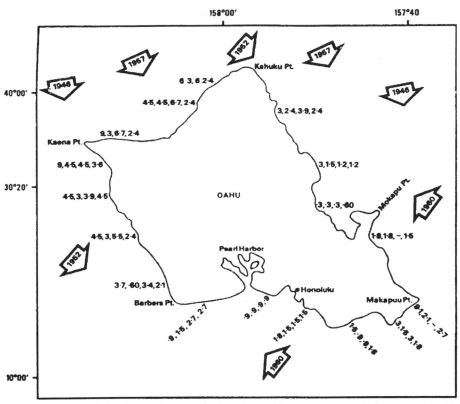

Fig. 12. Run-up heights (m) and approach directions of four tsunamis at Oahu, Hawaii: 1946, 1952, 1957, and 1960 [diagram in Murty, 1977, from Cox and Mink, 1963].

necessarily record maximum amplitude of breaking tsunami, which can be much greater than for the sheltered instrumental record. In Apia Harbor, for example, wave heights of up to 5 m were reported [Keys, 1963].

An earthquake located in the plate boundary zone offshore of the west coast of Nicaragua in 1992 generated a tsunami which had severe impacts on the continental coastline. Tsunamis also propagated across the Pacific and were recorded at numerous island locations (Figure 13), where instrumented amplitudes reached 0.82 m at Easter Island and 1.12 m at Galapagos Island (Figure 14).

The most lethal tsunami of record in the Asia-Pacific area was generated by the eruptive explosion of Krakatau in 1883, from which waves up to 40 m high killed 36,000 Javanese and Sumatran islanders many hundreds of miles from the volcanic source [Simkin and Fiske, 1983].

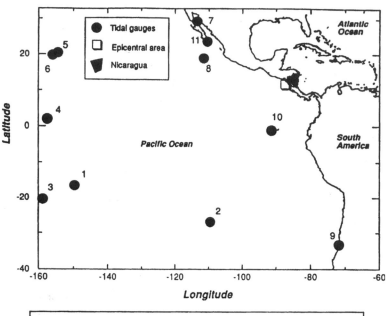

Station	Height [cm]	Arrival time
1. Papeete, French Pol	6	
2. Easter Island, Chile	82	246 07:51 (U.T.)
3. Rarotonga, Cook Island	10	
4. Christmas Island, USA	--	
5. Kawaihae, Hawai, USA	4	
6. Hilo, Hawai, USA	11	
7. La Liberdad, Mexico	16	246 03:29 (U.T.)
8. Socorro Island, Mexico	29	246 04:59 (U.T.)
9. Valparaíso, Chile	11	
10. Galapagos Island, Ecuador	112	246 02:38 (U.T.)
11. Cabo San Lucas, Mexico	28	

Fig. 13. Peak-to-trough heights at tide gauges of tsunami generated by the 1992 Nicaraguan earthquake [from Baptista *et al.*, 1993].

Mediterranean

The eastern Mediterranean, Aegean, Adriatic and Ionian sea coasts are the most affected by tsunami, a reflection of the many earthquake source areas for these events. An earthquake on Crete in AD 365 caused tsunami that affected the entire eastern Mediterranean and killed thousands of people [Papazachos and Dimitriu, 1991].

A volcanic explosion may have been the source of the most destructive tsunamigenic event in the Mediterranean, when the island of Santorini (Thera) blew up about 1470 BC, possibly giving rise to the biblical legend of the Great Flood and contributing in part to the end of Minoan civilization [Smith and Shepherd, 1993].

The epicenter of the largest recent tsunami was between Amorgos and Astypalea islands in the southern Aegean on July 9, 1956. Tsunami height reached 25 m on the southern coast of Amorgas Island, 20 m at the north coast of Astypalea Island and 2.6 m on the east coast of Crete. Damage was substantial and three persons were reported drowned on Kalimnos [Papazachos and Dimitriu, 1991].

Caribbean

This seismically and volcanically active area is ringed by tectonic plate boundaries (see Hendry, Chapter 12), and potential for tsunami generation is substantial.

The 1867 Virgin Islands tsunami generated waves with 9 m run-up at St. Croix. The tsunami was observed throughout the eastern Caribbean causing damage as far south as St. Georges Harbor in Grenada. The 1918 Virgin Islands tsunami produced wave heights up to 6.1 m at Pt. Agujereada, and 1.2 m at Krum Bay, St. John, and accounted for 40 of the 47 tsunami-related deaths in Puerto Rico and the U.S. Virgin Islands. The five tsunamis of record for the Virgin Islands sub-region all have a local source [Lander

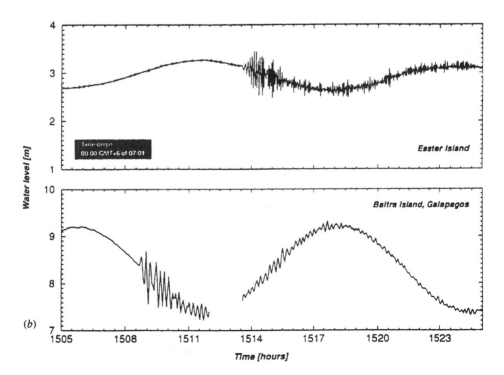

Fig. 14. Tide gauge records of the 1992 Nicaraguan tsunami at two island locations [from Baptista *et al.*, 1993]. See Figure 13 for location.

and Lockridge, 1989]. Earthquakes may trigger slope failures: major failure scars have been detected by side-scan sonar off the north coast of Puerto Rico (Hendry, Chapter 12), which, in turn, may generate tsunami.

In Jamaica, tsunami were produced during the great 1692 Port Royal earthquake, where waves in the inner harbor were up to 2 m high. Tsunami may have been generated by the seismic event or subsequent liquefaction and slumping of land on which the town was built [Hendry, 1979]. Initial withdrawal of water from within the harbor was also reported. There are also reports of tsunami 1-2 m amplitude on the north coast during the destructive 1907 earthquake.

External sources also generate significant tsunami in the Caribbean. For example, the Lisbon earthquake of 1755 caused 2-m high waves in Antigua and Barbados and in Martinique anecdotal evidence suggested that the waves were as high as houses.

Concern over the source of the next major Caribbean tsunami centers on the submarine volcano Kick 'em Jenny, located 8 km off the north coast of Grenada. The volcano has erupted at least 11 times historically, making it the most active in the Lesser Antilles (see Hendry, Chapter 12). Smith and Shepherd [1993] have estimated tsunami travel time and wave run-up for several types of energy release (Figures 15 and 16), and the peculiar propagation characteristics in the area place all islands of the eastern Caribbean at risk from such an event.

Predictions

A tsunami warning system has been established in several Pacific Ocean countries since 1946. However, the time available to give the alarm is usually very short, especially in areas near the source of the event. Key components of the warning system are the seismogram network, which detects the epicenter, depth and energy of an earthquake, and the regional tide gauge network, which provides data on the propagation and amplitude of the tsunami and allows for provision of timely warnings throughout the ocean.

Implications of tsunami for coastal area management are considerable. Vertical and horizontal tsunami run-up can cause sudden, catastrophic inundation and flooding in coastal areas. Apart from direct physical damage to natural areas, tsunami waves cause major loss of life and destruction or damage to property and infrastructure.

While warning systems allow for evacuation, physical damage can still be substantial. Re-location of homes and businesses and formation of buffer zones along the coast can be effective, but construction against tsunami may not. This was the case on the small Japanese island of Okushiri after an earthquake on July 12, 1993 when tsunami run-up heights between 15 and 30 m overtopped a seawall, causing at least 120 deaths [EOS, 1993].

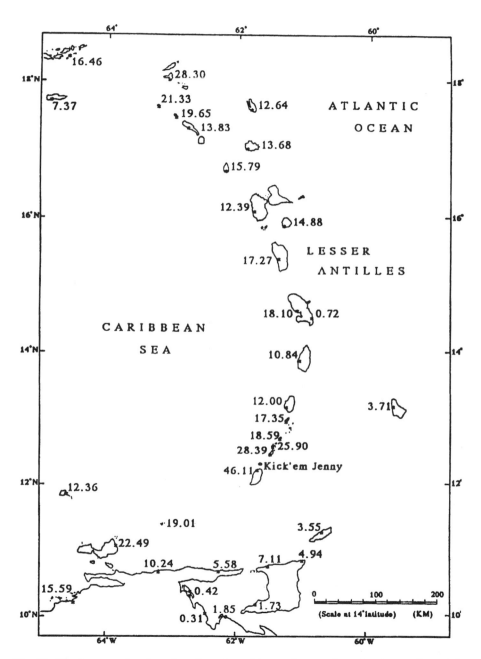

Fig. 15. Final run-up elevations for a "worst-case" tsunami-generating volcanic eruption at Kick 'em Jenny, in meters. This is equivalent to a Krakatau-like eruption, with a VEI of 6 [from Smith and Shepherd, 1993].

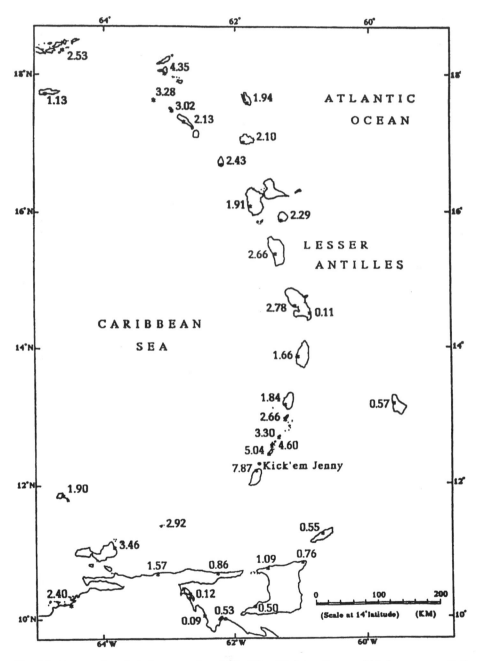

Fig. 16. Run-up elevations in meters for a VEI of 3 at Kick 'em Jenny, which is a more realistic eruption scenario with the volcano in its present state [from Smith and Shepherd, 1993].

The historical incidence of major tsunami events in the Caribbean and the potential for a catastrophic tsunami-generating event at Kick 'em Jenny submarine volcano indicate a crucial need for a similar warning system in the Caribbean. Routine measurement of the horizontal motion of plates is now possible using the Global Positioning System (GPS) [*e.g.*, Dixon, 1993], and should be also considered in any modern warning system.

Implications for Integrated Coastal Management

Day-to-Day Management

Management of ocean and land resources affected by sea level, tides, storm surge, and tsunamis is essentially an issue of prediction. At the highest frequencies (*q.v.* Figure 2) it is atmospherically driven events such as hurricanes and storms that require immediate attention for public welfare. Incoming waves and swell from distant storms can cause significant damage to coastal structures: with adequate warning, destruction from such causes can be ameliorated significantly. This certainly has been the case with storm surge (Chapter 8) and tsunamis as well, and requires the full attention of coastal area management charged with public safety.

At diurnal and semidiurnal tidal frequencies, much of the necessary prediction is routinely provided by national ocean services, either domestic or foreign. For some small island nations, tidal constants can be extrapolated to locations near the control station site, but this is not universally true. At sites where a significant investment will be made, it is essential to obtain tide observations simultaneously with those at the nearest control station and to determine correction factors for both tidal range and timing. The accuracy of prediction is controlled greatly by the quality of data generated by *in situ* measurement and monitoring of these processes. Nations that do not have operating tide gauges (control stations) should consider installation to be of highest priority.

Mean sea level (MSL) as defined by the U.S. National Ocean Service (NOS) is the average of hourly tidal heights over a specific 19-year period [NOS, 1988]. MSL is a critical engineering datum because it is the reference surface used by most countries to determine geodetic vertical control networks from which land elevations are referenced. The United States, Canada, and Mexico recently established the North American Vertical Datum of 1988 which provides a common vertical reference network for all elevations in North America [Zilkoski *et al.*, 1992]. There are a number of other datums determined from tide observations that provide decision-makers with information required for effective management of coastal resources. Tidal datums such as mean high water and mean lower low water are typically used by maritime nations for nautical charts and to define regulatory, jurisdictional, and marine boundaries. There is great complexity in dealing with boundaries, setbacks, erosion, and accretion due to differences of opinion and approach to what is represented as MHW, a problem which is exacerbated in many small island states where, due to rugged interior topography, there is a concentration of development and population directly at the land/ocean interface.

On a more futuristic vein, the rapidly developing ocean numerical circulation models will give management a new and highly relevant capability. For example, mesoscale ocean eddies with time scales of the order of one month and sea level height scales of the order of 30 cm are known to pass certain islands [Maul et al., 1992] with some regularity. Knowing that such an event is approaching, coastal managers will be in a position to better plan the effect on water resources and, if the eddy is forecast to pass during preparation for a tropical storm, heightened public awareness can be broadcast. Modeling of coastal and ocean currents has many direct management applications, ranging from protection of coasts from oil spills to determining optimum fishing strategies to variable chlorination expenditures in sewage treatment effluents to locations for sewage outfalls.

Decadal-Scale Planning

Assuming that the number of tropical storms and tsunamis remain relatively constant in the future and tides are not affected by engineering projects, the next most important management problem with which this chapter is concerned is sea level rise. To illustrate the issue, the record from Key West, Florida, is once again recalled. Maul and Martin [1993] report a sea level series at Key West dating back to 1846, which may be the longest record from a small island. Figure 17a shows a map of the island and the record of monthly mean sea level anomalies, 1846-1992. For scale purposes, Key West island is about 5 nautical miles long (~9 km) in the east-west direction shown.

Figure 17b is the record of mean annual RSL adjusted to a common datum with the residual anomalies being depicted by a dot. It shows that since the mid 1840's, water level with respect to the fixed benchmarks on land has risen about 30 cm (~1 foot). The linear least squares trend line is 2.1 mm/yr ± 0.1 mm/yr (± 1 ε); cf. Table 1. The linear correlation coefficient (r) 95% confidence interval for these data is 0.84>r>0.80 with the r^2 line explaining about 67% of the variance. The order zero RSL signal at Key West is a linear rise over the last 147 years with no statistically significant evidence of accelerated rise as might be expected from global warming.

It was of interest to determine the effect of the 30 cm rise on the coastline. To do so, a chart of the island from the 1851 Annual Report of the Superintendent of the U.S. Coast Survey was scaled to the modern (1991) coastline, both of which are also shown in Figure 17a. The modern coastline is drafted as a solid line surrounding the ca. 1850 coast. Simple application of the Bruun Rule would lead to an expected coastline regression of ~30 m with this sea level rise but, as can be seen, the area of Key West island has grown over the years rather than shrunk. The cause has been the extensive bulkheading and the creation of artificial land to support the 10-fold increase in human population. While the land increased during these years, the mangrove forest has virtually disappeared. Litz and Shinn [1991] demonstrated the likely shrinkage of land area in the Florida Keys assuming no human intervention, using established "global" scenarios for sea-level rise during the next century.

Fig. 17a. Chart of Key West, Florida, *ca.* 1850 (soundings in feet) with coastline *ca.* 1991 superimposed (solid line); inset shows monthly mean sea level anomaly from 1846-1992.

Although the sea level at Key West has risen, the tides have remained rather constant. Thus, mean high water and mean low water have both risen by about the same 30 cm. Therefore, the vertical datum for mapping, charting, coastal area management, setback regulations, *etc.* has also changed. If sea level rises faster in the next century, as many climatologists predict, changing vertical datums must be considered in the decadal-scale decision-making process. It must be reiterated here (*q.v.* Figures 4-9) that any change in sea level is very site specific, and in the case of larger islands (say $\geq 10,000$ km^2) it can vary from one side of the island to the other. There are dangers of over or under-interpretation of effects if assumed rates of sea-level change are extrapolated between sites, and problems in taking the IPCC projections and applying them without appropriate local correction. In low-lying coastal areas, however, or island nations with low elevations, application of even the simplified IPCC "common methodology" for vulnerability assessment in relation to sea level may provide an important trigger for development and implementation of a coastal zone management plan [Resource Analysis and Delft Hydraulics, 1993].

Ongoing Global Programs

Many agencies in the United Nations system have programs concerned with issues considered herein. The Intergovernmental Panel on Climate Change (IPCC) of the World Meteorological Organization is active in estimating future climate and sea level, response strategies, and assessment. The Regional Seas Programme of the United Nations Environment Programme has assumed a leadership role in studying the

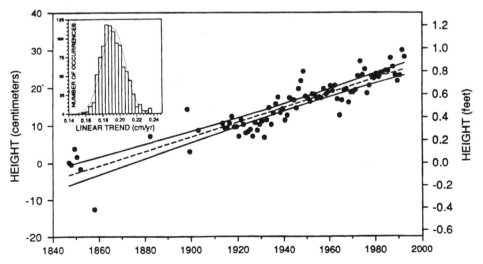

Fig. 17b. Annual mean sea level at Key West, Florida (dots); inset is the bootstrap analysis [Maul and Martin, 1993] showing the linear trend in sea level and its statistical distribution function.

implications of climatic change on the ecosystems and socioeconomic structure of marine and coastal areas on regional levels. The Intergovernmental Oceanographic Commission of UNESCO supports the Global Sea Level Observing System (GLOSS), as well as regional Groups of Experts on Ocean Processes and Climate and the International Tsunami Warning System. These agencies are but three of many concerned with forecasting environmental dangers.

Since sea level rise has center stage in the concerns of many small island developing states, the global nature of this issue needs further discussion. First, to summarize the degree of scientific uncertainty surrounding global predictions of future sea level, Figure 18 has been drawn. Based on the publication date of each estimate, Estevez (pers. comm.) plotted predictions of sea level rise to the year 2100. The heavy vertical lines in Figure 18 are used when the estimate was actually for 2100; the light vertical lines are used where the author's estimate was for a year before 2100 and it has been extrapolated to 2100; where the letter "B" is freestanding, the author did not estimate a range (high to low).

There are several interesting things to notice about the data in Figure 18. First, note that the estimates of the early to mid 1980's were for a very large rise in global sea level, compared to those of more recent times. Second, the range of the rise is generally smaller in the 1990's than earlier. Third, and perhaps most important, these predictions are for the "global" rise; there are many compelling arguments to think that on a regional basis, the change in sea level will be very different [e.g. Jeftic et al., 1992; Maul, 1993]. All these estimates are exclusive of vertical land motion, and may have very little to do with the RSL experienced at a single site (q.v. Figure 9). At climatologists' current understanding of the rate of greenhouse gas production by humankind projected to the year 2100, the "best" estimate is somewhat less than 50 cm for the global rise at the end of the next century.

Although three activities of United Nations agencies have been highlighted in this chapter, the overall number is far greater. This creates a particular problem for small island developing states in that the environmental entities of their countries are easily overwhelmed with information and experts. The mechanisms for regionalization are already established within most international organizations, governmental and non-governmental, but in many cases the regional boundaries are different. Agreement on common regional boundaries and integrated focus of assistance to small island developing states would create a cost/time effective means of communicating global program results and translating them into concrete results. The continued proliferation of programs and agencies is ineffective at best and non-productive at worst.

Research and Monitoring

At a recent conference on ocean circulation in the Intra-Americas Sea, the question was asked "What would you most like to see from this research effort?" One answer was that

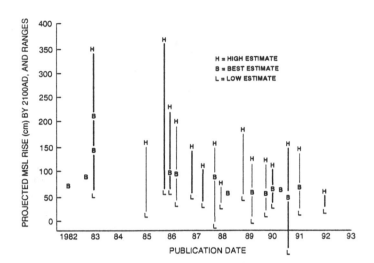

Fig. 18. Predictions of global sea level rise according to the date of publication (redrawn and updated from E. Estevez, Mote Marine Laboratory, pers. comm.).

there be at least one Ph.D. candidate in the environmental educational pipeline at all times from each small island developing state. Such an utopian circumstance may be unreasonable today, but it strikes at the heart of the issue of "research and monitoring." Effective monitoring is necessary to determine quantitatively single issues such as sea level change, but its coupling to an effective research agenda translates the routine measurement of say, hourly tidal heights, into the enlightening determination of vertical datums and their future prediction. These datums are of direct socioeconomic importance and crucial for decision-making for integrated coastal area management.

Almost every small island developing state has some national meteorological expertise, and thus they participate in the global weather research programs even if indirectly. This, in general, is not true in oceanography. Global research programs, such as the GLOSS Program on sea level, are all well developed and only need regionalization to provide direct and immediate relevance on a nation-by-nation basis. Participation in such highly pertinent research and monitoring efforts such as GLOSS should not await grants from external sources; it should be integral to national policy.

Conclusions

Summary

It is impossible to characterize the effects of sea level, tides, storm surge, and tsunamis on small islands in a general sense; that perhaps is the central message of this chapter.

Sea level is rising at some islands; falling at others. Tides are large at some small islands; almost non-existent at another. Tsunamis are a very real and omnipresent danger for one coast of a small island, and of little concern at another coast of the same island. The complexity of each issue requires that generalizations be avoided.

Needs for the Future

Perhaps the most frustrating aspect of quantifying sea level, tide, and tsunami effects on specific small islands is the lack of data. For the immediate future, the overwhelming need is to establish *and maintain* a water level/weather monitoring network and to participate in the global research community. Such a commitment is essential to a wide range of future activities, both social and economic, and it should be of such a national priority that the infrastructure is established internally.

Regional consortia within existing institutional frameworks are envisioned as being a cost effective approach for the necessary acquisition and deployment of instrumentation, training and education including active research partnerships, and for data and information exchange. Where warning systems are non-existent (such as a tsunami warning network for the Caribbean), existing systems should be expanded to include such regions. Mapping and charting should be an ongoing priority for improved storm surge prediction, as well as for economic and coastal area management requirements. A national vertical datum, using an updated mean sea level determination, is central to sustained economic development and should be given national priority in every small island developing state. Where practical, a geodetic quality Global Positioning System (GPS) receiver facility should be established for the direct measurement of vertical land motion near tide gauges.

Regional numerical ocean circulation models that explicitly calculate sub-tidal frequency sea level variations should be encouraged and developed. Such models have practical applications for many purposes that not only include the subject considered herein. Accurate forecasts of "ocean weather" should be given the same high priority for small islands as meteorological forecasts are given. As an intermediate step, calculations from existing basin-scale numerical ocean circulation nowcasts should be acquired and studied for potential application to coastal area management problems of immediate concern. PC-based socioeconomic models using numerical predictions of sea level change should become a common tool for quantitative management decisions [Engelen *et al.*, 1993].

Last, but certainly not least, is the need for education. Developed countries need to learn the many practical lessons that can be gleaned from the already existing traditional and cultural knowledge base in small island developing countries. Similarly, new technology needs to be introduced from the developed countries, and a balanced appreciation of the level of scientific uncertainty in global change predictions needs to spread from scientists to policy makers and environmental managers. A sincere ongoing mutual exchange of knowledge and a renewed appreciation of scholarship should be the most lasting legacy of any program, national, regional, or global.

Acknowledgments. We wish to express our appreciation to Dr. Ernest Estevez for permission to update and publish Figure 18. Several of the figures were redrawn by Mr. David Senn, and the photography was by Mr. Andrew Ramsay. Linda Pikula facilitated access to many of the publications quoted. Many other colleagues too numerous to mention freely gave of their time for this endeavor.

References

Andrews, J. T., Glaciation and sea level: A case study, in *Sea Surface Studies: A Global View*, edited by R. J. N. Devoy, pp. 95-126, Croom Helm, London, 1987.

Andrews, J. T., and G. H. Miller, Holocene sea level variations within Frobisher Bay, in *Quaternary Environments: Eastern Canadian Arctic, Baffin Bay, and Western Greenland*, edited by J. T. Andrews, pp. 585-607, Allen and Unwin, London, 1985.

Andrews, J. T., J. T. Buckley, and J. H. England, Late-glacial chronology and glacio-isostatic recovery, Home Bay, East Baffin Island, Canada, *Geol. Soc. Amer. Bull., 81*, 1123-1148, 1970.

Baltzer, F., Datation absolue de la transgression holocène sur la côte ouest de Nouvelle-Calédonie sur des échantillons de tourbes à palétuviers, interprétation néotectonique, *C.R. Acad. Sci. Paris, D, 271*, 2251-2254, 1970.

Baptista, A. M., G. R. Priest, and T. S. Murty, Field survey of the 1992 Nicaragua tsunami, *Mar. Geod., 16*, 169-203, 1993.

Belknap, D. F., B. G. Andersen, R. S. Anderson, W. A. Anderson, H. W. Borns, Jr., G. L. Jacobson, J. T. Kelley, R. C. Shipp, D. C. Smith, R. Stuckenrath, Jr., W. B. Thompson, and D. A. Tyler, Late Quaternary sea-level changes in Maine, in *Sea-Level Fluctuation and Coastal Evolution*, edited by D. Nummedal, O. H. Pilkey, and J. D. Howard, pp. 71-85, *Soc. Econ. Paleontol. Mineral.*, Spec. Publ. 41, Tulsa, 1987.

Bloom, A. L., Late Quaternary sea level changes on south Pacific coasts: A study in tectonic diversity, in *Earth Rheology, Isostasy, Eustasy*, edited by N. A. Mörner, pp. 505-516, Wiley, New York, 1980.

Bloom, A. L., and N. Yonekura, Coastal terraces generated by sea-level change and tectonic uplift, in *Models in Geomorphology*, edited by M. J. Woldenberg, pp. 139-154, Binghamton Symp. Geomorphol. Int. Ser., 14, Allen and Unwin, Boston, 1985.

Brandsma, M., D. Divoky, and L.-S. Hwang, Response of small islands to long waves, Tetra Tech. Inc., California, for U.S. Atomic Energy Commission, Nevada, 90 pp., 1974.

Cox, D. C., and J. F. Mink, The tsunami of May 23, 1960 in the Hawaiian Islands, *Bull. Seismol. Soc. Am., 53*, 1191-1209, 1963.

Digerfeldt, G., and M. Enell, Paleoecological studies of the past development of the Negril and Black River Morasses, Jamaica, Dep. Quat. Geol. Lund and Pet. Corp. Jam., Kingston, 145 pp., 1984.

Digerfeldt, G., and M. D. Hendry, An 8,000 year Holocene sea-level record from Jamaica: Implications for interpretation of Caribbean reef and coastal history, *Coral Reefs, 5*, 165-169, 1987.

Dixon, T. H., GPS measurement of relative motion of the Cocos and Caribbean Plates and strain accumulation across the Middle America Trench, *Geophys. Res. Lett., 20*(20), 2167-2170, 1993.

Dudley, W. C., and M. Lee, *Tsunami*, Univ. of Hawaii Press, 132 pp., 1988.

Emery, K. O., and D. G. Aubrey, *Sea Levels, Land Levels, and Tide Gauges*, 237 pp., Springer-Verlag, New York, 1991.

Engelen, G., R. White, and I. Uljee, Exploratory modeling of socio-economic impacts of climatic change, in *Climatic Change in the Intra-Americas Sea (Chapter 16)*, edited by G. A. Maul, pp. 350-368, Edward Arnold Publ., London, 1993.

EOS, Tsunami devastates Japanese coastal region, *EOS, Trans. Am Geophys. Union, 74*, 417-432, 1993.

Fairbanks, R. G., A 17,000-year glacio-eustatic sea level record: Influence of glacial melting rates on the Younger Dryas event and deep-ocean circulation, *Nature, 342*, 637-642, 1989.

Feller, C., M. Fournier, D. Imbert, C. Caratini, and L. Martin, Datations ^{14}C et palynologie d'un sédiment tourbeux continu (0-7 m) dans la mangrove de Guadeloupe (F.W.I.) - Résultats préliminaires, in Symp. Int. Evolution des Littoraux des Guyanes et de la Zone Caraïbe Méridionale Pendant le Quaternaire, pp. 69-79 (9-14 Nov. 1990), Résumés, ORSTOM, Cayenne, 1990.

Harris, R. A., Manual of tides - Part IVa: Outlines of tidal theory, Appendix No. 7, in *Report of the Superintendent of the Coast and Geodetic Survey from July 1, 1899 to June 30, 1900*, pp. 545-677 + 39 plates, Gov. Print. Office, Washington, D.C., 1901.

Hendry, M. D., A study of coastline evolution and sedimentology: The Palisadoes, Jamaica, Ph.D. thesis, Univ. of the West Indies, 232 pp., 1979.

Hendry, M. D., Sea level movements and shoreline changes, in *Climatic Change in the Intra-Americas Sea*, edited by G. Maul, 389 pp., Edward Arnold Publ., London, 1993.

Hine, A. C., S. W. Snyder, and A. C. Neumann, Coastal plain and inner shelf structure, stratigraphy, and geologic history: Bogue Banks area, North Carolina, Final Rep. to N.C. Sci. Technol. Comm. (quoted by Belknap *et al.*, 1987), 1979.

Ives, J. D., Deglaciation and land emergence in northeastern Foxe Basin, *Geogr. Bull., 21*, 54-65, 1964.

Jeftic, L., J. D. Milliman, and G. Sestini (Eds.), *Climatic Change and the Mediterranean*, 673 pp., Edward Arnold Publ., London, 1992.

Keys, J. G., The tsunami of May 22, 1960, in the Samoa and Cook Islands, *Bull. Seismol. Soc. Am., 53*, 1211-1227, 1963.

Kjerfve, B., Tides of the Caribbean Sea, *J. Geophys. Res., 86*(C5), 4243-4247, 1981.

Lander, J. F., and P. A. Lockridge, *United States Tsunamis 1690-1988*, 265 pp., National Geophysical Data Center, NOAA, Boulder, CO, 1989.

Lighty, R. G., I. G. Macintyre, and R. Stuckenrath, R., *Acropora palmata* reef framework: A reliable indicator of sea-level in the western Atlantic for the past 10,000 years, *Coral Reefs, 1*, 125-130, 1982.

Litz, B. H., and E. A. Shinn, Paleoshorelines, reefs and a rising sea: South Florida, USA, *J. Coast. Res., 7*, 203-230, 1991.

Løken, O. H., Postglacial emergence at the south end of Inugsuin Fjord, Baffin Island, N.W.T., *Geogr. Bull., 7*(3-4), 243-258, 1965.

Maul, G. A. (Ed.), *Climatic Change in the Intra-Americas Sea*, 389 pp., Edward Arnold Publ., London, 1993.

Maul, G. A., and D. M. Martin, Sea level rise at Key West, 1846-1992: America's longest instrument record? *Geophys. Res. Lett., 20*(18), 1955-1959, 1993.

Maul, G. A., D. V. Hansen, and N. J. Bravo, A note on sea level variability at Clipperton Island from GEOSAT and in-situ observations, in *Sea Level Changes: Determination and Effects*, pp. 145-154, Geophysical Monograph 69, IUGG Volume 11, 1992.

Miyata, T., Y. Maeda, E. Matsumoto, Y. Matsushima, R. P. Rodda, and A. Sugimura, Emerged notches and microatolls on Vanua Levu, Fiji, in *Sea-Level Changes and Tectonics in the Middle Pacific*, pp. 67-76, Rep. HIPAC Proj. in 1986 and 1987, Univ. Tokyo, 1988.

Montaggioni, L. F., Holocene reef growth history in mid-plate high volcanic islands, *Proc. 6th Int. Coral Reef Symp.*, Australia, vol. 3, pp. 455-460, 1988.

Murty, T. S., Seismic sea waves: Tsunamis, Department of Fisheries and the Environment, Ottawa, Canada, Bull. 198, 337 pp., 1977.

Nagano, O., F. Imamura, and N. Shuto, A numerical model for far-field tsunamis and its application to predict damages done to aquaculture, *Natural Hazards, 4*, 235-255, 1991.

Neumann, C. A., Quaternary sea-level data from Bermuda, *Quaternaria, 14*, 41-43, 1971.

NOS (National Ocean Service), *Geodetic Glossary*, 274 pp., U.S. Department of Commerce, NOAA/NOS, National Geodetic Survey, Rockville, MD 20852, 1986.

Nunn, P., Sea-level changes during the past 8,000 years in Fiji, Tonga and western Samoa: Implications for future coastline development, in Workshop on Coastal Processes in the South Pacific Island Nations, Lae, Papua New Guinea, 1-8 October 1987, *SOPAC Tech. Bull., 7*, 79-90, 1991.

Papazachos, B. C., and P. P. Dimitriu, Tsunamis in and near Greece and their relation to the earthquake focal mechanism, *Natural Hazards, 4*, 161-170, 1991.

Parke, M. E., O_1, P_1, N_2 models of the global ocean tide on an elastic earth plus surface potential and spherical harmonic decompositions for M_2, S_2, and K_1, *Mar. Geod., 6*(1), 35-81, 1983.

Pirazzoli, P. A., *World Atlas of Holocene Sea-Level Changes*, 300 pp., Elsevier Oceanography Series, 58, Elsevier, Amsterdam, 1991.

Pirazzoli, P. A., and L. F. Montaggioni, The 7,000 year sea-level curve in French Polynesia: Geodynamic implications for mid-plate volcanic islands, *Proc. 6th Int. Coral Reef Symp.*, Australia, vol. 3, pp. 467-472, 1988a.

Pirazzoli, P. A., and L. F. Montaggioni, Holocene sea-level changes in French Polynesia, *Palaeogeogr., Palaeoclim., Palaeoecol., 68*, 153-175, 1988b.

Pond, S., and G. L. Pickard, *Introductory Dynamical Oceanography*, 329 pp., Pergamon Press, Oxford, 1983.

Pugh, D. T., *Tides, Surges, and Mean Sea Level*, 472 pp., Wiley, Chichester, NY, Brisbane, Toronto, Singapore, 1987.

Redfield, A. C., Postglacial change in sea level in the western North Atlantic Ocean, *Science, 157*, 687-692, 1967.

Resource Analysis and Delft Hydraulics, How to account for impacts of climate change in integrated coastal zone management, World Coast Conference Organizing Committee, The Hague, The Netherlands, 40 pp., 1993.

Rodda, P., Visit to Western Samoa with the HIPAC Team, in *Sea-Level Changes and Tectonics in the Middle Pacific*, pp. 85-90, Rep. HIPAC Proj. in 1986 and 1987, Univ. Tokyo, 1988.

Simkin, T., and R. S. Fiske, *Krakatau 1883: The Volcanic Eruption and its Effects*, Smithsonian Institution Press, Washington D.C., 1983.

Smith, M. S., and J. B. Shepherd, Preliminary investigations of the tsunami hazard of Kick 'em Jenny submarine volcano, *Natural Hazards, 7*, 257-277, 1993.

Soloviev, S. L., Tsunamis in the Pacific in 1969-1974, in *Tsunami Evolution from Origin of Run to the Shore*, edited by S. L. Soloviev, pp. 75-87, Radio I svyaz', Moscow, 1982.

Sugimura, A., Y. Maeda, Y. Matsushima, P. Rodda, and E. Matsumoto, Lobau lowland, Viti Levu, Fiji, in *Sea-Level Changes and Tectonics in the Middle Pacific*, pp. 59-65, Rep. HIPAC Proj. in 1986 and 1987, Univ. Tokyo, 1988a.

Sugimura, A., Y. Maeda, Y. Matsushima, and P. Rodda, Further report on sea-level investigation in Western Samoa, in *Sea-Level Changes and Tectonics in the Middle Pacific*, pp. 77-84, Rep. HIPAC Proj. in 1986 and 1987, Univ. Tokyo, 1988b.

Suguio, K., L. Martin, and J. M. Flexor, Quaternary sea levels of the Brazilian coast: Recent progress, *Episodes, 11*, 203-208, 1988.

Woodworth, P. L., The Permanent Service for Mean Sea Level and the global sea level observing system, *J. Coast. Res., 7*, 699-710, 1991.

Yonekura, N., T. Ishii, Y. Saito, Y. Maeda, Y. Matsushima, E. Matsumoto, and H. Kayanne, Holocene fringing reefs and sea-level change in Mangaia Island, southern Cook Islands, *Palaeogeogr., Palaeoclim., Palaeoecol., 68*, 177-188, 1988.

Yonekura, N., Y. Matsushima, Y. Maeda, and H. Kayanne, Holocene sea-level changes in the southern Cook Islands, in *Sea-Level Changes and Tectonics in the Middle Pacific*, pp. 113-136, Rep. HIPAC Proj. in 1981, 1982 and 1983, Kobe Univ., 1984.

Zilkoski, D. B., J. H. Richards, and G. M. Young, Results of the general adjustment of the North American Vertical Datum of 1988, *Surv. Land Inform. Syst., 52*(3), 133-149, 1992.

7

Practical Aspects of Physical Oceanography for Small Island States

Judith Wolf

Abstract

The problems to be solved for a small island state in obtaining physical oceanographic data in a cost-effective manner are addressed. Examples are taken from experience in Trinidad and Tobago.

Introduction

For many small island states, the coast and surrounding seas are an important resource. Such problems as coastal defense from erosion and flooding, the development of beaches and coastal land for tourism, and the protection of the environment from pollution urgently need to be addressed. Typically, however, the movement of water nearshore and the susceptibility to extreme events such as storms are not well understood. There is likely to be little data available and, even where there is some, the skills necessary to interpret that data may be lacking locally. Yet, planners are increasingly demanding information on various environmental parameters including physical oceanographic data. Therefore, it seems appropriate to address the question of how to obtain data in the most cost-effective way, with limited local resources. Further benefits may be obtained by a better understanding of the physical oceanography of the surrounding seas, such as improved exploitation of fisheries, sea-bed mining or oil exploration, and better navigational information for pilots.

Small Islands: Marine Science and Sustainable Development
Coastal and Estuarine Studies, Volume 51, Pages 120–131
Copyright 1996 by the American Geophysical Union

These problems are discussed, with particular reference to the situation of Trinidad and Tobago, with illustrations from projects undertaken by the Institute of Marine Affairs (IMA) (see Figure 1). The solutions obtained are likely to be relevant in other small island states. The demands for oceanographic data for environmental impact analysis and coastal engineering are increasing. Typical projects include port development, oil spill contingency planning, protection of marine parks, beach protection, and renourishment. Transport of hazardous cargo requires more precise information for navigation. In general, the problems of small islands are likely to be similar, and generic solutions can be found. Most of the work is coastal: global-scale phenomena are best addressed by Large-Scale international experiments; however, these may have implications for small islands and collaboration is desirable. An attempt is made here to identify suitable techniques and list the basic equipment required.

Data Collection

The main parameters to be measured are sea level, wave height, and currents. These are inputs required for design of coastal defense structures and calculations of sediment transport and beach processes [U.S. Army Corps of Engineers, 1984], as well as being the fundamentals for any study of physical oceanography. Together with measurements of temperature and salinity, these data can lead to an understanding of the mechanisms in operation, which can enable better longer-term predictions to be made. This can pay for the extra initial investment in data collection. Although small island states cannot generally indulge in the luxury of "pure" research, some progress can usually be made with careful planning of applied research. It is important that wherever possible client-oriented research allows the data also to be used for further academic research.

A certain amount can be achieved by very low cost methods, often requiring visual observation. The disadvantages include poor space and time coverage, as well as being highly labor intensive. Modern solid-state digital recording instruments, on the other hand, are expensive initially, although they are becoming relatively cheaper with improvements in electronic technology. However, they can collect several orders of magnitude more data than visual methods. Therefore, if the economic barrier can be crossed to achieve this type of data collection, there are many benefits. Observing extreme (and hence, rare) events requires long time series of automatically recorded data. Having this type of equipment also makes it possible for a small group of oceanographers to contribute to international collaborative exercises.

The problems for small islands will be primarily coastal and, in general, ocean-going research vessels will only be available during collaborative exercises. Most work must be done from small boats.

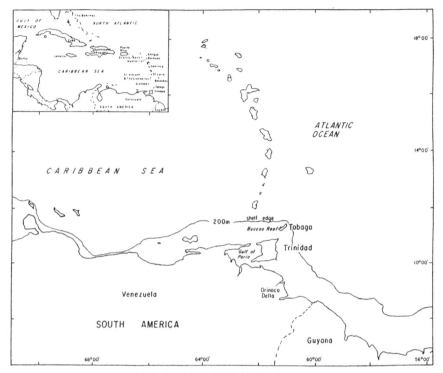

Fig. 1. Location of Trinidad and Tobago.

Tides and Sea Level

The most useful data are hourly observations. Basic data can be collected using a tide staff. Usually the longest this would be monitored on an hourly basis would be 25 hours for initial station establishment or datum transfer. With one month of data a tidal analysis can be carried out and residuals may be used to investigate storm surges, for example. Longer term records are necessary in order to study seasonal trends, and mean sea level requires many years of records from an established tide gauge with a good bench mark datum. Since low-lying island states could be at risk from sea-level rise, it would seem important for the establishment of at least one good tide gauge at a site not susceptible to land movements [Pugh, 1987]. Other records can usefully be added using a temporary tide gauge installed for at least one month.

Waves

Most historical wave data available at island sites will be statistics from visual observations, i.e., ship report data. At an oceanic island, this data, usually gathered over several degrees of latitude and longitude, would give a good representation of the deep water waves arriving at the island. This may be less useful at an island surrounded by a wide continental shelf where it would be advisable to apply refraction models to bring the waves nearer to shore. Another problem relates to the coverage of observations. An island off the main shipping routes would not generally get much useful data. Statistical data is also available from satellites which now give good global coverage. Data can be usefully obtained from wave models such as that running at the European Center for Medium Range Weather Forecasting (ECMWF) in real time [Jansen and Komen, 1987], which may fill in gaps not covered by observations such as in the Southern Ocean, and may be used after satisfactory validation. However, such data is not usually available, especially in developing countries, in real time.

Wave data is a rather expensive commodity, yet the impact of waves on exposed coasts can be severe. Directional wave measurement devices are particularly expensive. Perhaps the most inexpensive wave measurement device is a bottom pressure gauge deployed in shallow water (ideally less than 10 m deep but outside the surf zone), suitable for nearshore wave measurements where wave direction is not required. Visual observations of waves can be made easily and quite reliably by a trained observer and, given a long enough time series at reasonable intervals (ideally daily) representative of the climatic variations, may be very useful. This has now been carried out for several years in Grenada and Dominica [Cambers, 1993]. Significant wave height, period, and direction can be estimated for local wind-sea and swell. It is most useful to record wind speed and direction simultaneously.

Currents

The traditional moored current meter is still the best way to obtain substantial amounts of current data. It has limitations in spatial coverage, as it is not possible to deploy enough instruments simultaneously to properly study the spatial variability of currents. Ideally, more than one current meter should be deployed on each mooring if there is a likelihood of a change in current with depth, especially if stratification of temperature or salinity is observed. The particular problems of current meter deployment will be different depending on whether the island lies on a continental shelf or not. Deep ocean moorings are very expensive and usually only justified by large scale international experiments. Local deployments nearshore, e.g., in harbors or bays, can be carried out from small boats if necessary with no lifting equipment and the minimum of cost for mooring gear.

An inexpensive way of acquiring current data can be the use of drogues. These will give a quick overview of transport paths, particularly useful for dispersion studies. Previously, drogue tracking was conducted by using shore theodolite stations or sextant observations of landmarks. This was very labor intensive and limited in offshore extent. Recently, the IMA has acquired two GPS (global positioning system) survey grade receivers. In stand-alone mode these can achieve an accuracy of 30-100 m in absolute position. This is reduced in relative position terms and, if two GPS receivers are used with postprocessing differential correction, the accuracy improves to about 2 m. A single receiver is now very cheap. Differential correction is an order of magnitude more expensive, being about the same cost as a recording current meter. GPS has other applications in the surveying of coastlines and offshore position fixing [Santamaria *et al.*, 1990].

Dye-tracking is most useful for dispersion studies which are often required for planning effluent disposal.

Other Measurements

Other parameters which are required include temperature and salinity. These may be measured with direct-reading probes, from water bottle samples or with a CTD (conductivity-temperature-depth) instrument. The cost of a CTD is rather high; direct-reading probes are limited in depth and of variable accuracy. Acquisition of a CTD marks the commitment to more strategic oceanographic research. More sophisticated current measuring devices are also more appropriate for deep ocean work such as a ship-mounted ADCP (acoustic Doppler current profiler), *e.g.*, Smith and Morrison [1989].

Remote Sensing

Due to the ubiquity of PC hardware, the technology of remote sensing becomes available to the developing nations. Real time data may not yet be available, since this requires access to a satellite receiving station. A substantial initial investment in training in the use and interpretation of satellite images is required, plus a significant amount of programming effort or expensive software. This aspect is likely to be underestimated relative to the comparatively cheap hardware required for image processing. In fact, there is a long learning curve, and experience is required to maximize the potential of this very powerful tool. In principal, virtually every oceanographic parameter can be measured from space. An application in the Caribbean is given by Müller-Karger *et al.* [1989].

Remotely sensed data requires ground-truthing which may be out of reach for the small-scale user. Some field data is essential, however, particularly where the data is calibrated using algorithms developed for different sea and climate conditions.

Numerical Models

Simple models are now commonplace and very cost effective. The simplest model is the depth-averaged, finite difference model, suitable for modeling tides and the essentials of wind-driven flow. The variation of current with depth can be included as a post-processing stage. Models are very useful for studying extreme events [Flather *et al.*, 1991; Cooper and Thompson, 1989; Kjerfve *et al.*, 1986].

Case Studies at the IMA

The IMA has faced a few problems in instrument deployments. Marine fouling is a serious problem for long-term moorings, particularly in the Gulf of Paria, due to working in shallow tropical waters with high levels of nutrients. Most instruments have been developed and extensively deployed in temperate climates. The manufacturers do not have solutions to the fouling problem. The IMA does not have acoustic release systems as yet so deep deployments are not possible. However, shallow water moorings may be deployed off small boats, *e.g.*, 8 m open fishing boats ("pirogues") in depths accessible to divers. Loss of equipment probably due to interference by fishermen is a problem. Many small islands will have a large population of artisanal fishermen to whom anything in the sea is fair game. There is an unresolved debate as to whether it is advisable to use a surface marker or not and instrument losses are inevitable. Insurance is probably a necessity since losing a single piece of equipment may be more devastating than for a large institution with many items. The largest boat available is a 12.5 m launch which is capable of 25-hour stations but not extended periods at sea.

Gulf of Paria Model

A two-dimensional numerical model of the Gulf of Paria (5 km grid size) has been set up using a general purpose model program based on Wolf [1991]. This has been validated using existing tide gauge data and tidal streams. Further work with the model will allow the study of wind-driven circulation and the mean flow due to the Guiana Current. There is a need for more data, especially on currents, but this is gradually being acquired through other projects. It is hoped that there will be many applications for the model, e.g., in navigation and oil spill trajectory modeling. One product will be a current atlas for the Gulf of Paria using vectors generated by the model such as in Figure 2. The model development so far has all been carried out on a 386-PC with no other expenses, which demonstrates the cost benefits of modeling, as well as the ability of models to maximize the use of existing data.

Buccoo Reef Study

A year-long physical oceanographic study of the Buccoo Reef marine park area has just been completed [Wolf *et al.*, 1993]. This included repeated field measurements to study the seasonal variations. A model of the Buccoo Reef area has been constructed with a

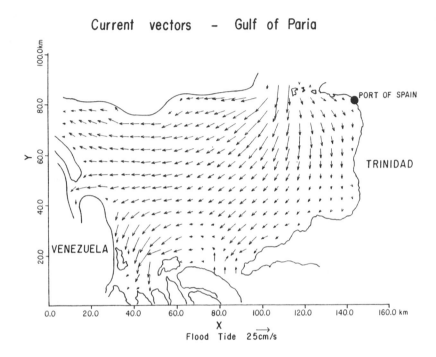

Fig. 2. Gulf of Paria model current vector plot.

100 m grid. The study area is shown in Figure 3. Using the model and measurements hand-in-hand, an assessment of the tidal and residual circulation patterns has been completed and flushing times estimated. Moored current meter deployments have allowed the acquisition of a wealth of data which will generate further research. Some problems that were faced included the cost of traveling and the logistical problems of transporting equipment from Trinidad to Tobago, the lack of availability of spares and even batteries.

The benefits of a new digital tide gauge were immediate. After months of struggling with outdated mechanical chart-recording instruments which had a tendency to malfunction, and the difficulty of changing charts regularly, no satisfactory sea-level data had been obtained. Three days after receipt of a SUTRON data logger and an IMO pressure sensor the tide gauge was deployed in Buccoo Reef. One month later the data was recovered onto a laptop PC and the following day a tidal analysis had been completed. At the IMA some effort has been put into digitization of analogue chart records from older type tide gauges. The effort requires many man-hours compared to the ease of use of a modern digital tide gauge which can also be set at a higher frequency than hourly, allowing detection of short period seiches and the elimination of aliasing in the hourly data.

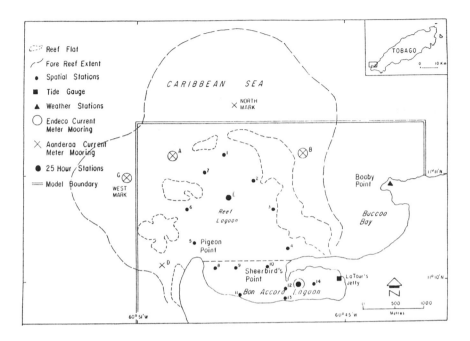

Fig. 3. Map of the Buccoo Reef physical oceanography project.

An attempt was made to use local observers to gather regular visual measurements of waves. A training exercise was carried out with the Reef Patrol Officers of the Tobago House of Assembly and one month of good data was obtained. Due to various logistical problems, unfortunately, they could not carry on. However, this is still something to be encouraged for the future.

East Coast of Trinidad Oil Spill Contingency Planning

A multidisciplinary pilot project is underway to identify the uses of remote sensing data to the oil spill contingency plan for the east coast of Trinidad. This includes ecologists and oceanographers identifying sensitive shorelines, mapping land-use, and studying the oceanographic conditions off the east coast. This study is highlighting the amount of effort required to use satellite imagery effectively.

Equipment Pool and Staffing

A basic minimum of instrumentation is recommended for efficiency. Ideally, this should be shared by several small islands by establishing a marine research center which could maintain and deploy the equipment as necessary. A core group of three oceanographers, a data analyst and two instrument technicians plus a few trainees is identified by Swamy [1992]. It is essential to train and retain trained staff. This is best done by providing sufficient incentive and facilities for the scientists to carry on some personal research, as well as carrying out contract work. It must be emphasized that training alone is not sufficient.

A core of essential equipment is listed in Table 1. It will be seen that there is an emphasis on current meters, the sine qua non of physical oceanography. This equipment requires specialist servicing and deployment, and the preliminary routine data analysis can also be done by instrument technicians. Extra items such as direct-reading instruments, water bottles, *etc.*, are not listed, but may be assumed.

Alleng and Joseph [1993] discuss the benefits of global and regional integration and discuss some of the institutions and projects involved in the Caribbean.

TABLE 1. Minimum Equipment Requirements

Instrument Description and Number	Available at IMA
1 profiling current meter	Yes
6 recording current meters	Yes
1 digital tide gauge	Yes
1 automatic weather station	Yes, but not digital
1 shallow water CTD	No (on order)
1 portable echosounder	Yes
1 directional wave recorder	Yes
1 GPS (preferably differential)	Yes
1 laptop PC	Yes
1 portable fluorometer	No (on order)

Data Analysis

Most data analysis can now be done on a PC, which brings it within the reach of everyone. The skills needed to carry out basic analysis are not too demanding. Some general purpose software is available, although most oceanographic institutes still produce software to their own requirements. Much of this is available, although perhaps at a price. Solid state instruments usually come with a software package for preliminary data analysis, calibration and plotting.

The most useful programs are a spreadsheet package, statistical packages, plotting libraries, FORTRAN, BASIC, and C compilers. A minimum level of programming skill is a must for modern oceanographers, since the most common problem with the purchase of a "black box" program is that it will not be properly used or not used at all.

Caribbean and Global Perspectives

The problems of the eastern Caribbean states, in the Lesser Antilles, are somewhat different to Trinidad and Tobago. Trinidad is located on a relatively wide continental shelf, close to the mainland, and has an adjacent semi-enclosed sea: the Gulf of Paria. Tobago is very near to the continental shelf edge on its northern side and more exposed to the Atlantic conditions. Many islands, *e.g.*, St. Lucia, are steep-sided, of volcanic origin, with narrow or non-existent continental shelves. The applications of physical oceanography are mainly in coastal management and port and harbor development. A pressing problem is the erosion of beaches which are critical to tourism [Deane, 1987]. Most development is on or very close to the coast and could be susceptible to sea-level rise. The projected rise in sea level could impact many small island states. Hoozemans *et al.* [1993] list the areas of the world most at risk. Of the top 10 most vulnerable coastal areas the Indian Ocean small islands come fifth, the Caribbean islands are seventh, and the Pacific Ocean small islands are ninth.

Some Large-Scale data collection exercises have tended to concentrate on the Caribbean and the flow through the Lesser Antilles passages or the tropical and equatorial Atlantic. Some examples of Large-Scale exercises have been CICAR (Cooperative Investigation of the Caribbean and Adjacent Regions), CORE (Caribbean Oceanographic Resources Exploration), STACS (Subtropical Atlantic Climate Studies), and FOCAL/SEQUAL (Francais Ocean Climat en Atlantique Equatorial/Seasonal Response of the Equatorial Atlantic). Little work has been done indigenously and generally these exercises do not address the most pressing problems for small islands. However, there are issues for which an understanding of Large-Scale current patterns can be important, *e.g.*, petroleum pollution is found on the windward coasts of many Caribbean islands [Atwood *et al.*, 1987].

The urgent problems which have a direct economic impact tend to be in nearshore circulation and beach processes. These concerns are, no doubt, common to many tropical islands.

Discussion

The solution for small island states seems to be the establishment of regional centers for marine science rather than each attempting to acquire expensive technology. This will allow local knowledge to inform the scientific work and skill transfer will benefit local inhabitants. It is considered more cost-effective to purchase a pool of relatively sophisticated modern equipment rather than rely on primitive labor-intensive techniques of visual observation, although these can be useful, particularly for coastal work. Generally, the equipment required for deep ocean work will be beyond the reach of most small island states; however, much useful work in coastal oceanography can be tackled locally.

It must be emphasized that there is a general lack of physical oceanographic data. The developed countries are now showing concern about environmental problems which developing countries, including small islands, could yet avoid if the lesson is learned. However, without data on which to make informed decisions irrevocable damage might be done by poorly planned development just as it has been in the developed countries. The acquisition of good data rapidly is more important than finding a cheap solution which may not produce results for a long time. A data base of sufficiently long time series data can be reused many times, repaying the initial investment, whereas small amounts of site-specific data may never be used again.

References

Alleng, G. P., and B. R. Joseph, Implications of technical and legislative harmonization for integrated coastal zone management in the wider Caribbean region, *Coastal Zone '93, Proceedings, 8th Symposium on Coastal and Ocean Management*, 1092-1107, 1993.

Atwood, D. K., F. J. Burton, J. E. Corredor, G. R. Harvey, A. J. Mata-Jimenez, A. Vasquez-Botello, and B. A. Wade, Petroleum pollution in the Caribbean, *Oceanus, 30*(4), 25-32, 1987.

Cambers, G., Beach stability and coastal erosion in the eastern Caribbean islands--a regional program, *Coastal Zone '93, Proceedings, 8th Symposium on Coastal and Ocean Management*, 870-882, 1993.

Cooper, C., and J. D. Thompson, Hurricane-generated currents on the outer continental shelf, 1. Model formulation and verification, *J. Geophys. Res., 94*(C9), 12,513-12,539, 1989.

Deane, C., Coastal erosion and accretion in the Caribbean Lesser Antilles, *Coastal Zone Management of the Caribbean Region: A Status Report*, CSC Technical Publication No. 227, 138-147, 1987.

Flather, R. A., R. Proctor, and J. Wolf, Oceanographic forecast models, in *Computer Modeling in the Environmental Sciences*, edited by D. G. Farmer and M. J. Rycroft, pp. 15-30, Clarendon Press, Oxford, 1991.

Hoozemans, F. M. J., M. J. Stive, and L. Bijlsma, A global vulnerability assessment: Vulnerability of coastal areas to sea level rise, *Coastal Zone '93, Proceedings, 8th Symposium on Coastal and Ocean Management*, 390-404, 1993.

Jansen, R. A. E. M., and G. J. Komen (Eds.), *WAM Newsletter, 2*, 4 pp., KNMI de Bilt, Netherlands, 1987.

Kjerfve, B., K. E. Magill, J. W. Porter, and J. D. Woodley, Hindcasting of hurricane characteristics and observed storm damage on a fringing reef, Jamaica, West Indies, *J. Mar. Res., 44*(1), 119-148, 1986.

Müller-Karger, F. E., C. R. McClain, T. R. Fisher, W. E. Esaias, and R. Varela, Pigment distribution in the Caribbean Sea: Observations from space, *Progr. Oceanogr., 23*, 23-64, 1989.

Pugh, D. T., *Tides, Surges and Mean Sea Level*, 472 pp., John Wiley and Sons, 1987.

Santamaria, R., S. Troisi, and L. Turturici, Marine applications of GPS, *Mar. Geod., 14*(1), 13-20, 1990.

Smith, O. P., and J. M. Morrison, Shipboard acoustic Doppler current profiling in the eastern Caribbean Sea, 1985-1986, *J. Geophys. Res., 94*(C7), 9713-9719, 1989.

Swamy, G. N., A physical oceanographic research plan for the Caribbean, with special reference to Trinidad and Tobago, Institute of Marine Affairs Internal Report, 1992.

U.S. Army Corps of Engineers, *Shore Protection Manual*, vols. 1 and 2, 1984.

Wolf, J., A unified framework for water quality modeling in shallow seas, Proudman Oceanographic Laboratory Report No. 19, 43 pp., 1991.

Wolf, J. A., F. S. Teelucksingh, D. Neale, N. Gopaul, F. Charles, and B. Greenidge, Final report: Physical oceanographic and bathymetric surveys of Buccoo Reef, Tobago, Caribbean Conservation Association, Institute of Marine Affairs Technical Advisory Services Report, 144 pp., 1993.

8

Design Values of Extreme Winds in Small Island States

Calvin R. Gray

Abstract

Records of surface wind speed data recorded at Norman Manley and Sangster International Airports in Jamaica since 1951 are analyzed using different methodologies with a view towards selecting the most appropriate for extreme winds associated with tropical cyclones. Annual maximum 1-min wind speed from 1885 to 1988, associated with tropical cyclones in the region, are then analyzed to establish design values of extreme winds in small island states.

The mixed Fretchet type extreme value distribution [after Thom, 1968] is used for the annual series of maximum monthly average wind speeds. The Fisher-Tippet Type II, which is an exponential transformation of the Type I distribution, is applied to the annual series of maximum 1-min winds associated with tropical cyclones; the Lieblein Fitting Technique is also applied to negate the effect of the moments giving poor estimates. The estimates obtained from the analysis of maximum 1-min winds indicate that, on an average, at least once every 50 years a 1-min sustained wind speed of 168 mph is likely to be equaled or exceeded anywhere within the path of tropical cyclones. These statistics and their interpretation are of utmost importance since investigation of numerous cases of building damage due to wind action has shown that, while some failures were undoubtedly due to defects of workmanship, many cases of damage resulted from an under-estimate of wind forces in the design of buildings and other structures.

Small Islands: Marine Science and Sustainable Development
Coastal and Estuarine Studies, Volume 51, Pages 132–145
Copyright 1996 by the American Geophysical Union

Introduction

Air in motion is what we call wind and is the result of conversion of potential energy of the atmosphere into kinetic energy, mainly through the work of pressure forces. The ultimate energy source is, of course, the Sun.

The local wind, at a particular place and height, shows considerable variation in strength and direction. The traveling large scale weather systems or local mesoscale circulations create a framework within which local factors determine the actual wind conditions. Some of the most important factors influencing the air flow are:

1. Turbulence, *i.e.*, local irregular and random motions;
2. Properties of the underlying surface such as varying degrees of roughness: smooth seas or rough urban areas;
3. Day or night: heating or cooling of the underlying surface;
4. Topographic features and local small-scale obstacles, buildings, trees, ridges and valleys;
5. Outside disturbances, *e.g.*, down-drafts from thunderstorms.

The prediction of wind-induced effects for design purposes begins with the definition of the local characteristics of the wind climate.

Small island states in the tropics experience trade winds and by day this flow adds vectorially with the local thermal breezes, the resultant being the sea breeze and its night-time counterpart known as the land breeze. In addition, migratory frontal systems also penetrate into the tropics, producing wind speeds usually from a direction opposite to the trade flow.

Small island states in the tropics are bombarded also by some of the most violent weather disturbances in the world, *i.e.*, tropical cyclones. It is these systems that produce extreme wind speeds of the highest magnitude. The problem of extreme winds and their effects on structures is obviously one of considerable magnitude. It is a problem abetted by a lack of site specific, reliable, extended data bases that are appropriate for the well-known statistical analyses for building design purposes.

In this paper, the appropriateness of existing data is examined, and the data is synthesized to generate design values and their recurrence interval for small island states. In synthesizing the data to obtain the suggested design values, the underlying assumption is that the choice and format of the input data for the statistical model determines the accuracy and reliability of the output statistics. It is these output statistics which guide engineers and designers of buildings and/or low-cost housing in small island states to withstand the extreme winds associated with the mature form of a tropical cyclone (see Table 1).

TABLE 1. A Tropical Cyclone Intensity Scale*

Intensity Level	Wind Speed (mph)	Description
1	74-95	No real damage to building structure. Damage primarily to unanchored homes.
2	96-110	Some roofing material, door and window damage to buildings. Considerable damage to exposed unanchored homes.
3	111-130	Some structural damage to small residences and utility buildings with a minor amount of curtain-wall failures. Unanchored homes are destroyed.
4	131-155	More extensive curtain-wall failures with some complete roof structure failure on small residences.
5	155+	Major tropical cyclone. Complete roof failure on many residences and industrial buildings. Some complete building failures with small utility buildings blown over or away.

*A scale from one to five based on the tropical cyclone's present intensity which gives an estimate of the potential damage to building structures. This scale is adopted from the Saffir/Simpson Hurricane Scale (SSH), and winds refer to the maximum sustained 1-min wind speed.

Data

Wind direction and wind speed data recorded at the Norman Manley and Sangster International Airports are acquired using Munro wind equipment; the output from the sensors at 10 m are sent to a strip-chart recorder (anemograph). The response time of the instrument is between 2 and 5 sec; hence, gust speeds are taken as a 3-sec gust. From the anemograph chart, the data is reduced to allow the tabulation of mean wind direction, mean wind speed and the maximum gust for each hour of the day. Thereafter, the monthly average wind speed is computed.

It is to be noted that the highest monthly averages, both at Norman Manley and Sangster International Airports, were not associated with tropical cyclones. Similarly, not all annual extreme gusts were associated with tropical cyclones. As a matter of fact, at Manley, only in 1951, 1980 and 1988 were the extreme gusts associated with tropical cyclones, whilst at Sangster, it was only in 1988. It must be borne in mind, therefore, that for any given monthly average speed and for any given value of extreme gust, depending on the time of the year, there are at least three separate and distinct wind populations: speeds associated with tropical cyclones; speeds associated with migratory frontal systems; and speeds associated with the normal trade flow.

In order to ensure homogeneity of the data series, it is imperative that the data sample be from a "single" population; otherwise, adjustments must be made to ensure the statistical estimates will be valid estimates of the population parameters. Indeed, with respect to the data used in Figures 1-4, they are not from a single population. With respect to the instruments and instrument exposures at the Norman Manley and Sangster International

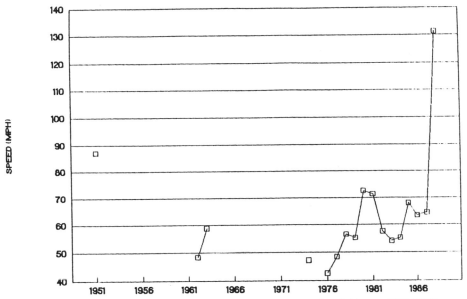

Fig. 1. Annual extreme gust (mph), Norman Manley International Airport, 1951-1988.

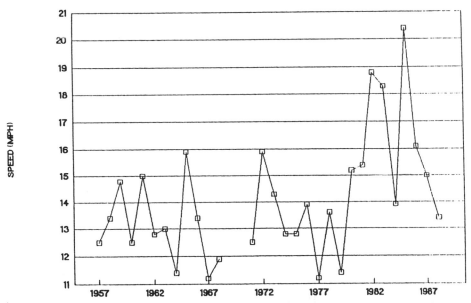

Fig 2.   Maximum monthly average speed (mph), Norman Manley International Airport, 1957-1988.

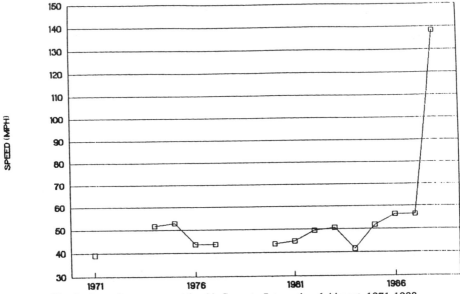

Fig. 3. Annual extreme gust (mph), Sangster International Airport, 1971-1988.

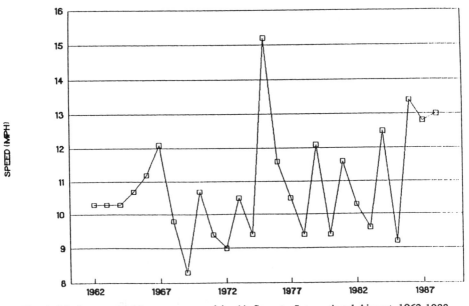

Fig. 4. Maximum monthly average speed (mph), Sangster International Airport, 1962-1988.

Airports, there has been no changes during the data period, and the accuracy of the instruments has remained well within the required limits.

The annual extreme 1-min wind speed associated with tropical cyclones in the North Atlantic Ocean, Gulf of Mexico, and the Caribbean Sea is used to generate an annual series. This data series may contain estimated values, especially prior to the advent of organized aircraft surveillance in 1944.   Since this era, the combination of reconnaissance aircraft, weather satellite, radar, and more accurate surface wind recording equipment have ensured continuous, reliable wind speed values. This data base is quite representative for small island states in the entire tropical region around the globe. This data was reduced from a print-out of the hurricane data (HURDAT) tape obtained from the National Hurricane Center in Miami, Florida. It is to be noted that the extreme 1-min wind speed refers to the maximum sustained 1-min wind at a height of 10 m or 32 ft (surface wind). Figures 1-8 are plots of the data series used in this paper.

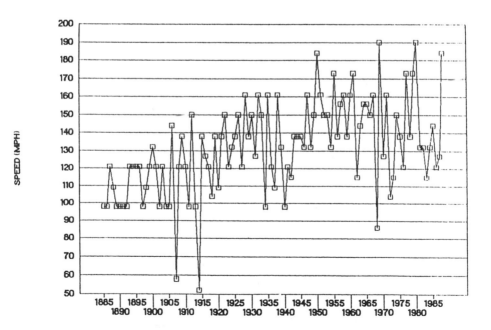

Fig. 5. Annual extreme 1-min speed (mph), Atlantic Ocean, Gulf of Mexico, and Caribbean Sea, 1885-1988.

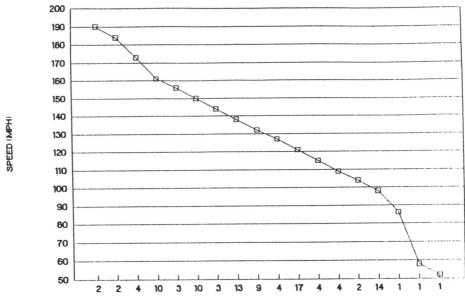

Fig. 6. Variability of annual extreme 1-min speed (mph), Atlantic Ocean, Gulf of Mexico, and Caribbean Sea, 1885-1988.

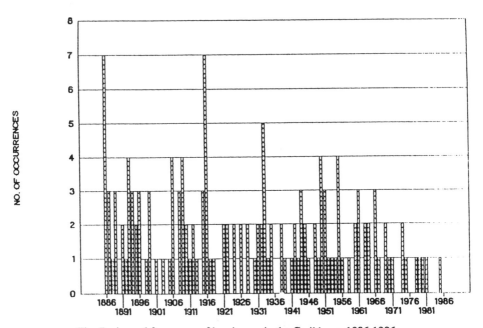

Fig. 7. Annual frequency of hurricanes in the Caribbean, 1886-1986.

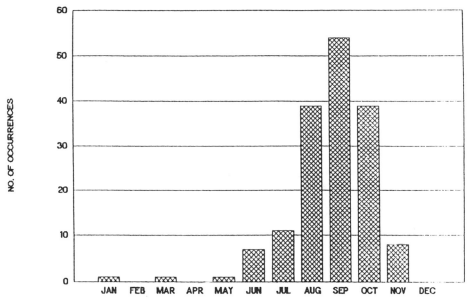

Fig. 8. Monthly frequency of hurricanes in the Caribbean, 1886-1986.

Extreme Value Distribution

Fisher-Tippet Type I

A commonly used distribution of extreme values (annual series) is the Fisher-Tippet Type I distribution, which has been widely applied by Gumbel [1958] and often bears his name. In this method,

$$VTr = V + K*Sv$$

and

$$K = (YTr - Yn)/Sn,$$

where VTr denotes the magnitude of the event reached or exceeded on an average once in Tr years; V is the mean value; Sv is the standard deviation; K is a frequency factor and is a function of the sample size, n, and the return period, Tr, of the event; Yn is the reduced mean; Sn the reduced standard deviation; and YTr, the reduced variate, is related to the return period by

$$YTr = -(0.83405 + 2.30259 * \log (Tr/(Tr - 1))).$$

Fretchet Extreme Value Distribution

The quasi-universal extreme wind distribution developed by Thom [1968] is also utilized in this paper. This mixed extreme value distribution function is given by (Table 2)

TABLE 2. Definition of Parameters used in Estimating Extreme Wind Speeds

Parameter	Definition
Va	Highest maximum monthly average speed
Vm	Fastest mile speed = Va × 1.12
G	1.09 * Vm + 8.0
Vg	Recorded, mean annual extreme gust
Vt	Recorded, annual extreme tropical cyclone 1-min wind
Std	Standard deviation
B	Scale parameter, function of Vm
f	Mean number of tropical storms and hurricanes per year
P(t)	Probability of tropical cyclone
P(e)	Probability of an extra-tropical system

$$G(v) = P(e)*F(v) + P(t)*F(v),$$

$$F(v) = \exp{(-(v/B)^{-9})},$$

$$F(v) = \exp{(-(v/B)^{-4.5})},$$

$$P(e) = 1 - P(t),$$

where P(e) and P(t) are the probability of an annual extreme wind being produced by an extra-tropical and a tropical storm, respectively. The relationships between B and Vm, the highest average monthly mean wind speed, and between P(t) and f, the mean number of tropical storms per year, established by Thom [1968], are:

$$B = (347.5*Vm + 364.5)^{0.5}$$

for speeds of up to 18 mph, and

$$P(t) = 1/(1 + 99 \exp{(-3.0*f)}).$$

Fisher-Tippet Type II with the Lieblein Fitting Technique

The Type II distribution has been found most useful in fitting extreme winds. Since the Type I distribution on the logarithms is a Type II distribution, in this paper the Type I distribution is fitted to the logarithms of the annual extreme winds, in this instance, the annual extreme 1-min wind speed. The Type I distribution function is given by:

$$F(v) = \exp\left(-(v/x^2)^{-y}\right)$$

where x and y are the scale and shape parameters. As with most other skewed distributions, the moments give poor estimates of the parameters. Lieblein [1954] has provided a method of fitting the Type I distribution, which gives much better estimates. The Lieblein Fitting Technique involves carefully maintaining the original time order of the homogeneous data series and dividing it into suitable sub-groups for the computations. Order statistics weights developed by Lieblein [1954] are utilized in the computations. For minimum values or lower extremes, the magnitude order arrangement is from high to low values. For further details of the methodology, those interested are referred to H. C. S. Thom, *Some Methods of Climatological Analysis* [1968].

Analysis

In the Gumbel [1958] method, the mean, V, and standard deviation, Sv, of the annual series of extreme gusts are computed and used as the input parameters. Only 17 and 14 years of data are available from Norman Manley and Sangster International Airports, respectively, and in both cases the data period is not continuous. Further "outliers" are evident in both samples; hence, the means and standard deviations won't be realistic. Extrapolation of this data to obtain estimates for return periods of 50 or 100 years is also unrealistic.

In the Thom [1968] method, the highest maximum monthly average speed, Va, for the data period is used. This value is multiplied by a factor of 1.12 to obtain the fastest mile speed, Vm. This factor is an imported one and, as such, should be treated as a possible error source, until it is verified for our wind regime. The fastest mile speed is used to develop the scale parameter, B. This scale parameter is applicable for speeds of up to 18 mph; from the Manley data the value of Vm is 22.8. For this method, 30 and 27 years of data from Norman Manley and Sangster International Airports, respectively, were used; however, the data periods were not continuous.

For the method with the Lieblein Fitting Technique (FTLFT), 104 years of continuous data were analyzed. The input parameter is the actual annual extreme 1-min wind associated with a tropical cyclone. The data series represents a sample from a single population. No outliers are evident in the data series, and its length is adequate to facilitate reliable estimates with return periods of up to 200 years.

Summary and Discussion

The results are given in Tables 3, 4, and 5. All estimates of wind speeds are minimum values or lower extremes. That is, these values are likely to be equaled or exceeded, on

TABLE 3. Fastest Mile Estimates (mph) for Selected Return Periods (years)

Site	Data Period	Years of Data	Analyst	Method	Return Period (years) 5	10	25	50	100	200
Manley	1950-1962	12	Shellard	Thom	--	71	83	93	105	117
Manley	1959-1974	14	Evans	Thom	--	83	96	107	119	132
Manley	1957-1987	29	Gray	Thom	85	95	111	126	143	164
Manley	1957-1988	30	Gray	Thom	85	95	111	126	143	164
Sangster	1962-1987	26	Gray	Thom	72	80	93	105	119	138
Sangster	1962-1988	27	Gray	Thom	72	80	93	105	119	138

TABLE 4. Estimates of 3-sec Gust Speed (mph) for Selected Return Periods (years)

Site	Data Period	Years of Data	Analyst	Method	Return Period (years) 5	10	25	50	100	200
Manley	1950-1962	12	Shellard	Thom	70	85	98	110	123	136
Manley	1959-1974	14	Evans	Thom	83	98	113	125	138	152
Manley	1957-1987	29	Gray	Thom	101	112	129	145	164	187
Manley	1957-1988	30	Gray	Thom	101	112	129	145	164	187
Manley	1951-1987	16	Gray	Gumbel	70	78	88	96	103	111
Manley	1951-1988	17	Gray	Gumbel	83	97	115	129	142	156
Sangster	1962-1987	26	Gray	Thom	86	95	109	122	138	158
Sangster	1962-1988	27	Gray	Thom	86	95	109	122	138	158
Sangster	1971-1987	13	Gray	Gumbel	54	58	63	67	71	75
Sangster	1971-1988	14	Gray	Gumbel	78	96	118	135	151	167

TABLE 5. Estimates of 1-min Wind Speed (mph) for Selected Return Periods (years)

Site	Data Period	Years of Data	Analyst	Method	Return Period (years) 5	10	25	50	100	200
Tropics	1885-1988	104	Gray	FTLFT	149	156	163	168	171	175

an average, at least once within the specified recurrence interval. The values in Table 3 are for the fastest mile. An imported factor, a 3-sec gust factor, G, given by

$$G = 1.09 * Vm + 8.0,$$

where Vm is the fastest mile speed, is applied to the values in Table 3 to obtain the 3-sec estimates in Table 4 for Thom's [1968] method. This factor should be treated as a possible error source until it can be verified.

For the Gumbel [1958] method, estimates in Table 4 are taken as 3-sec gust estimates since the instrument yielding the input data has a mean response time of 3 sec. The estimates computed by Shellard [1972] and Evans [1976] using Thom's [1968] method are seen as a first approximation since the data lengths are very small, and in the case of Evans, the data period is discontinuous. For the 50-year return period, Shellard [1972] has an estimate of 115 mph and Evans [1976], an estimate of 125 mph.

In Gumbel's [1958] method, the 3-sec gust estimates prior to Gilbert are meaningless. The very small sample sizes and discontinuous data periods are, no doubt, contributing factors. When the extreme winds associated with Gilbert are included in the analyses, the estimates increased and their magnitude were of a higher order than those indicated by either Shellard [1972]or Evans [1976] who used Thom's method. The main problem here, in my opinion, is the lack of sufficient reliable data from the appropriate population.

From Table 5, it is seen that this method yields an estimated 1-min wind speed of 168 mph at a recurrence interval of 50 years. This is saying that for any small island state in the tropics , a 1-min wind speed of 168 mph is likely to be equaled or exceeded, on an average, at least once every 50 years.

The data length of 104 years, although small, is significantly greater than what is currently available for any of the other distributions utilizing wind speeds from a mixed population. For all practical purposes, the mean of the sample (131.4) is equal to the median (132), and the standard deviation is only 26.0 bearing in mind that the data ranges from 52 to 190 mph. The variability of the annual extreme 1-min speeds and the appropriateness of the data for the wind climate are certainly not a shortcoming of this data series. The advantage of this 104 years of data is that it is from a single, real population of wind speeds associated with tropical cyclones, and the data period is continuous.

Recommendations

In the absence of a longer and more reliable data base, it is being recommended that the 50-year return value of 168 mph should be adhered to, especially in designing buildings that will have important post-disaster functions, *e.g.*, emergency centers and shelters, hospitals, schools, churches, *etc*. In other critical areas, *e.g.*, government buildings, *etc.*, on no account should a design value of less than 155 mph be used. A design value of 155 mph is the critical value associated with hurricanes, typhoons, and cyclones classified as "major."

The suggested design values are also appropriate for non-island states since the population sample used to generate these estimates is pertinent to all countries in the path of tropical cyclones. Irrespective of where the small island state is located in the tropics, in choosing design values within the suggested limits for a specific site, reliable correction factors will have to be applied for: tropical cyclone frequency and exposure; degree of gustiness, particularly in urban areas; and the location and height of the design in question. It may also be necessary to take into account other meteorological and non-meteorological factors.

With respect to using a 1-min wind, it is recognized that the longer the time interval over which the wind speed is averaged, the lower the indicated peak wind speed will be. Therefore, the calculated design loads will thus depend upon the averaging time used to determine the design wind speeds. This introduces the cost-benefit factor. The engineering institutions are, therefore, strongly urged to cooperate in developing and documenting statistics relating to this area with the tropical cyclone frequency and exposure factor being a key input.

Acknowledgments. I would like to thank Mr. Alfrico Adams of SMADA Consultants, Ltd. in Jamaica for his help in providing most of the reference material. Thanks also to the Director and staff of the National Hurricane Center in Miami for providing a computer listing of HURDAT. The useful comments from all the parties mentioned were indeed truly beneficial.

References

Caribbean Community Secretariat, Georgetown, Guyana, Caribbean uniform building code, Part 2: Structural design requirements, Section 2: Wind load, 1986.

Davenport, A. G., P. N. Georgio, and D. Surry, A hurricane wind risk study for the eastern Caribbean, Jamaica and Belize with special consideration to the influence of topography, University of Western Ontario, Faculty of Engineering Science, London, Ontario, Canada, BLWT-SS31-1985, 1985.

Evans, C. J., Design values of extreme winds in Jamaica, *National Meteorological Service, Jamaica,*1976.

Georgio, P. N., Design wind speeds in tropical cyclone-prone regions, University of Western Ontario, Faculty of Engineering Science, London, Ontario, Canada, BLWT-2-1985, 1985.

Georgio, P. N., and A. G. Davenport, Estimation of the wind hazard in tropical cyclone regions, University of Western Ontario, London, Ontario, Canada.

Georgio, P. N., A. G. Davenport, and B. J. Vickery, Design wind speeds in regions dominated by tropical cyclones, *J. Wind Eng. Industr. Aerodyn., 13,* 139-152, 1983.

Georgio, P. N., D. Surry, and A. G. Davenport, Codification of wind loading in a region with typhoons and hills, University of Western Ontario, London, Ontario, Canada.

Gray, C. R., History of tropical cyclones in Jamaica, 1886-1988, *National Meteorological Service, Jamaica,* 1988.

Gumbel, E. J., *Statistics of Extremes,* Columbia University Press, 375 pp., New York, 1958.

Lemilin, D. R., D. Surry, and A. G. Davenport, Simple approximations for wind speed-up over hills, *J. Wind Eng. Industr. Aerodyn., 28,* 117-127, 1988.

Maul, G. A. (Ed.), *Climatic Change in the Intra-Americas Sea,* 389 pp., United Nations Environment Programme, Edward Arnold Publ., London, 1993.

Molina, M., Gray, C. R., Frequency distribution of hurricanes and tropical storms in Jamaica, 1900-1980, *National Meteorological Service, Jamaica*, 1986.

Neuman, C. J., G. W. Cry, E. L. Caso, and B. R. Jarvinen, Tropical cyclones of the North Atlantic Ocean, 1871-1980, U.S. Government Printing Office, Washington D.C., 186 pp., 1981.

Powell, M. D., and P. N. Georgio, Response of the Allied Bank Plaza Tower during Hurricane Alicia (1983), *J. Wind Eng. Industr. Aerodyn.*, *26*, 231-254, 1987.

Shellard, H. C., Extreme wind speeds in the commonwealth Caribbean, *J. Barbados Assoc. Profes. Eng.*,1972.

Surry, D., and A. G. Davenport, Modeling the wind climate: An overview, University of Western Ontario, London, Ontario, Canada.

Thom, H. C. S., Some methods of climatological analysis, World Meteorological Organization, WMO No. 199, TP. 103, Technical Note No. 81, Geneva, Switzerland.

Thom, H. C. S., New distribution of extreme winds in the United States, *J. Struct. Div., Amer. Soc. Civil Eng.*, 94(ST7), 1787-1801, 1968.

U.S. Department of Commerce, NOAA-National Hurricane Center, Computer listing of tropical cyclones of the North Atlantic Ocean, Gulf of Mexico, and Caribbean Sea, 1885-1988, 50 pp., 1988.

Vickery, B. J., On the estimation of extreme speeds in mixed wind climates, University of Western Ontario, London, Canada.

Workshop Proceedings, Development of improved design criteria for low-rise buildings in developing countries to better resist the effects of extreme winds, Manilla, Philippines, November 14-17, National Science Development Board, 1973.

9

A Real-Time System for Forecasting Hurricane Storm Surges Over the French Antilles

Pierre Daniel

Abstract

A depth-averaged numerical storm-surge model has been developed and configured to run on a personal workstation to provide a stand-alone system for forecasting hurricane storm surge. Atmospheric surface pressure and surface winds are derived from an analytical hurricane model that requires only hurricane positions, central pressures, and radii of winds. The storm-surge model was tested in hindcast mode on three hurricanes which gave significant surges over Guadeloupe and Martinique during the last 15 years. This model could be used for other small islands in the Caribbean.

Introduction

Storm surge is the elevation of water generated by strong wind-stress forcing and a drop in atmospheric pressure. For most small islands in the Caribbean, storm surge results from the passage of hurricanes. Destruction within coastal communities is caused by a combination of surge-induced flooding and wind damage. It is important to take such events into account when planning for sustainable development. The use of numerical models for the prediction of storm surge is a well-established technique and forms the basis of such operational prediction systems as the SLOSH model [Jelesnianski et al., 1992] over the United States coast, the BMRC storm surge model [Hubbert et al., 1991] over the Australian coast, and the North Sea models reviewed by Peek et al. [1983].

Small Islands: Marine Science and Sustainable Development
Coastal and Estuarine Studies, Volume 51, Pages 146–156
Copyright 1996 by the American Geophysical Union

The advent of powerful workstations has opened up the possibility of direct use of dynamical models in operational centers. The second WMO International Workshop on Tropical Cyclones [WMO, 1990] recommended stand-alone systems to forecast tropical cyclone storm surges. Such a system was developed for the French Antilles (Martinique and Guadeloupe). A brief description of the storm-surge model and numerical solution is given in the next section. Atmospheric forcing is then described and a few case studies are analyzed.

Storm Surge Model

Although baroclinic effects have a significant influence on deep ocean circulation over long time scales, the main short-term variations in ocean circulation, particularly on a continental shelf, are due to surface wind stress, surface pressure, and the tides. As a result, baroclinic effects can be neglected for prediction of ocean circulation over periods of a few days on the continental shelf. Hence, a depth-integrated model has been adopted for storm surge prediction. The model is driven by wind stress and atmospheric pressure gradients. It solves the non-linear, shallow-water equations, written in spherical coordinates:

$$\frac{\partial U}{\partial t} = f \cdot V - \frac{g}{R \cdot \cos\varphi} \cdot \frac{\partial \eta}{\partial \lambda} - \frac{1}{\rho \cdot R \cdot \cos\varphi} \cdot \frac{\partial P_a}{\partial \lambda} - \left(\frac{U}{R \cdot \cos\varphi} \cdot \frac{\partial U}{\partial \lambda} + \frac{V}{R} \cdot \frac{\partial U}{\partial \varphi} \right) + \left(\tau_{sx} - \tau_{bx} \right) + A_H \cdot \nabla^2 U$$

$$\frac{\partial V}{\partial t} = -f \cdot U - \frac{g}{R} \cdot \frac{\partial \eta}{\partial \varphi} - \frac{1}{\rho \cdot R} \cdot \frac{\partial P_a}{\partial \varphi} - \left(\frac{U}{R \cdot \cos\varphi} \cdot \frac{\partial V}{\partial \lambda} + \frac{V}{R} \cdot \frac{\partial V}{\partial \varphi} \right) + \frac{1}{\rho \cdot H} \cdot \left(\tau_{sy} - \tau_{by} \right) + A_H \cdot \nabla^2 V$$

$$\frac{\partial \eta}{\partial t} = -\frac{1}{R \cdot \cos\varphi} \left[\frac{\partial}{\partial \lambda} (U \cdot H) + \frac{\partial}{\partial \varphi} (V \cdot H \cdot \cos\varphi) \right]$$

where λ is the east longitude (positive eastward), φ is the north latitude (positive northward), U and V are components of the depth-integrated current, η is the sea surface elevation, H is the total water depth, f is the Coriolis parameter, P_a is the atmospheric surface pressure, τ_{sx}, τ_{sy} are the components of surface wind stress, τ_{bx}, τ_{by} are the components of bottom frictional stress, ρ is the density of water, g is the gravitational acceleration, A_H is the horizontal diffusion coefficient (2000 m^2/s), and R is the radius of the earth.

These equations are integrated forward in time on an Arakawa C-grid [Mesinger and Arakawa, 1976] using a split-explicit finite difference scheme. The numerical solution scheme is described in detail in Hubbert et al. [1990], together with a stability analysis. The bottom stress is computed from the depth-integrated current using a quadratic relationship with a constant coefficient of 0.002.

At coastal boundaries the normal component of velocity is zero. At open boundaries a gravity wave radiation condition [Pearson, 1974] is used. Tides can be modeled but are not included since the major forecasting requirement is for surge heights above local tides. The bathymetry used in the forecast system has a latitude and longitude resolution of 1 min (Figure 1).

Atmospheric Forcing

The primary data requirements for modeling storm surge are accurate surface winds and atmospheric pressure fields, in particular, the vicinity of maximum winds. These fields are inferred from the analytical-empirical model of Holland [1980]. An advantage of this approach is that the model can be used in a stand-alone mode.

For hurricane forecasting, the French Antilles Weather Service relies heavily upon advisories issued by the Miami, Florida National Hurricane Center (NHC). These advisories provide analyses and forecasts with hurricane positions, central pressure and, for each quadrant (northeast, southeast, southwest, northwest), the radii of 34 kt, 50 kt, and 64 kt wind speeds.

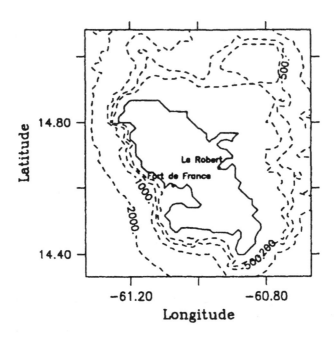

Fig. 1. Model domain for Martinique with bathymetry (m). Relevant place names are marked.

The pressure field is derived as follows [Holland, 1980]:

$$P = P_c + (P_n - P_c) \exp[-(r_m/r)^b],$$

where P is the atmospheric pressure at radius r, P_c is the central pressure, P_n is the environmental pressure defined here as the climatological mean for the region and month calculated from ten years of European Center for Medium Range Weather Forecasts (ECMRWF) analysis (1982-1992), r_m is the radius of maximum winds, and b is the scaling on the profile shape.

$$b = \rho_a \exp(1) \, v_m^2/(P_n - P_c),$$

where v_m is the maximum wind and ρ_a is the air density. The azimuthal wind component is estimated [Holland, 1980] by

$$v = \{b(r_m/r)^b \, (P_n - P_c) \exp[-(r_m/r)^b]/\rho_a + r^2 f^2/4\}^{1/2} - rf/2$$

where f is the Coriolis parameter.

The radius of maximum winds, r_m, is calculated for each direction in order to fit the wind profile to the advisories' wind speed radii. In cases where the asymmetry is unknown, an asymmetry is included by adding the hurricane translation to the symmetric field and rotating the field so that the maximum wind is 70° to the right of the direction of hurricane motion [Shapiro, 1983]. The radial wind field is constructed by rotating the flow to a constant inflow angle of 25° outside the radius of maximum winds [Shea and Gray, 1973].

Surface wind stress is computed using the quadratic relationship

$$\tau_{sx} = C_d \, \rho_a \, (u^2 + v^2)^{1/2} u$$

and

$$\tau_{sy} = C_d \, \rho_a \, (u^2 + v^2)^{1/2} v,$$

where u and v are the horizontal components of wind velocity 10 m above the sea surface and C_d is the drag coefficient. For wind speeds below 25 m s^{-1}, C_d is given by the expression [Smith and Banke, 1975]:

$$C_d = (0.63 + 0.066 \, (u^2 + v^2)^{1/2}) \times 10^{-3}.$$

For wind speeds above 25 m s^{-1}, the dependence of C_d on wind speed is reduced and expressed as

$$C_d = (2.28 + 0.033 \, ((u^2 + v^2)^{1/2} - 25.0)) \times 10^{-3}.$$

Operating Procedure

A typical procedure would be as follows. The user provides hurricane positions, central pressures, and radii of winds at any time (typically every 3 h for 24 h). The user also is prompted to provide an arbitrary number of stations for time series display of surge heights. A temporal interpolation is made to provide hurricane parameters at each time step. The hurricane model and surge model are then run for the required forecast period (typically 24 h). The output is the hourly forecast of surface winds and sea-level pressure fields (Figure 2), hourly sea levels (above the astronomical tide) (Figure 3) and current fields (Figure 4), maximum surge field (Figure 5), and station time series with a 1-min resolution (Figure 6). A 24-h forecast can be carried out on a workstation in a few minutes. This system enables an investigation of multiple forecast scenarios to be made in real time.

Numerical Simulations

The three following simulations were made using trajectory and intensity data provided by the Miami, Florida National Hurricane Center [Jarvinen, 1988]. Table 1 shows observed and modeled maximum storm surge during the passage of Hurricanes Hugo, Allen, and David.

TABLE 1. Observed and Modeled Maximum Storm Surge Magnitudes

Hurricane	Location	Observed Elevation (m)	Model Elevation (m)
Hugo	Pointe Fouillole	>0.70	1.48
Hugo	Pointe à Pitre (marina)	1.50	1.48
Hugo	Baie-Mahault	2.50	2.48
Hugo	St. Francis	1.50	1.41
Allen	Le Robert	0.59	0.53
David	Pointe Fouillole	0.37	0.25

Hurricane Hugo

Hurricane Hugo (1989) was one of the most devastating hurricanes of the last decade in the Antilles region. It crossed the island of Guadeloupe on September 17, 1989 (Figure 7). A 24-h simulation started at 18 UTC on September 16, 1989 and continued to 18 UTC on September 17, 1989.

Only one tide gauge, located at Pointe Fouillole near the Pointe à Pitre marina in an area of complex bathymetry, is available on the island (Figure 8). The storm surge (total elevation minus predicted astronomical tide) at this location is not known with accuracy.

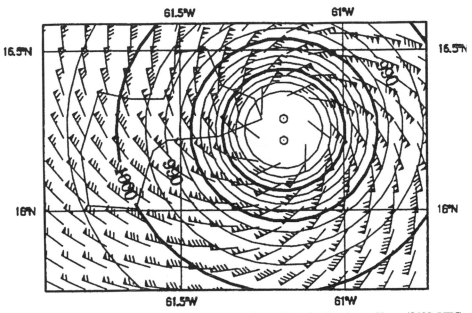

Fig. 2. Surface pressure (5 hPa contours) and surface winds for Hurricane Hugo (0400 UTC, September 17, 1989).

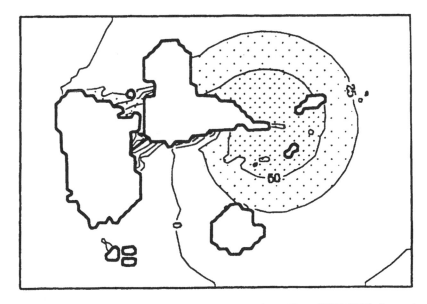

Fig. 3. Sea surface elevations (25 cm contours) for Hurricane Hugo (0400 UTC, September 17, 1989).

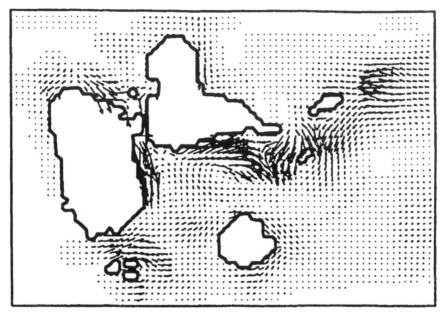

Fig. 4. Depth-integrated currents for Hurricane Hugo (0400 UTC, September 17, 1989) (scale: 1 cm = 1 m/s).

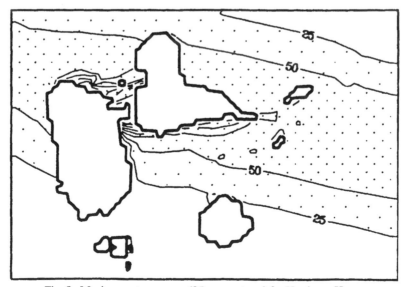

Fig. 5. Maximum storm surge (25 cm contours) for Hurricane Hugo.

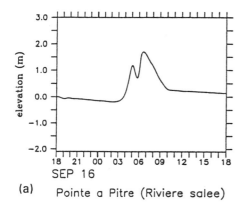

(a)   Pointe a Pitre (Riviere salee)

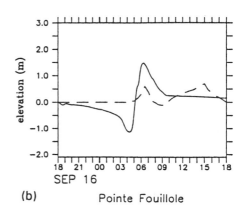

(b)   Pointe Fouillole

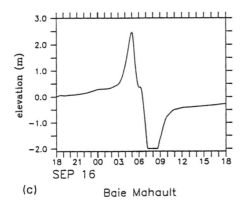

(c)   Baie Mahault

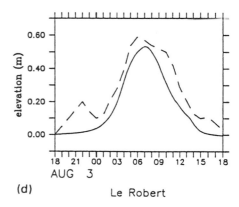

(d)   Le Robert

Fig. 6.   Model (solid line) and observed (broken line) storm surges at Point à Pitre (a), Pointe Fouillole (b), Baie-Mahault (c) for Hurricane Hugo and Le Robert (d) for Hurricane Allen.

In fact, the recording paper on the tide gauge was not wide enough and the pen left the paper when the surge reached 70 cm. Two waves can be seen on the hydrograph: one at 6.30 UTC and a second at 14 UTC on September 17. Other observations complete this record: the Pointe à Pitre marina pontoons rose up to 1.50 m; at St. Francis a 1.50 m surge was estimated; and at Baie-Mahault the waterline of a stranded 1,600-tonner was 2.50 m above the average sea level [SMIRAG, 1990].

Figure 3 shows sea surface elevation, and Figure 4 shows depth-integrated currents. Offshore, the maximum elevation was in the eye of the hurricane. An amplification appears near the coast, first in the Grand cul de sac Marin with northerly winds and then in the Petit cul de sac Marin with southerly winds (Figure 5).

Figure 6a-c shows the time series of elevation at three locations around the island. At Pointe Fouillole the timing of the peak surge coincided with the observed time of the first wave, but the model peak surge was stronger. Hence, this magnitude fits the observed surge at Pointe a Pitre marina, 400 m away from Pointe Fouillole. At Baie-Mahault the model surge reached a maximum of 2.48 m, coincident with the estimated peak surge of 2.50 m;   at St. Francis a 1.41 m model surge is close to the 1.50 m observed surge. At Pointe à Pitre, two waves can be seen. The first one comes from the north through the Rivière salée (Rivière salée is a narrow shallow water passage between Grand cul de sac Marin and Petit cul de sac Marin (Figure 8)); the second one comes from the south when southern winds are blowing. A higher resolution in this region of complex bathymetry should improve the results.

Hurricane Allen

Hurricane Allen (1980) passed through St. Vincent passage south of the island of St. Lucia on August 4, 1980 (Figure 7). A 24-h simulation started at 18 UTC on August 3 and continued to 18 UTC on August 4. Only one tide gauge is available on the island of Martinique, located at Le Robert on the east coast (Figure 1). The storm surge at this location reached a maximum of 59 cm [SMIRAG, 1980], which coincided well with the 53 cm model peak surge (Figure 6d).

Hurricane David

Hurricane David (1979) passed through the Martinique passage between Dominica and Martinique on August 29, 1979 (Figure 7). David produced storm surges over both Guadeloupe and Martinique [SMIRAG, 1979]. A simulation was made only for Guadeloupe because of the lack of observations over Martinique, although the surge was significant over this island. A 24-h simulation started at 12 UTC on August 29 and continued to 12 UTC on August 30. The tide gauge at Pointe Fouillole indicated the storm surge was 37 cm; the model storm surge at this location was 25 cm.

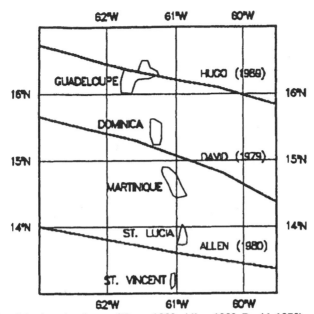

Fig. 7. Tracks of the three hurricanes (Hugo, 1989; Allen, 1980; David, 1979) used in this study.

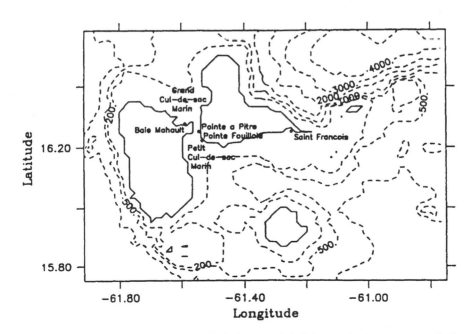

Fig. 8. Model domain for Guadeloupe with bathymetry (m). Relevant place names are marked.

Conclusion

These three studies have shown that the model can accurately simulate storm surge generated by hurricanes in proximity to Guadeloupe and Martinique. This forecast system is now used for operational storm surge forecasting in Météo-France, Direction Interrégionale Antilles Guyane. The system can be used in real-time as a hurricane approaches an island.   An alternate procedure would be to prepare an atlas of precomputed surges based on hurricane climatology. The next step is the adaptation and installation of this model for the islands of St. Martin and St. Bartholomew. It could also be adapted for other small islands in the Caribbean.

Acknowledgments. This research has been partially supported by the Direction de la prevention des pollution et des risques, Delegation aux Risques Majeurs (DRM).

References

Holland, G. J., An analytical model of the wind and pressure profiles in hurricanes, *Mon. Wea. Rev., 108,* 1212-1218, 1980.

Hubbert, G. D., L. M. Leslie, and M. J. Manton, A storm surge model for the Australian region, *Q. J. Roy. Meteorol. Soc., 116,* 1005-1020, 1990.

Hubbert, G. D., G. J. Holland, L. M. Leslie, and M. J. Manton, A real-time system for forecasting tropical cyclone storm surges, *Wea. Forecast., 6,* 86-97, 1991.

Jarvinen, B. R., C. J. Neumann, and M. Davies, A tropical cyclone data tape for the North Atlantic basin: Contents, limitation and uses, NOAA Tech. Memo., NWS, NHC-22, 22 pp., 1988.

Jelesnianski, C. P., J. Chen, and W. A. Shaffer, SLOSH: Sea, lake and overland surges from hurricanes, NOAA Tech. Rept., NWS-48, 79 pp., 1992.

Le cyclone Allen, SMIRAG, 1980.

Le cyclone David, SMIRAG, 1979.

L'ouragan Hugo, Service Meteorologique Interregional Antilles Guyane (SMIRAG), 1990.

Mesinger, F., and A. Arakawa, Numerical methods used in atmospheric models, GARP Publ. Ser. No. 17, ICSU/WMO, Geneva, Switzerland, 1976.

Pearson, R. A., Consistent boundary conditions for numerical models of systems that admit dispersive waves, *J. Atmos. Sci., 31,* 1481-1489, 1974.

Peek, H. H., R. Proctor, and C. Brockmann, Operational storm surge models for the North Sea, *Cont. Shelf Res., 2,* 317-329, 1983.

Shapiro, L. J., The asymmetric boundary layer flow under a translating hurricane, *J. Atmos. Sci., 40,* 1984-1998, 1983.

Shea, D. J., and W. M. Gray, The hurricane's inner core region: Symmetric and asymmetric structure, *J. Atmos. Sci., 30,* 1544-1564, 1973.

Smith, S. D., and E. G. Banke, Variation of the sea surface drag coefficient with wind speed, *Q. J. Roy. Meteor. Soc., 101,* 665-673, 1975.

WMO, Proc., WMO Second International Workshop on Tropical Cyclones (ETC-II), WMO/T No. 83, WMO, Geneva, Switzerland, 1990.

10

Geography of Small Tropical Islands: Implications for Sustainable Development in a Changing World

Orman E. Granger

Abstract

Small tropical islands, although physically diverse and ranging from rugged mountainous terrain of volcanic antecedents to low-lying, flat coralline reefs, are the products of similar tectonic activities and geologic structures from the Cretaceous to Recent times. They can be classified according to their location and mode of formation into: islands of trench/arc systems, oceanic islands, and islands associated with continental plate dynamics. There may be combinations of types in the same geographic location. The crustal instability that underlies their genesis is the source of economic strength for some but of disasters that threaten their viability as developing independent states. Their economic bases are their varied terrestrial and oceanic bioproductivity systems which are not generally robust. In a few cases, that base may include some mineral deposits of economic value: gold, silver, copper, nickel, and bauxite. In addition to socio-economic constraints on sustained development such as hegemonic marginalization, serious balance of payments problems, diseconomies of scale, limitations on natural resources base, dependence on a very narrow range of generally uncompetitive export products and foreign aid, environmental constraints to development face these small islands and island states: a limited supply of fertile soils, unreliable water supply that may become more restricted and even less reliable with global warming, geologic and meteorologic disasters that include volcanic eruptions, earthquakes, hurricanes and typhoons, floods and droughts that periodically, and perhaps more frequently in the next 50 years, completely destroy the already shaky economic base. With global warming and its anticipated oceanic impacts including

Small Islands: Marine Science and Sustainable Development
Coastal and Estuarine Studies, Volume 51, Pages 157–187
Copyright 1996 by the American Geophysical Union

increased storm intensities, coastal and estuarine inundations, salinization of coastal aquifers, changes in the spatial and temporal distribution of rainfall, increased temperatures and hence increased evapotranspiration, these constraints will be exacerbated. Solutions are not easy, but present and expected constraints on sustained development must be taken into consideration now in preparation for an increasingly vibrant socioeconomic future. More importantly, small tropical islands must attempt to define the nature and scope of sustained development on their own terms and in keeping with their cultural and social presuppositions rather than externally imposed expectations and definition.

Introduction

Article 121 of Part VIII of the International Convention on the Law of the Sea defines an island "as a naturally formed piece of land surrounded by water on all sides, emerging above the surface of the sea at the highest tide, capable of sustaining human habitation or economic life on its own and whose dimensions are smaller than those of a continent." Islands may be distinguished from continents in many ways. Continents, by virtue of their mass tend to generate their own spatial patterning, their own biological and physical environments, and their own climates. Islands, with the exception of the very large ones such as New Guinea, Borneo, Sumatra, Hispaniola, Madagascar, and Sri Lanka, on the other hand, are subject to and cannot materially modify the natural hydroclimate because of their volume. As a first approximation, when a mountainous mass of more than 1500 m of average height extends over more than 20,000 km^2, it is large enough to generate its own hydroclimate effects and, therefore, it enters the continental category despite its insularity. Many attempts to specify threshold criteria for separating insular land masses into islands and continentalized islands exist. The criteria have generally been arbitrary and include land area and morphology, population size, and gross disposable product [Dommen, 1980; Jalan, 1982; Dolman, 1982; Doumenge, 1983; Non-Aligned Movement, 1983; UNCTAD, 1971].

There is a consensus among geographers, ecologists, demographers, anthropologists, and development economists that tropical islands have a distinct character even though the component characteristics of that distinctiveness have not been precisely defined. First, there is the issue of size, although there is serious doubt about its usefulness as an analytic and prescriptive concept. There are hundreds of thousands of pieces of tropical land territory of subcontinental size ranging from sandbanks and huge rock protrusions through transitional sizes such as Trinidad, Mauritius, Jamaica and Sri Lanka to extensive masses of land such as Madagascar, Borneo, Sumatra, Cuba, Hispaniola, Puerto Rico, and New Guinea that are generally identified as islands. They differ in geology, geomorphology, ecology, and climate. As political, social, economic, and cultural systems, they exist as single isolated land masses or as archipelagos that are spread over wide ocean spaces. Unlike the comparatively uniform and homogeneous continental masses, the universe of islands is comprised of an infinity of diverse entities.

However, the locations of islands and island chains on the earth are associated with similar sets of geologic conditions.

A large number of islands arose from the sea as volcanoes in areas of tectonic instability such as mid-oceanic ridges and spreading zones and subducting tectonic plate boundaries, or as atolls arising from volcanic activities. Others arose as a result of one or more eustatic changes, sedimentation, and subsequent emergence. Figure 1 shows the locations of the major islands and island groups of the world. Because their origins are associated with instability in the earth's crust, islands are susceptible to extreme hazards associated with that instability such as high incidence of volcanic eruptions, high seismicity that is, however, not statistically different from continental areas and unstable ground.

In what follows, we confine ourselves to a consideration of small tropical islands and archipelagos within the tropics. The tropics is that region of the earth bounded in the north by the Tropic of Cancer and in the south by the Tropic of Capricorn or between 23.5°N latitude and 23.5°S latitude. In some of the discussions that follow we will deviate from those boundaries to include the Bahamas. The Mediterranean islands have been excluded, although some of them such as Malta have features analogous to those of tropical islands. The existence of large oceanic masses relative to the mass of the islands, the large amount of solar radiation, and moisture input at these low latitudes together with the atmospheric dynamics and tectonic history there have created characteristic natural environments that make small tropical islands as a genre quite homogeneous. Consequently, the natural environments in the outer islands of Papua-New Guinea are strikingly similar to those of islands of the eastern Caribbean. To facilitate discussion of the geography of these small tropical islands we will divide them into the islands of the North Atlantic-Caribbean region (Figures 2a and 2b), those of the Pacific Ocean (Figure 3), and those of the Indian Ocean (Figure 4) when such divisions will enlighten the discourse. The discussion will include the distribution of tropical islands, their geology, geomorphology and tectonics, climate and soils, biogeography, their susceptibility to natural disasters, and their economic geographies. We will then use a synthesis of these in our discussion of sustainable development and examine the viability of these small tropical islands in the context of trace gas-induced global warming and environmental change.

Geology and Tectonics

Small tropical islands may be divided into three groups based on their geologic origin and tectonic evolution: islands of trench/arc systems, oceanic islands, and islands associated with continental plate dynamics.

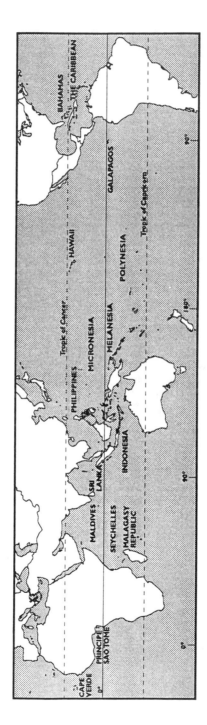

Fig. 1. Tropical islands of the world.

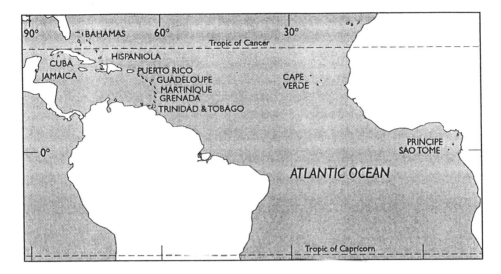

Fig. 2a. Tropical islands of the Atlantic and Caribbean.

Fig. 2b. Caribbean islands.

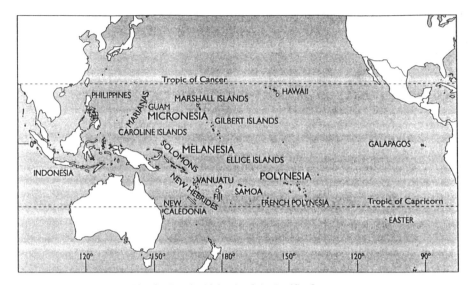

Fig. 3. Tropical islands of the Pacific Ocean.

Islands of Trench/Arc Systems

In lithospheric subduction zones, often arcuate chains of volcanic islands associated with a deep-sea trench exceeding 6,000 m in depth are produced. These volcanic islands that are the result of extrusions of metamorphosed substrata are structurally complex. The extruded materials are usually acidic and produce explosive eruptions of varied ejecta (*e.g.*, Pelée in Martinique and Soufriere in St. Vincent and the many volcanoes of southeastern Papua-New Guinea). In these islands volcanism is still prevalent, as are earthquakes that accompany volcanic eruptions or that occur by themselves. In the Caribbean region (Figure 2b) there have been notable volcanic eruptions: Mt. Pelée in Martinique in 1902 that destroyed the town of St. Pierre killing 29,000 people within seconds; the St. Vincent Soufriére eruptions in 1812 in which 75 people perished, in 1902 when 1,565 people died and again in 1979 when 20,000 people were evacuated from the northern areas of St. Vincent to relief centers whose maintenance cost was around EC$90,000 a day or the equivalent of the entire budgeted current expenditure for 1979 [Granger 1990].

The relief of volcanic islands are usually dissected (*e.g.*, Dominica, Grenada, St. Vincent, Martinique, and St. Lucia in the eastern Caribbean) and erosion produces monoliths of resistant lava (*e.g.*, the Pitons in the eastern Caribbean and the towering plugs of the New Ireland and New Britain chains of northeastern Papua-New Guinea). Volcanic ash, pumice, and breccia erode quickly to a fertile but highly permeable soil. Successive explosive eruptions reorganize and redistribute strata including older

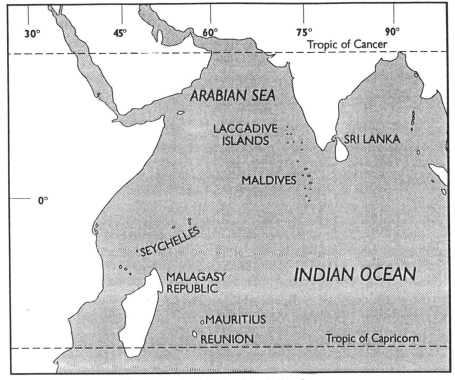

Fig. 4. Tropical islands of the Indian Ocean.

lithospheric materials that may contain veins of valuable mineral resources (gold, copper, nickel, and silver).

The inner walls of trenches accumulate marine and volcanic sedimentary deposits which become calcareous islands when anticlinal uplift occurs. The resulting relief is not unlike that of raised atolls although there may not be any phosphatic deposits or the contrast between fringing reefs and lagoon basins. Anticlinal uplift may produce a coastline topography that is very indented, providing these islands with good harbors.

The Caribbean island-arc and trench are typical of the island-arcs of the world with minor exceptions. It represents the locus of downwelling seafloor. Uplift averages 1.0 mm/yr over the entire island-arc shelf platform. Earthquake foci are distributed in a dipping plane and dip-slip motion of shocks indicates that the inside arc is over-riding the oceanic plate. But dip-slip may not be the predominant motion throughout the arc as seen in the dominantly compressive interactions with the oceanic plate in the Lesser Antilles while the present volcanic quiescence in the Greater Antilles and the character of the observed structure suggest strike-slip interactions. Seismicity indicates that the Caribbean plate is moving relative to its borders at a rate of 2.1 cm/yr eastward relative to the Atlantic plate. The catastrophic effects of earthquakes in the Caribbean are ever present and reside in the destruction of buildings for which there is no generally adopted construction codes, the generation of destructive landslides on the markedly bedded clayey sub-strata, and coastal devastation by tsunamis. There is hardly a town here that has not experienced a seismic disaster sometime in its history.

The geologic history of the larger Caribbean islands (Greater Antilles) can be divided into four phases which partly overlap each other and which are not simultaneous everywhere: (1) a Cretaceous period with much submarine volcanism and little or no supply of sialic detritus, (2) a late Cretaceous or early Paleogene phase of foldings, serpentine mobilization, quartz diorite intrusion, thrusting, and a limited amount of non-calcareous turbidite sedimentation accompanied by volcanic activity, (3) a Tertiary phase of volcanic quiescence, of widespread subsidence and carbonate deposition, some folding and considerable faulting, and (4) a younger Tertiary to Recent phase characterized by the formation of the present morphology and submarine topography [MacGillavry, 1970]. The Lesser Antilles forms a classical island-arc with a volcanic inner arc and a non-volcanic outer arc. The geologic history of this arc is entirely Tertiary. Volcanic activity of the inner arc succeeded such activity in the Greater Antilles. The Caribbean seafloor was stable at the same time that its margins were being severely tectonized. Three different categories of relief of structural origin may be distinguished in the region: (1) the large local relief faultblock mountains of the Greater Antilles, the Leeward islands and Trinidad and Tobago; (2) the volcanic mountains of the islands of the inner arc; and (3) the low and uniform limestone plateaus of the orogenic foreland and small pockets within the fold mountains of the greater part of Cuba, the islands of the outer arc and the Bahamas [Donnelly, 1968].

Chains of trench/arc system islands are stretched along the length of the western rim of the Pacific basin and in association with the American and Caribbean continental plates. The small islands of the eastern Caribbean exemplify the ordered structure in which an active internal volcanic arc and an external calcareous arc are juxtaposed [Granger, 1990, pp. 11-25], providing the existing contrast between the natural potential of neighboring islands, some of which are exclusively volcanic (Grenada, St. Vincent, St. Lucia, Martinique, Dominica, Montserrat, St. Kitts, Nevis, St. Eustatius, Saba, *etc.*) while others are exclusively calcareous (Barbados, Marie Galante, Antigua, Barbuda, Anguilla, *etc.*). Twin island Guadeloupe is the only island in which both types of formation (volcanic Basse-Terre and calcareous Grand-Terre) are found together [Granger, 1990].

The island-arc systems of the southwest Pacific are essentially Tertiary and came to occupy their present position well after the Early Cretaceous. While all the Pacific islands are basically of volcanic origin they can be placed into three physical-geologic categories: (1) complex serpentine formations, *e.g.*, Papua-New Guinea, Solomon Islands and New Caledonia (Figure 3); (2) volcanic structures with significant relief, *e.g.*, the Samoas and Rarotonga in Cook Islands; and (3) coral atolls, *e.g.*, Tokelau, the Northern Cook Islands, Wallis and Futuna, and the Marshalls (Figure 3) [Coleman,

1973]. The islands in the first category are large, rugged, and mountainous and have a variety of morphologies and natural resources. The mountains and rugged terrain foster fragmentation of population and diversity of socio-cultural and language groups. The islands of the second category are also physically diverse, being characterized by rugged mountain ranges, river systems, and some mineral resources but have smaller land masses. The third category are tiny, low-lying, except for raised coral atolls such as Niue, and flat. These lack terrestrially-based resources of any significance.

Oceanic Islands

These islands are formed when isolated summits of large volcanoes linked to mid-oceanic ridges or to lateral fracture zones protrude above ocean levels. These volcanoes are formed when the basaltic material underlying the major ocean basins rise to the surface as flattened cones. This basic lava when eroded forms rich soils. However, some volcanic rocks, especially extruded basalt, retraction fissures formed during solidification in those rocks, and tunnels formed by sub-surficial lava flows under solidified basaltic crust, create water resource problems in some islands due to the porosity and high infiltration capacity of the rocks. The basaltic rocks in this type of oceanic volcano do not generally contain minerals of economic value. The coastal areas of these islands attract coral reefs if the hydro-biochemical conditions are favorable. The corals may grow to considerable thickness if after volcanic emergence, the island subsides at a rate equal to or less than the growth rates of the reef complex. In the Fiji group, for example, the majority of the geologic formations belong to the late Tertiary but in the Lomaiviti Islands and in Taveuni there is evidence of Pleistocene vulcanicity with successive basaltic flows and ash fall well into Recent times [Bayliss-Smith *et al.,* 1988]. Intermittent earthquakes in 1953 between Kadavu and Viti Levu and in 1977 north of Taveuni that appeared to have affected ground water levels indicate continuing seismic activity into the present.

Oceanic islands may further be divided into five categories:

Simple volcanic islands of recent formation: These are usually small in size, have steep inaccessible coasts because of the short span of coastal denudation and have biological populations that are limited both in the variety of species and in the number of individuals in each species. There has not been enough time to form natural communities.

Old complex volcanic islands: These have been built up by several successive eruptions over the greater part of the Tertiary age. The relief has been smoothed by denudation and deep rich volcanic soils have been formed. Coastline is regular and coral formations, initially fringing reefs, have been built up. Agricultural development of the originally fertile land has been sustained.

Volcanic islands with lagoons and barrier reefs: Due to subsidence and the growth of coral formations, these islands are surrounded by barrier reefs that enclose lagoons. Because of the faster growth of corals on the windward side of islands, the reefs there are thicker and soon form continuous barriers, with parts composed of consolidated conglomerates that remain above high water level. On the leeward side, the reefs are thinner and discontinuous due to the absence of breaking waves which should transport and distribute nutrients and rapidly oxygenate the water. Breaches in these reefs enable exchanges of water between lagoons and the ocean. Islands in this group have rich volcanic soils that have supported intense farming and the biological resources of the reefs and lagoons have provided high quality protein. Where subsidence has been great enough to produce lagoons of great depth, the leeward openings have been valuable for shipping. Fringing reefs and the lagoons they enclose have provided great potential as artisanal fisheries and as tourist attractions.

Atolls: Atolls are raised discontinuous reefs enclosing lagoons. Like barrier reefs of high islands, atolls are asymmetrical. The size of atolls vary according to the extent and depth of the bases but the portion above sea level is usually limited to a few square kilometers. The hypercalcic soils of these islands are poor because they are in the early stages of decomposition of young conglomerates and in some cases are made poorer by significant amounts of magnesium carbonate in addition to calcium carbonate. Infiltration capacity is very high resulting in the perennial problem of lack of water. On bigger islands of higher relief, this water problem is relieved by Gyben-Hertzberg lenses but sustained yield principles in water resource management have been necessary to ensure a reliable supply of water.

The biogeography of these islands is characterized by small plant and animal populations and a limited number of species. There are numerous herbaceous plants and shrubs and the subspontaneous vegetation includes ironwood, screwpine and the coconut palm which has been the only exploitable subsistence crop and commercial commodity (copra). Other crops including various taros, sweet potatoes and a few vegetables and fruits have been grown on these islands. Atolls have exceptional developmental potential predicated on the biological resources of their lagoons. Some, such as those in the Tuamotu archipelago in French Polynesia, have developed aquaculture (pearl oysters).

Raised atolls: When subsidence is replaced by uplift, atolls emerge slightly, the water exchange channels dry up, fresh water fills the lagoons and the outer rim emerges totally (Niau in the Tuamotu archipelago). With further emergence, networks of interlinked basins appear in the lagoons (Christmas Island) and these support large colonies of birds resulting in extensive phosphate deposits (Tamaiva). If emergence continues yet further to tens of meters, a fragmented Karst landscape emerges and a residual red clay soil formed by decalcification fills the Karst pockets to provide cultivatable but not very productive areas. Some raised atolls are 60-70 m above the sea around their rims, yet still have central depressions and stepped sides suggesting two or more stages of emergence (*e.g.,* Nauru, Niue, *etc.*). Water problems abound because of the high infiltration of the substrate. However, phosphate deposits have supported industrial complexes, high standards of living and have provided the capital resources for investments in new industries when the deposits are depleted as in the republic of Nauru or the converse in the case of Ocean Island in Kiribati and Makatea in French Polynesia.

The spatial distribution and relative size of oceanic islands reflect their geological evolution. In the Atlantic and Indian Oceans, islands are isolated and have a characteristic size when lateral ridges run off the central axis of the mid-ocean ridge. In the Pacific, the alignment of archipelagos depends on the major fractures such that volcanic formations tended to rise in the southeast, giving rise to higher and larger islands and subside in the northwest where most of the atolls are to be found. The Samoan Islands is an exception in that the spatial and size patterns are reversed.

Islands Associated with Continental Plate Dynamics

Continental plate island systems are more extensive and complex than those of trench/arc systems and oceanic island systems. Here, a succession of sedimentary deposits are laid down on stable cores of metamorphic crystalline masses built from consolidated plate fragments. Ancient powerful tectonic movements resulted in complex structures with upwellings and effusion of deep ultrabasic materials. Because of the varied bases and their large areal extent, these islands have a greater and more diverse potential for agricultural development than either oceanic or island-arc islands. They are more likely to have mineral resources which have built up over their long geological history, *e.g.,* hydrocarbons in Trinidad; nickel and other metals in New Caledonia; gold, manganese and copper in Fiji. The islands on the northern Seychelles shelf, for example, although belonging in this group have features similar to those of other systems, though they are more stable. They stand on portions of emerged outcrops of a continental plate least prone to tilting and are emerged volcanic features linked with fractures at its boundary. In the Pacific, New Guinea, the Solomon Islands, New Hebrides, New Caledonia, and Fiji are in this group. They are mountainous and, for the greatest part, heavily forested, comprising more than 95% of the total land area in the tropical Pacific and containing over three-quarters of the population.

Environments of Tropical Islands

Climate and Water Resources

The major controls on the climates of small tropical islands are the same all over the world. However, the physical configuration and size of the ocean basins together with the distribution of islands within a basin produces other controls that make the climates of small tropical islands differ in several important respects from one region to another.

In the Caribbean region the islands are all situated north of the equator (Figure 2b); in the Indian Ocean, the Maldives and Laccadive Islands are in the northern tropics while the Seychelles, Comoros, Mauritius, Reunion and Malagasy Republic are south of the equator (Figure 4); in the Pacific Ocean, islands have a much wider latitudinal distribution (Figure 3). Among the major controls are large inputs of solar radiation throughout the year and, because of the amount of precipitable water in the atmosphere, small longwave radiational losses. Hence, air and sea surface temperatures tend to be high and the warm ocean moderates the temperatures of islands at various time scales. This maritime moderating effect is the result of the ocean-atmosphere coupling which transfers large amounts of latent and sensible heat throughout the year with the former being the major driving force of hurricanes and typhoons. The amount of solar radiation incident on the surface is greatest during the low sun period when cloud cover and precipitable water are least. The second major control is the movement of the Inter-Tropical Convergence Zone (ITCZ) as the thermal equator migrates seasonally. This migratory motion influences both the temporal and spatial distribution of rainfall and the intensity and height of the Trade Wind Inversion that in turn influences the formation of tropical storms, hurricanes, and typhoons. The position and extent of migration of the ITCZ is different in the Atlantic from that in the Pacific. The third major influence are the waves in the upper Easterlies and their control of surface and mid-tropospheric convergence, on which convective activities depend, and divergence which is responsible for long dry spells that sometimes decimate island agricultural systems. These wave-like perturbations vary in size and intensity and are most frequent in summer when they may occur as often as every three to five days.

In the Pacific and, to some extent, the Atlantic, characteristic distributions of coherent pools of warm ocean water give rise to particular climate phenomena (El Niño) that affect the global climate systems and the biological productivity of the ocean including the fisheries that are very important in the nutrition and economy of small islands. Associated with these sea surface temperature perturbations and exemplified in the western Pacific is an aperiodic change in the velocity of the Trade Winds and a reversal of the direction of the winds due to oscillations of pressure centers there (Southern Oscillation). These oscillations occur alternatively in each hemisphere and between the eastern and western parts of that sphere with a highly variable periodicity. Together, these two coupled ocean-atmosphere effects is the ENSO phenomenon which is known to affect the variability of precipitation over large regions of the world by its effects on convective activity including the strength and duration of the monsoons, and on the distribution of hurricanes and typhoons and hence floods, droughts, and coastal disasters associated with them.

Among tropical islands there is a wide range of climates from the "ever wet" ones to climates with low rainfall and long dry seasons. There are even desert islands such as Curaçao and Aruba in the Caribbean. Close to the equator where everyone expects it to be very wet, there is, in the central Pacific, for instance, an extensive area of deficient rainfall. The interannual variability of rainfall can be considerable and unpredictable, while annual reliability is directly proportional to rainfall amounts. Tropical rainfall is usually of high intensity although high degrees of intensity, over 60 mm an hour, are commonly short lived. Local falls of 50-70 mm in one hour can be expected once every two years. Nevertheless, even very short periods of high intensity rainfall can have deleterious effects on water supplies, soils, and crops. Even more so are the intense 1 in 100 year rainfall episodes from slow-moving tropical depressions which con precipitate 20mm/hr. for as long as 100 consecutive hours.

Tropical island temperatures are more striking in their uniformity than in their magnitudes. In the lowland areas the mean monthly range is rarely more than 2.8°C, but temperature range increases with distance from the equator. Diurnal temperature ranges are larger than annual ones. The bio-productive systems of tropical islands make use of these thermal resources to ensure the highest degree of continuity in the biological cycle.

Differences in relative humidity patterns between the equatorial lowlands and their drier counterparts are important for human comfort and agriculture through evapotranspirational influences. Diurnal variations are considerable despite the oceanicity of islands and relative humidities are usually lowest in the early afternoon and highest in the early morning and just before afternoon showers when they reach 100%.

The main climatic factors relevant to this study of tropical island sustainable development are rainfall, temperature, solar radiation intensity, and their duration. Other derivative factors such as humidity, evapotranspiration rates, soil moisture amounts, and rates of change are also important. In tropical island environments rainfall rather than temperature determines seasonality and it is the amount and timing of rainfall that is crucial for the all-important bio-productive systems that form the economic backbone of most of these islands. However, we must be careful about treating climate as an abstraction because there are limitations to using average rainfall, temperature, and similar statistical specifications of weather variables for agricultural and other environmentally-based planning. The actual meteorological conditions occurring in any particular year, especially in their extreme form, over a strictly limited area are of greater significance than the theoretical abstractions of climate. Later we will see that weather phenomena in their extreme manifestations can be the bane of development in small tropical islands.

At first glance, it seems paradoxical that tropical islands should suffer from difficulties in water supply: after all, they are hot, wet lands where bioproductivity is possible year round without irrigation. Yet, water is probably the most important single determinant of tropical land use because the moisture supply is commonly unsatisfactory--either too much or insufficient. In practice, however, the unequal distribution of rainfall throughout the year is a problem in islands affected by monsoons and those in which a distinct and sometimes prolonged dry season occurs or where the moisture balance between rainfall and evapotranspiration losses are critical. Difficulties in water supply in bioproductivity systems are due to the characteristics of rainfall--seasonality, variability and high intensity; to evapotranspiration rate and its variability; to hydrologic changes and to topographic influences on the spatial distribution of rainfall [Granger, 1983]. But the significance of these factors depends on geological and edaphic conditions: porosity and permeability of the substrata, topography, soil, and vegetal cover.

Finally, the development and utilization of water resources may depend as much upon economic, social, and political factors as upon the amount of available water. Different crops have widely differing moisture requirements in amount, distribution, and reliability. Water shortage may also be simply an expression of unsuitable methods of cultivation. Suitable methods of cultivation must also be predicated on the types and fertility of the soils. On rainy islands, excessive moisture and high temperature permit weathering of rocks to depths of 10-15 m below the surface, where a thick layer of clayey red and yellow latosol tends to develop as the zonal soil type. These soils are usually leached of plant nutrients and when cleared of vegetation may prove to be infertile. On these wet islands, azonal soil types derived from the weathering of recently deposited alluvium and young volcanic ejecta are formed. These soils have not yet been leached of valuable plant nutrients and minerals and are, therefore, premier agricultural lands. In the wet-dry islands too, the soils are various but are not as deeply weathered or leached as those of the wetter islands. On drier islands an alkaline, little leached and red brown to black, quite fertile soils containing an abundance of lime are found. Different moisture regimes result in different weathering patterns and resultant clay structure. In St. Vincent in the Caribbean, for example, where the soil is continuously moist, the clay consists of allophane with high aluminum at times to the level of toxicity and moderate amounts of iron. In areas where the dry season is short to moderate the allophane still dominates but the soils are low in aluminum. Where there is a marked and intense dry season the clays are dominated by halloysite and kaolinite or montmorillonite. In this case the soil may contain high levels of iron [Limbird, 1988]. It is apparent from the foregoing that geologic, climatic, and geomorphic factors in small tropical islands exert significant influence on the edaphic and hydrologic environments and by extension on the water resources and bioproductive systems. These influences, if not adequately accounted for in the process of planning sustainable development strategies, can become very significant constraints.

Because of this climate diversity among islands, it is more enlightening to look closely at two of the regions delimited earlier in this discussion than to try to elucidate the myriads of different climates that exist among all small tropical islands or island states. We will, therefore, concentrate our climate discussion on the islands of the Caribbean region and those of the Pacific Ocean.

Regional Climates of the Caribbean Region

Caribbean temperatures are much less variable than precipitation. The annual temperature range lies between 1.9°C in Trinidad and 5.5°C in Cuba. The period of greatest warmth is the wet season when cloud cover and high atmospheric humidity accentuate the greenhouse effect [Granger, 1985]. Monthly mean diurnal temperature range is significantly larger than mean annual range. The values lie between 6.1°C in Curaçao to 11.1°C in Haiti and averages 8.2°C over the region. Very extreme temperatures are rare in the Caribbean due to the moderating effects of the sea and the very high level of evapotranspiration. No station in the region has a mean maximum that exceeds 36°C. In Trinidad the average temperature (1951-1970) is 26.7°C varying from 25.0°C in January to 27.2°C in May. The annual range is 1.9°C and the mean diurnal range is 9.4°C. In the Bahamas, mean maximum and minimum temperatures are respectively 23.9°C and 16.1°C. Night-time minima rarely fall below 15.5°C in the southern islands; in the north, however, temperatures below 10.0°C may occur after the passage of a cold front when dry polar continental air temporarily replaces the normal maritime tropical air. The lowest temperature recorded in the Grand Bahamas is 6.1°C [Granger, 1982].

In the Greater Antilles, terrain and elevation effects affect spatial distribution of temperature significantly. In Cuba, average temperatures range from 21.1°C in January to 27.8°C in August. In the mountainous areas, temperatures may fall below 10.0°C. From Guadeloupe to St. Lucia mean annual temperature ranges between 25.0°C and 27.8°C. In Guadeloupe and Martinique there is a distinct cool season from November to February-March.

The most important climatic element in the Caribbean is rainfall and, although the region is always influenced by maritime air masses, the spatial and temporal variations are surprisingly pronounced, ranging from semi-aridity of Curaçao, Aruba, St. Maarten, the Cul de Sac region of the Dominican Republic to the very heavy rainfall of the windward side of the more mountainous islands. Seasonal divisions in the region are based on precipitation distribution and variation. The climate year, therefore, is divided into a dry season lasting roughly from January to May and a wet season lasting roughly from June to December. The beginnings and endings of these seasons vary with latitude.

There are differences from island to island, yet it is possible to group islands on the basis of the general configuration of their annual rainfall regimes. The islands between Puerto Rico and Guadeloupe (Figure 2) have maxima in June and July and again in September and October. Minima occur in February, in March or in both but not in January. Because of their topographic configurations, Dominica, St. Vincent and Martinique have a comparatively long rainy season on their leeward coasts lasting roughly from May to December with a peak in July and August. The windward stations have a wet season of similar length but with the maximum in November. The islands from St. Lucia to Trinidad experience a broad rainfall minimum in March and April and while the time of maximum shows some interannual variability, two peaks are discernible, one in July and August and another in October and November. [Granger, 1985]. In Curaçao and Aruba, the broad minimum in rainfall occurs in March and April and the maximum in November; everywhere the rainfall totals are comparatively low.

In many Caribbean islands terrain and elevation influence the spatial distribution of rainfall. A transect across the mountains of Jamaica from north to south along longitude 76°25'W shows an increase in mean annual rainfall from 3,000 mm on the north coast to 8,000 mm at the crest, decreasing again to 1,800 mm on the southeast coast. Camp Jacob in Guadeloupe, 5.4 m elevation, reports 3560 mm of rain on 274 rain days annually whereas Neufchateau at 250 m upstream reports 4414 mm annually. In the Dominican Republic, the annual rainfall varies from as much as 203 cm on the windward side of mountainous areas to 127 cm in the interior lowlands of the Cibao and 145 cm on the southern or leeward lowlands. Topographic influences can be significant in smaller islands with mountainous backbones that are almost normal to the prevailing winds. In St. Lucia, coastal lowlands average 129 cm against 197 cm annually in the mountains. In Martinique, while most of the island receives and annual average of around 152 cm, Monte Pelee in the mountains receives 1,000 cm. A similar situation occurs in Dominica where the eastern mountain slopes receive between 508 cm and 762 cm; Shawford at 152 m elevation receives 470 cm but along the west coast the annual mean is around 178 cm (1941-1970 Climo) [Granger, 1985].

Here again topographic configurations of tropical islands interacting with processes in the atmospheric and hydrologic environments affect the spatial distribution of rainfall and hence of water supply resulting in some areas within islands having substantial deficits and others, large surpluses. Even in small islands, development plans cannot be spatially generic but must be regionalized to accommodate these physical environmental differences; alternatively costly and sometimes elaborate water storage and transfer systems have to be constructed with resources that have alternative and often more pressing uses, to equalize water availability across the island.

Regional Climates of the Pacific

Throughout the Pacific, sea level temperatures are remarkably steady averaging between 25.0°C and 29.5°C all year. Close to the equator the annual range is less than 1.6°C. The moderating influences of the warm expansive ocean is so pronounced that on all but the largest islands annual ranges of temperature rarely exceed 4.5°C. For the same reason, temperatures seldom rise above 35°C or fall below 16°C. The thermal equator remains north of the geometric equator in the eastern Pacific but migrates far to the south of it during the southern summer in the western Pacific. This gives rise to some spatial differentiation in this apparently monotonous climatic system. Consequently, two major climatic zones can be distinguished: an equatorial oceanic belt extending approximately ten degrees on both sides of the equator except on the southern side of the eastern Pacific. This zone is characterized by constant high temperature (always above 25°C) and humidity (always above 80%), frequently unreliable rainfall, and freedom from typhoons and hurricanes. On both sides of this belt lies the tropical oceanic zone that has an eastern and western subzone. Here, temperatures show slight seasonal variation throughout, but the regular trade winds and the generally moderate rainfall of the eastern zone give way, in the western zone, to a seasonal alteration of "monsoon" and trade winds, heavier but more variable rainfall, and seasonal but irregular incidence of typhoons and hurricanes.

Rainfall distribution in the Pacific is both spatially and temporally variable. Here latitude is the factor most responsible for some islands being dry while others wet, although on the whole, there is less rainfall on low, flat islands than in high mountainous ones. There is an equatorial zone where predominantly convective showers in sharp thundery downpours occur, variation is large and the highest variations coincide with the equatorial dry belt of the central Pacific where prolonged droughts of more than two months duration may alternate with many months of excessive rainfall (Ocean Island with an average of 200 cm recorded 38.0 cm and 308 cm, respectively, in two consecutive years). On either side of this zone rain arises from disturbances associated with the movement of the ITCZ. The rainfall in this "tropical" zone is more seasonal and tends to follow the sun. We may divide the Pacific islands into the following groups: dry islands with low and variable rainfall; wet islands that are strongly influenced by the dynamics and movement of the ITCZ; islands with alternating wet-dry seasons including those susceptible to typhoons and those in which monsoon effects are significant.

In the high islands, as in the Caribbean, rainfall is orographic, especially where island mountain ranges lie directly across the path of prevailing winds. In these cases, windward slopes tend to receive a higher annual total than leeward slopes. This results in sharp discontinuities in vegetation and soil distributions across mountain transects although the effect is partially observed in seasonal reversals of winds which may be expected as in the Fiji Islands. In some Melanesian islands, notably in New Caledonia and the tail of Papua-New Guinea, the main axial mountain ranges parallel the direction of the prevailing winds so that the coastlands on both sides have a low rainfall (260 cm annually in most parts).

Hurricanes, Typhoons, and Tropical Cyclones in Small Tropical Islands

Tropical cyclones are an integral part of the general circulation of the troposphere in low latitudes and of the climate of most tropical islands. They serve to transport accumulated energy and momentum together with precipitable water mass poleward. Because vorticity is essential to their formation, they do not develop at latitudes closer than five degrees to the equator where the Coriolis parameter is too small to provide the necessary cyclonic rotation. Some tropical islands in the Pacific basin, therefore, are not affected by them. These cyclones, typically 100 km to more than 1500 km in diameter at maturity, derive their kinetic energy from the latent heat of condensation. They are generally associated with waves in the deep upper Easterlies and appear to require weak vertical shear. Once formed, they tend to move westward and may turn northeastward at about 30-35°N and south latitude. Tropical cyclones are designated hurricanes in the tropical Atlantic and eastern Pacific, typhoons in the western Pacific, and cyclones in the Indian Ocean when their strongest sustained surface winds (averaged over one minute) exceed 33 m s^{-1}; with winds between 17 m s^{-1} and 33 m s^{-1} they are called tropical storms. Winds in a hurricane increase toward the center where the surface pressure is usually below 950 mb, and sustained wind velocities that reach values between 50 m s^{-1} and 110 m s^{-1} near the center. Apart from the high winds, other destructive features of hurricanes and typhoons include torrential rains over wide areas that lead to instantaneous and severe flooding and storm surge heights of 5 to 20 m above normal high tide levels resulting in coastal inundations; all of these lead to deaths and extensive damage to commercial and industrial infrastructure and other property. The atmospheric dynamics governing these storms are similar in all parts of the tropics. There may be differences in intensity and geographic movements due to the shape and size of the oceanic basin, the level of sea-surface temperature, and the position of warm and cold ocean currents. Therefore, further discussion of these storms will focus on hurricanes in the greater Caribbean region.

Caribbean hurricanes have been extensively studied by Riehl [1954], Tannehill [1956], Simpson [1973], Saffir [1977], Neumann et al. [1981], and Simpson and Riehl [1981]. They form over the ocean where surface temperature is in excess of 26.7°C, travel on average about 15-30 km hr^{-1}, and may be very erratic in their movement. The energy of an average hurricane is equal to 5×10^5 atom bombs of the type used on Nagasaki, almost all of it coming from the release of the latent heat of condensation. Between 1871 and 1980, the average duration of tropical cyclones in the region was about eight days with a modal duration of six days [Neumann et al., 1981]. In that same period, the average occurrence was 4.9 per year although the mean occurrence was not temporally stationary. If we take periods of equal length, Caribbean cyclones averaged eight annually before 1900, six between 1900 and 1930, and 10 between 1931 and 1960 (Figures 5a-5e). Since 1951, the average has decreased slightly and, although the causes of the fluctuations are not entirely known, there is evidence to suggest that oscillations in sea surface temperature in both the Atlantic and eastern Pacific Oceans are key factors [Wendland, 1977]. The comparatively larger number of ENSO occurrences in that period (1957-1958, 1963, 1965, 1969, 1973, 1976-1977, 1982-1983, 1986, etc.) may have also been a determining factor in their temporal distribution since there appears to be an inverse relationship between Caribbean hurricane occurrences and ENSO [Shapiro, 1982]. What is significant in terms of human habitation and the viability of the region for sustained development is that since 1960 the intensity of hurricanes has increased, giving rise to some of the most devastating storms ever experienced in the region, e.g., David, Allen, Gloria, Gilbert, Hugo, Andrew, etc.

The damage potential of hurricanes can be assessed through the Saffir/Simpson scale. Hurricanes above category 3 on that scale can result in loss of life and substantial environmental destruction. In the eastern Caribbean alone there were at least three hurricanes of category 5, six of category 4, and six of category 3 in the period 1951-1980 for an average of one hurricane every two years capable of causing death and destruction traversing the Lesser Antilles [Granger, 1990].

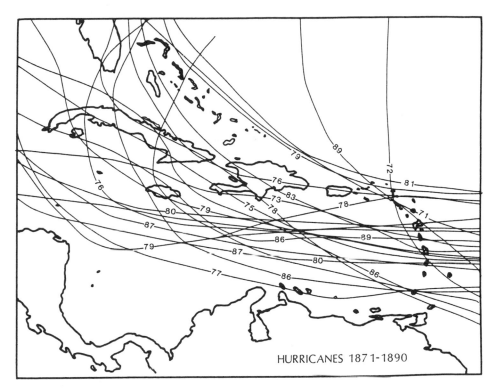

Fig. 5a.   Tracks of some hurricanes that have affected the Caribbean during 1871-1890 [modified from Neumann *et al.*, 1978].

Of the estimated U.S. $40 billion a year in damages caused by natural disasters globally, 40% is due to floods, 20% to tropical cyclones, and 15% to drought. The impact of these hazards expressed as proportions of GNP is 20-30 times higher in poor than in rich countries and among poor than rich in these countries [Kates, 1980]. A convoluted set of circumstances make it difficult to effect the kind of precise allocation of damages suggested in the above estimates in the Caribbean region or in many tropical regions for that matter. Paradoxically, reduced hurricane hazards lead to increased drought hazard, whereas more hurricanes tend to increase damage from floods and the risk of damage due to landslides.

If the expected greenhouse-gas induced global warming and sea level changes materialize as most climatologists expect, the implications for the Caribbean region and for most small tropical islands could be serious. The indications from models and simulations of tropical circulation patterns and of hurricanes are that the intensity of hurricanes will increase by up to 60% [Emmanuel, 1987], if the GCM derived equilibrium temperature to 2030 is   realized. Already in the Caribbean, temperature changes are consistent with those for the Northern Hemisphere. The decade ending 1986 was the warmest in the last 60 years as was the pentad 1982-1986. Some of the warmest years in the record of most stations are the same ones identified in hemispheric studies [Jones *et al.*, 1986, 1989; Karl and Jones, 1989; Hansen and Lebedeff, 1987]. For most stations in the region, monthly mean annual and seasonal temperatures have been increasing at between 0.15°C and 0.47°C per decade, a rate that is an order of magnitude greater than that of the whole hemisphere. The variability in the data set is quite large so that trends are masked by the interannual variability. A cautious extrapolation of the signals from the regional data suggest, however, that increases in the annual and seasonal temperature of between 0.75°C and 2.3°C over the next 50 years can be expected [Granger, 1991].

Between 1931 and 1986, annual rainfall in the Caribbean region increased between 0.2 and 1.0 cm yr^{-1} for islands south of 15°N but decreased between 0.3 and 0.6 cm yr^{-1} between 15°N and 20°N. These changes have occurred in both the rainy and the dry seasons. Such changes have been small compared to the total annual rainfall on most

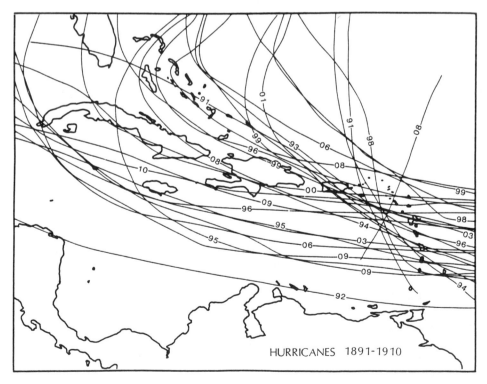

Fig. 5b. Tracks of some hurricanes that have affected the Caribbean during 1891-1910 [modified from Neumann *et al.*, 1978].

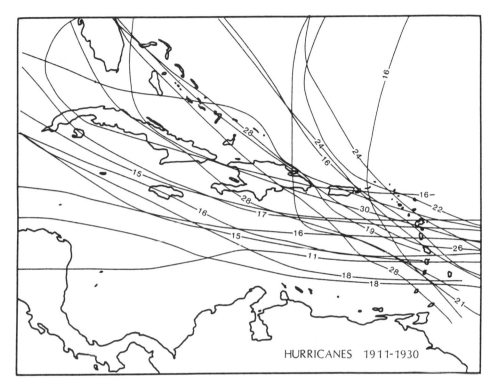

Fig. 5c. Tracks of some hurricanes that have affected the Caribbean during 1911-1930 [modified from Neumann *et al.*, 1978].

islands. In the Caribbean, a wide range of magnitudes and even directions of sea level changes over the past 60 years exist because of tectonic instability, sediment accumulation and compaction. The noise in the data make the extraction of spatial signals difficult. In recent years regional relative sea level rise estimated from tide gauge trends show a 3.2 mm yr^{-1} increase and estimates are that over the next 40 years the rise will be between 15 and 20 cm greater than the worldwide average because of natural subsidence and subsidence due to petroleum and natural gas extraction, ground water pumping, and sediment compaction [Gable, 1987].

Putting all of these observed changes and the trends derived from them, we get a realistic basis for a climate scenario for the Caribbean region to the year 2030: mean temperatures 0.75°C to 2.0°C warmer than today, but given the uncertainties in the data and extrapolations from them, the mean could be as small as 0.5°C or as large as 2.3°C; over the ocean, slight transient increases in SST will continue in some regions with decreases in others, but these would be followed by a more widespread increase of between 2.3°C and 4.8°C by 2030. Such temperature changes are likely to alter the magnitude and probably the frequency of extreme events such as hurricanes, floods, and droughts and of extreme heat loads on humans; sea level rise will destroy coastal settlements by inundation and coastal erosion, affect coastal aquatic systems and induce salt water intrusions in aquifers leading to substantial retreat from coastal economic zones with deleterious effects on the tourism upon which most small tropical islands depend; all of these together will lead to changes in bioproductive systems both positively and negatively depending on the ecosystemic structure and the adaptability of the biota, and increased degradation of land and water quality. Such short-term changes in the frequency of extremes would produce a succession of natural disasters that will threaten the viability of many small island states everywhere in the tropics.

Small Tropical Islands and Sustainable Development

This chapter deals with small tropical islands in the context of sustainable development. Consequently, there are a number of important operative or defining terms here--

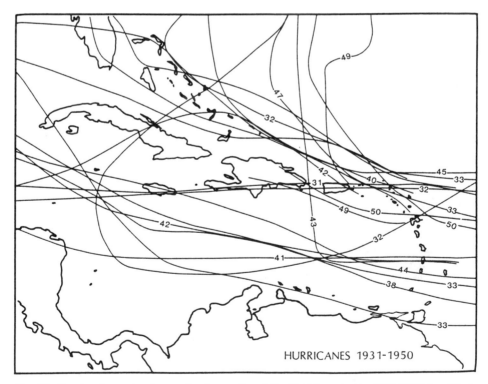

Fig. 5d. Tracks of some hurricanes that have affected the Caribbean during 1931-1950 [modified from Neumann *et al.*, 1978].

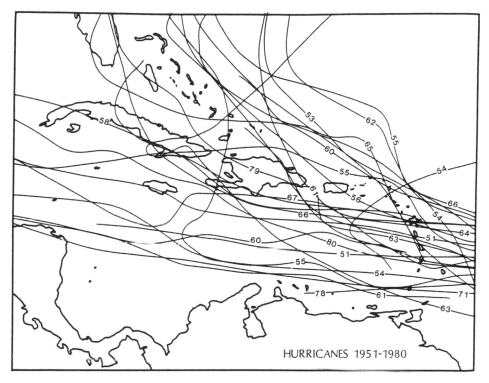

Fig. 5e. Tracks of some hurricanes that have affected the Caribbean during 1951-1980 [modified from Neumann *et al.*, 1978].

"islands," "small," "tropical," "development," and "sustainable"--that pose significant analytic problems. We have already dealt with "islands"; now for a look at "small."

Whether an island is small or not is more a matter of interpretation than of fact. The definition usually incorporates not only areal extent but also size and distribution of population and economy. There are some large islands with sparse populations and a comparatively limited economy that may be classified as "small," while others are limited but have population densities and economies, specified by their gross disposable products, GDPs, as large as or larger than some continental entities and are classified as "large." Some researchers have argued for various combinations of area and population. For some, "small" is 13,000 to 20,000 km^2 with populations less than 1.0-1.2 million [World Bank, 1983; Dolman, 1982; Doumenge, 1983; UNCTAD, 1971]. For others, a small island has a population of less than 400,000 and an arable land area of 700 to 2,500 km^2 [Jalan, 1982; Non-Aligned Movement, 1983]. Yet others have proposed a higher cut-off for area and population: 5-10 million population and 25,000 to 150,000 km^2 [Taylor, 1971; Demas, 1965; Kohr, 1977]. Many have questioned whether areal extent has any analytic or prescriptive value [*cf.* Selwyn, 1980]. In the context of development, however, the United Nations in many of its resolutions arising from discussions in UNCTAD since 1972 has identified smallness, remoteness, constraints on transportation and communication, great distances from market centers, limited internal markets, limited natural resources base, dependence on a few commodities for their foreign exchange earnings, heavy financial burdens and shortage of administrative personnel (UNCTAD Resolution 111(V), para. 1; Resolution 37/206 of 1982 and 138 (vi) of 1983) as developmental "handicaps." Not only does smallness head the list but several of the other constraints are directly related to small areal extent. Dommen [1980] compared a sample of small island countries with a corresponding sample of continental countries with respect to a number of social and natural characteristics and concluded that size was an important factor. Further, of the developing countries and territories with populations less than one million listed in the *UN Statistical Yearbook*, there are relatively few (16) that are not islands compared to 60 which are. A large proportion of these island states and territories are concentrated in the Caribbean and the South Pacific and, therefore, the considerable body of literature on these regions necessarily focuses on the problems of small size.

Truly small islands are those whose volumes are so small that oceanic insularity cannot be overcome. They usually have large coastline lengths to land area ratios. As island dimension grows, this ratio decreases. Insularity is further accentuated if there are numerous coastal indentations that lengthen the coastline and fragment emerged land areas. Martinique in the Caribbean has a land area of 1,080 km^2 and a coastline length of 360 km for an index of 1:3, whereas Tahiti in the Pacific with 160 km of coastline for an area of 1,040 km^2 has an index of 1:6.5. Once the ratio falls to 1:25 (1 km of coastline for 25 square kilometers of area) continentality affirms itself becoming more pronounced as the ratio falls further [Doumenge, 1983]. Socio-cultural factors and their ultimate expression are very important in defining the insular characteristics of islands. The more numerous and powerful the links with the outside world, the less pronounced will be insularity while geographic, socio-economic and political isolation increase the characteristics of insularity no matter what the size. Castro's Cuba is a case in point while the political divisions and the socio-economic depression of the Lesser Antilles greatly increases relative insularity of islands that are close to each other. Political partitioning, as in Hispaniola and New Guinea, can create or encourage characteristics of insularity.

The concept of insularity in the sense of boundedness and isolation has always been closely associated with small islands but it should not be overworked. Imperialism and colonial exploitation contaminated small islands with alien cultures, inappropriate technologies, and diseases that wiped out whole populations and cultures. More recently, political and economic hegemony and new communications technology have all but obliterated geographic, social, cultural, and economic isolation. Resource exploitation or super power military strategic requirements have resulted in environmental despoliation and the displacement of populations in many remote islands. Technologically induced global warming and sea level rise to which small islands have contributed very little, if anything, are threatening to obliterate small islands and atolls so much so that some atolls in the Maldives and the Seychelles are already being evacuated because of rising sea water. In short, the world's remotest islands have been drawn into the problematic repercussions of continental industrialization without any benefits to themselves despite their geographic isolation.

We are still left with "development" and "sustainable." There has been almost universal adoption of the phrase "sustainable development." The concept was important in the 1972 UN Conference in Stockholm on the Human Environment; it was central in the World Conservation Strategy published in 1980 [IUCN, 1980] and is the foundation of the report of the World Commission on Environment and Development in 1987 [Brundtland, 1987]. It has been extensively written about [Riddell, 1981; Sachs, 1979, 1980; Glaeser, 1984; Redcliff, 1987; Clark and Munn, 1987; Carim et al., 1987] and has been seriously considered in economic studies [Goodland and Ledec, 1984; Pearce, 1988a, 1988b; Turner, 1988]. The United Nations Environment Program (UNEP) and the International Union for Conservation of Nature and Natural Resources (IUCN) have backed sustainable development. This widespread embrace of sustainable development by widely disparate groups is not accidental in that it reflects the success of environmentalist pressures combined with renewed concerns about the global environment particularly trace gas-induced global warming and stratospheric ozone depletion. A pressing and important although probably unpopular question is how deep is the commitment to this apparent revolution; to what extent is "sustainable development" or "ecodevelopment" backed up by logical, coherent, theoretical concepts rather than convenient rhetoric.

In *Our Common Future* [Brundtland, 1987, p. 43], "sustainable development" is defined as "development which meets the needs of the present without compromising the ability of future generations to meet their own needs." This is a superficially attractive, beguilingly simplistic and high-sounding statement [O'Riordan, 1988, p. 29]. On the one hand, it could be used to demonstrate the relevance of the proper management of ecosystems, or as a concept, that it truly integrates environmental issues into development planning. On the other hand, development theorists and politicians use the label to suggest radical reform and liberal participatory approaches without providing the specifics of change or the political economy of reform. The unclear and overlapping meaning of sustainable development is the semantic and political confusion about what development itself means. In most cases development and conservation are defined and the concepts intermingled so as to assure their inevitable compatibility. For example, development is presented as "the modification of the biosphere and the application of human, financial, and both inanimate and animate resources to satisfy human needs and improve the quality of life [IUCN, 1980, para. 1.4], while conservation is the management of the human use of the biosphere to yield the greatest sustainable benefits. Consequently, conservation and development are mutually dependent [IUCN, 1980, para. 1.10; Brundtland, 1987, p. 3] and it is simplistic to attempt to deal with development and environmental problems without a broader perspective encompassing the factors that underlie world poverty and international inequity. From this it is clear that sustainable development must be global in scope and internationalist in formulation. It must both be ecocentrist as well as anthropomorphic.

The ideology of development rooted in mercantilism and economic liberalism outlines a process that reproduces an industrialized, urbanized, democratic and capitalist world in which economically successful, modern and affluent societies emerge [Goulet, 1971; Chilcote, 1984; Aseniero, 1985; Frank, 1987]. Here, development means the projects and policies, the infrastructure, flows of capital, transfers of technology and ways of doing things that imitate the so-called developed world--"the North." Yet the social and economic conditions of the majority of people in many developing and undeveloped countries worsened at the peak of development in the 1950's at the same time that adverse environmental impacts in terms of lost ecosystems and species, unreasoned exploitation of depletable natural resources for use in the industrialized world and air, land, and water pollution proceeded apace. Farmers in Haiti, Dominica, Grenada, St. Kitts/Nevis and the Philippine Islands farm eroded hillsides through necessity and not perversity, but degraded environments create poverty. These farmers have neither the freedom to stop degradation nor the opportunity to easily move elsewhere except as eco-refugees. Poverty and environmental degradation produce an inescapable trap and demonstrate the centrality of social, political, and economic issues in environment and sustainable development.

Sustainable development implies an anticipate-and-avoid, rather than a react-and-cure approach to development [IUCN, 1984]; it demands that we never engage in actions that reduce the potential of future generations to meet their needs; it should encourage activity that raises social welfare with the maximum of resource conservation and the minimum amount of environmental degradation allowable within given economic, social, and technical constraints [Brandt, 1983]. We have touched on some of the environmental constraints facing small tropical island states: infertile soils, unreliable water supply, geologic and meteorological hazards that include volcanic eruptions, earthquakes, land slides, hurricanes and typhoons, droughts, and floods.

The history of island development is replete with outsiders getting rich at the expense of islands and islanders. In Grenada, St. Vincent, Dominica, and St. Lucia in the eastern Caribbean, for example, the primacy of bananas in the 1960's as a replacement for plantation crops such as sugarcane, cocoa, coffee, and coconuts was supposed to ensure peasant autonomy through changes in land ownership and control, marketing modes, and wage-labor relationships. Bananas were a more viable export crop due to their resiliency after hurricanes which were more prevalent there in the period 1951-1980 than in the preceding 40 years, in addition to doubling as a food staple. It was to have signaled the triumph of peasant farmers over the landed gentry of the estates and the bondage of slave-like wage labor. It should have been the reverberation of Eric Williams' slogan "Masa day done." However, the monopolistic control of the banana market, price, and transportation by Geest Industries and later in Dominica, by a consortium of American investors rendered peasant autonomy more apparent than real. Geest Industries gets the profits while peasants bear the risks arising from climate disasters, crop diseases, and market and currency fluctuations [Granger, 1990]. Similarly, "industrialization by invitation" was the development clarion call in the Caribbean in the 1950's that resulted in environmental degradation due to resources exploitation, *e.g.*, bauxite in Jamaica and extensive capital export. The promise of economic diversification into tourism as a major means of foreign currency acquisition resulted in the financial aggrandizement of outsiders, providing a few menial jobs for islanders while expenses accrued to the islands through tax breaks and other financial incentives, air and water pollution and social problems [Granger, 1990].

Despite these drawbacks, tourism has become crucially important in the economies of literally hundreds of small tropical islands. There is no doubt that the leisure industry is a growth industry throughout the tropics and in islands in particular. As a result, tourism has become the predominant industry in hundreds of small tropical islands, and in many cases now, the only one, leading many of these islands to absolute dependence on that industry. There are environmentalists and most certainly politicians who view tourism as the panacea for tropical island economies under stress today. The new and rapid development of ecotourism in many small tropical islands worldwide attests to the confidence of conservationists that tourism and environmental conservation can co-exist to the economic and ecological enhancement of these small islands. Of course, the approach is of recent vintage and so the jury is still out on that question. Among the most noteworthy of the smallest of these tropical islands to become dependent on tourism as the backbone of their economy are Batam, Bintan and Biak in Indonesia; several islands in the central Pacific with new jet airports; some of the islands of the Bahamas, Caicos Islands, and Caman Brac in the Caribbean.

This push by small tropical island states to make tourism the mainstay of their economies should raise a number of substantial concerns by planners and policy makers. Tourism is very susceptible to environmental factors such as sea level rise, climate change and natural disasters. In the past decade, a series of devastating hurricanes and typhoons, and droughts and floods have caused insurance companies and in particular global re-insurers like Lloyds of London, to reevaluate their willingness to take heavy risks against disasters in small islands. Many Caribbean islands are presently being affected by this change in policy by insurers. Consequently, governments are being asked to underwrite development in tourism in the absence of commercial insurance, but not only are they reluctant to do so in the present financial climate but most of them do not have the means.

If we look at small island states today it is possible to infer that they are locked in a permanent state of crises of politics, economics, and environmental degradation--crises of development and environment. The former emphasize problems of debt, falling commodities prices, falling per capita food production, growing poverty, and growing socio-economic inequities and automatically raises questions about viability and size. In so far as viability is taken to mean a state's ability to balance its budget, to bring about political stability, economic development and social transformation, to maintain specified levels of public services, international representation, military establishment, *etc.,* all paraphernalia of a modern industrial state, rather than transforming and modernizing society not necessarily to conform to western concepts and ways of life, but rather in keeping with its own cultures, mores and value systems, so long will *ad hoc* approaches that do not incorporate discussions and understanding but treat symptoms rather than causes predominate to the detriment of small tropical islands.

Finally, in the preceding discussion we have used "small tropical island" as our geographic entity and raised the question as to whether it is a useful analytic category. It is apparent that when dealing with the concept of development we need more than territory or physical environment; we need social human organizations to provide the framework for governance and we need to examine it as an integrated study of population-environment interactions. Therefore, while it is possible to describe the geography of small tropical islands, when doing so in the context of sustainable development, we need to think of "states" or "countries," of human-use systems rather than structurally integrated objects called islands. The interconnectedness of people and places does not inhere within neatly bounded spatial units. There are many single islands that are states but many island states or countries, especially in the Pacific, are made up of many islands, typically an archipelago spread over vast areas of ocean space. The Maldives is more than 1000 islands spanning an arc more than 800 km long; the Seychelles is composed of more than 100 volcanic and coralline islands spread over a length of more than 1000 km; the 33 islands that make up Kiribati are scattered over 3.5 million km^2 of the Pacific Ocean; Fiji is made up of 320 islands, Tonga of 150 and Vanuatu of 80; French Polynesia has 130 islands spread over 5×10^6 km^2 of ocean. Island regions are vast. In the Pacific, the Solomons are separated from French Polynesia by 5,500 km or about the distance from Paris to New York. The Caribbean is compact by island standards, yet Jamaica to Barbados is the same distance as Toronto to Miami or London to Leningrad.

An island state may contain coral atolls with infertile soils and little more than subsistence agriculture together with volcanic islands with rich soils and a cash economy; significant differences in degree of urbanization, ethnic composition, social systems, cultural values, and even language. Whether as a country of tropical islands or as a tropical island country, they all face special sets of constraints to their sustainable development. We list them here subject to elaboration later: diseconomies of scale, limitations on natural resources base, dependence on a very narrow range of tropical agricultural products, distance from major markets resulting in high transport costs, serious balance of payments problems due to stagnating exports and rapidly growing volumes of imports, especially food and energy, very narrow range of local skills and heavy reliance on expatriates, limited access to capital markets and a heavy dependence on aid agencies and external institutions, proneness to natural disasters from hurricanes and typhoons, volcanic eruptions, earthquakes, droughts and floods, highly fragile ecology and very vulnerable physical environments and dependence on one or a few large foreign-owned companies operating under privileged terms [Granger, 1990]. Some of these constraints are almost insurmountable. There is very little many of these island states can do about their size or their resource endowment or about economies of scales but to come together economically or politically and exploit their comparative advantage. They can diversify their economies so as to promote capital formation and real local savings and away from narrow ranges of tropical agricultural products. In most of these states land is a non-renewable (fixed stock) resource, although it has a renewable capacity to support bioproductive systems indefinitely but only with good management.

References

Aseniero, G., A reflection on developmentalism: From development to transformation, in *Development as Social Transformation: Reflections on the Global Problematique*, edited by A. Aldo *et al.*, pp. 48-85, Hodder and Stoughton, Sevenoaks, for UN University, 1985.

Bayliss-Smith, T. P., R. Bedford, H. Brookfield, and M. Latham, *Islands, Islanders and the World: The Colonial and Post-Colonial Experience of Eastern Fiji*, 321 pp., Cambridge University Press, 1985.

Brandt, W., *Common Crisis North-South: Cooperation for World Recovery*, 174 pp., Pan, London, 1983.

Brundtland, H., *Our Common Future*, 383 pp., Oxford University Press, 1987.

Carim, E., G. Barnard, G. Foley, D. de Silva, J. Tinker, and R. Walgate (Eds.), *Towards Sustainable Development*, 112 pp., Panos Institute, London, 1987.

Chilcote, R. H., *Theories of Development and Underdevelopment*, 278 pp., Westview Press, Boulder, 1984.

Clark, W. C., and R. E. Munn (Eds.), *Sustainable Development of the Biosphere*, 491 pp., Cambridge University Press, Cambridge, 1987.

Coleman, P. J. (Ed.), *The Western Pacific: Island Arcs, Marginal Seas Geochemistry*, 675 pp., Crane Russak and Co. Inc., New York and University of Western Australia Press, 1973.

Demas, W. G., *The Economics of Development in Small Countries with Special Reference to the Caribbean*, 64 pp., McGill University Press, Montreal, 1965.

Dolman, A. J., The development strategies of small island countries: Issues and options, in *Small Island Countries, Regional Cooperation and the Management of Marine Resources*, 167 pp., The Hague, RIO Foundation, 1982.

Dommen, E. C. (Ed.), *Islands*, 216 pp., Pergamon Press, Oxford, 1980 (originally a special issue of *World Development*, *8*(12), 1980).

Dommen, E. C., Some distinguishing characteristics of island states, in *Islands*, edited by E. C. Dommen, 216 pp., Pergamon Press, Oxford, 1980.

Donnelly, T. W., Caribbean island-arcs in the light of the seafloor spreading hypothesis, *N. Y. Acad. Sci. Trans.* (Ser. 2), *30*(6), 745-50, 1968.

Doumenge, F., Viability of small island states, UNCTAD Document TD/B/950, 25 pp., 1983.

Emmanuel, K. A., The dependence of hurricane intensity upon climate, *Nature*, *326*, 483-485, 1987.

Frank, L., The development game, *Granta*, *22*, 231-243, 1987.

Gable, F., Changing climates and Caribbean coastlines, *Oceanus*, *30*(4), 53-56, 1987.

Glaeser, B. (Ed.), *Ecodevelopment: Concepts, Projects, Strategies*, 247 pp., Pergamon Press, Oxford, 1984.

Goodland, R. J., and G. Ledec, *Neoclassical Economics and Principles of Sustainable Development*, 100 pp., World Bank Office of Environmental Affairs, Washington, D.C., 1984.

Goulet, D., *The Cruel Choice: A New Concept in the Theory of Development*, 362 pp., Athanaeum, London, 1971.

Granger, O. E., Climatic fluctuations in Trinidad, West Indies and their implications for water resources planning, *Caribbean J. Sci.*, *17*(1-4), 173-201, 1982.

Granger, O. E., The hydroclimatonomy of a developing tropical island: A water resources perspective, *Ann. Assoc. Amer. Geogr.*, *73*(2), 183-205, 1983.

Granger, O. E., Caribbean climates, *Prog. Phys. Geogr.*, *9*(1), 16-43, 1985.

Granger, O. E., *Natural Disasters and Social Change: An Eastern Caribbean Perspective*, 271 pp., CRT Book Printing, Oakland, 1990.

Granger, O. E., Climatic change interactions in the greater Caribbean, *The Environ. Prof.*, *13*, 43-58, 1991.

Hansen, J., and S. Lebedeff, Global trends of measured surface air temperature, *J. Geophys. Res.*, *92*(13), 345-372, 1987.

IUCN, *The World Conservation Strategy*, 116 pp., UNEP, World Wildlife Fund, Geneva, 1980.

IUCN, *National Conservation Strategies: A Framework for Sustainable Development*, 121 pp., IUCN, Geneva, 1984.

Jalan, B., Classification of economies by size, in *Problems and Policies in Small Economies*, edited by B. Jalan, 275 pp., Croom Helm, London, 1982.

Jones, P. D., S. C. B. Raper, R. S. Bradley, H. F. Diaz, P. M. Kelly, and T. M. L. Wigley, Northern hemisphere surface air temperature variations, 1851-1984, *J. Climatol. Appl. Meteorol.*, *25*, 161-179, 1986.

Jones, P. D., P. M. Kelly, C. M. Godess, and T. R. Karl, The effect of urban warming on the northern hemisphere temperature average, *J. Climatol.*, *2*, 285-290, 1989.

Karl, T. R., and P. D. Jones, Urban bias in area-averaged surface air temperature trends, *Bull. Amer. Meteorol. Soc.*, *70*, 265-270, 1989.

Kates, R. W., Climate and society: Lessons from recent events, *Weather*, *35*(1), 17-25, 1980.

Kohr, L., *The Overdeveloped Nations*, 184 pp., Christopher Davies, 1977.

Limbird, A., Micro-environments and the need for localized soil testing in the tropics, in *The Tropical Environment*, Proc., Int. Symp. on Physical and Human Resources of the Tropics, edited by L. C. Nkenderin, 178 pp., Univ. of Calgary, 1988.

MacGillavry, H. J., Geological history of the Caribbean, *Knonkl. Nederlandse Akad. Wetensch. Proc. Ser. B*, *73*(1), 64-96, 1970.

Neumann, C. J., G. W. Cry, E. L. Caso, and B. R. Jarvinen, Tropical cyclones of the North Atlantic Ocean, 1871-1980, U.S. Dept. of Commerce, NOAA-National Climatic Center, Ashville, N.C., 174 pp., 1981.

Non-Aligned Movement, Final document of the non-aligned meeting of experts on small island developing countries, NAC/CONF. 7/EM/DOC .4/Rev. 2, 1983.

O'Riordan, T., The politics of sustainability, in *Sustainable Environmental Management: Principles and Practice*, edited by R. K. Turner, pp. 29-50, Westview Press, Boulder, 1988.

Pearce, D., The sustainable use of natural resources in developing countries, in *Sustainable Environmental Management: Principles and Practice*, edited by R. K. Turner, pp. 102-117, Belhaven, London, 1988a.

Pearce, D., Economists befriend the earth, *New Scientist*, 34-39, 1988b.

Redclift, M., *Sustainable Development: Exploring the Contradictions*, 221 pp., Methuen, London, 1987.

Riddell, R., *Ecodevelopment*, 218 pp., Gower, Aldershot, 1981.

Riehl, H., *Tropical Meteorology*, 392 pp., McGraw Hill Co., New York, 1954.

Sachs, I., Ecodevelopment: A definition, *Ambio, 8*(2/3), 113, 1979.

Sachs, I., *Strategies de l'ecodeveloppement,* 216 pp., Les Editions Ouvrieres, Paris, 1980.

Saffir, H. S., *Design and Construction Requirements for Hurricane Resistant Construction,* 20 pp., American Society of Civil Engineering, New York (Reprint 2830), 1977.

Selwyn, P., Smallness and islandness, in *Islands,* edited by E. Dommen, pp. 67-81, Pergamon Press, Oxford, 1980.

Shapiro, L. J., Hurricane climate fluctuations, Part II: Relation to Large-Scale circulation, *Mon. Weather Rev., 110,* 1014-1023, 1982.

Simpson, R. H., Hurricane prediction: Progress and problems, *Science, 181*(4103), 899-907, 1973.

Simpson, R. H., Hurricane prediction, in *Geophysical Predictions,* pp. 142, National Academy of Sciences, Washington, D.C., 1978.

Simpson, R. H., and H. Riehl, *The Hurricane and its Impact,* 398 pp., Louisiana State University Press, Baton Rouge, 1981.

Tannehill, I. R., *Hurricanes, Their Nature and History,* 308 pp., 9th Rev. Ed., Princeton University Press, Princeton, 1956.

Taylor, C. L., 1971; Statistical typology of micro-states and territories: Towards a definition of a micro-state, in *UNITAR, Small States and Territories,* Arno Press, New York, 1971.

Turner, R. K., Sustainability, resource conservation and pollution control: An overview, in *Sustainable Environmental Management: Principles and Practice,* edited by R. K. Turner, pp. 167-184, Westview Press, Boulder, 1988.

UNCTAD, Developing island countries, Report of the Panel of Experts, TD/B/443/Rev. 1, United Nations Publication Sales No. E. 74. II. D. 6, 1971.

Wendland, W. M., Tropical storm frequencies related to sea surface temperatures, *J. Appl. Meteor., 16*(5), 477-481, 1977.

World Bank, *World Development Report,* 146 pp., The World Bank, Washington, D.C., 1978-1983.

11

Small Island Geology: An Overview

Georges Vernette

Abstract

Small islands are generated by different processes. Some of them correspond to volcanic and tectonic activities and occur in plate boundaries (island-arc systems and ocean ridge islands) or far from plate boundaries (intraplate volcanic islands). On passive margins, islands resulting from folded or diapiric structures are related to sea level changes. Islands can also be constructed by hydrodynamic processes (barrier islands) and terrigenous input in deltas and estuaries. In tropical zones numerous islands are constructed by coral reefs.

Introduction

The structure, physical features and environmental attributes of small islands vary according to the interaction of geological, biological and meteorological processes. These processes vary in time and space and leave distinctive geological signatures. Consequently, in this paper it is possible to broadly group these processes and their products - the small oceanic islands - according to the major controls on their formation.

Islands Created by Volcanic and Tectonic Activities

Ninety percent of volcanic and tectonic activities is concentrated within, or adjacent to, plate boundaries. One of major consequences of these activities for general earth features is the formation of numerous topographic highs, some of them rising above of sea level as islands.

Small Islands: Marine Science and Sustainable Development
Coastal and Estuarine Studies, Volume 51, Pages 188–204
Copyright 1996 by the American Geophysical Union

Island Arc Systems Related to Subduction Zone

Island arc systems are among the most spectacular tectonic features of the earth and represent the sites where the ocean plate is subducted under another plate. Geomorphologically, the island arc systems exhibit the following characteristics:

1. Arcuate line of islands (Figure 1).
2. Prominent recent volcanic activity.
3. Deep seismicity (earthquake focus deeper than 70 km).
4. Deep trench on the ocean side (trench deeper than 6000 m).
5. Shallow seas on the continental side of the arc.
6. High heat flow on the continental side of the arc.
7. Gravity anomaly [Sugimura and Uyeda, 1973].

Island arcs are conspicuous features in the circum-Pacific, especially in the western sector where a tectonic and volcanic island belt can be clearly traced northward from New Zealand, through Melanesia, into Indonesia, the Philippines, Japan, Kuriles, Kamtchatka and eastward through the Aleutian Islands (Figure 1). More than 75% of the active and recent extinct volcanoes are located in the circum-Pacific belt. In the Atlantic Ocean, island arc systems occur in the Lesser Antilles Arc and Scotia Arc. In the Indian Ocean they occur only in Java and Sumatra. All of these island arc systems exhibit the convex side of the arc facing toward the ocean.

Island arc areas are marked by a high negative gravity anomaly of the order of -300 milligals (mgal) over the trench and a positive anomaly over the island arc itself. These anomalies show that isostatic equilibrium is absent in these areas. Specifically, these features indicate mass deficiency in the trench and mass excess in the island arc. This systematic arrangement indicates that all the islands and trench systems are caused by a common mechanism which corresponds to crustal subduction.

Subduction zone volcanism coincides with the focus of intermediate to deep seismicity paralleling the adjacent trenches. Subduction of oceanic crust creates stress to melt the volcanic magmas. The distribution of the volcanoes is restricted to the Benioff zone [Caron *et al.*, 1992]. The nature of the volcanism at these sites is explosive due to the highly volatile nature of the magma generated in the subduction zone (*e.g.*: Soufriere in St. Vincent, Pelée in Martinique and the many volcanoes of southeastern Papua, New Guinea). This leads to the production of a large pyroclastic material that expands on all neighboring regions, onto the adjacent shelf and carried by wind a vast distance from the source [Bardintzeff, 1993]. It is generally the only source of sediments for these islands. The circum-Pacific suite of volcanic rocks, displaying a broad spectrum of petrologic variation, is know as the "high alumina suite," and is generally marked by an abundance of andesite [Kenneth, 1982; Baker and Eggler, 1983]. The successive explosive eruptions generate strata of volcanic ash, breccia, including older lithospheric materials that may contain veins of valuable mineral resources (gold, copper, silver and nickel).

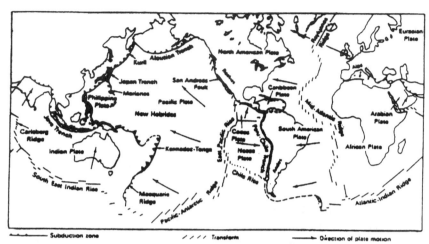

Fig. 1. Island arc systems and subduction zone.

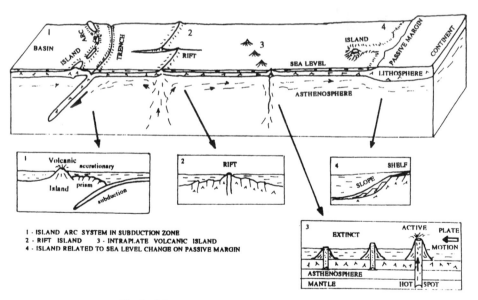

Fig. 2. Islands in different geological environments.

Morphologically, island-arc systems are a thousand to several thousand kilometers in length with a narrow width of only 200 to 300 km including the trench (Figure 2). Island arcs can occur as single or double arcs, as in Indonesia or in the Lesser Antilles, where "ancient" and "recent" arcs exist. Most island arc systems contain backarc or marginal basins between the island arc and the continent. Shelf topography on the ocean side of the island arcs is smooth and wider than shelf topography of the shallow sea side.

Seismicity indicates, for example, that the Caribbean plate is moving relative to the eastward at the rate of 2.1 cm/yr. The Lesser Antilles is the result of the tectonic stress against the Atlantic plate. The main volcanic phase of the Lesser Antilles are during Terciary [Butterlin, 1977]. The volcanic inner arc (Martinique, St. Vincent, St. Lucia, Grenada islands) is mountainous and juxtaposed by an external lowland calcareous arc (Anguilla, Antiqua, Barbuda, Marie Galante islands). These two characteristics are found together only in Guadeloupe where Basse-Terre is volcanic and Grand-Terre calcareous. The island arc system of southwest Pacific are also essentially Terciary, and basically of volcanic origin.

Islands Associated with Oceanic Ridges

Oceanic ridges form from the upwelling of material from the upper mantle (Figure 2). It is injected through tensional crustal fissures into a narrow zone only a few kilometers wide at the ridge axis. Volcanism is generally concentrated near the axis of a crustal rift or grabben (the central rift valley), which is contained by symmetrical hills which sometimes form islands, e.g., Iceland, Azores and Tristan da Cunha islands for the Mid-Atlantic Ridge, Galapagos islands in the Pacific ocean. The focus of seismicity is also concentrated near the surface [Williams and Birney, 1979]. Hydrothermalism is frequent and accompanied with iron and manganese oxide precipitates [Bougault et al., 1993]. These worldwide oceanic ridges along divergent plate margins correspond to the greatest source of magmatic rocks on the Earth's surface [Juteau, 1993].

The volcanic environment and the island construction correspond to lava sheet flows of basaltic material. Iceland, for example, is the most active volcanic region of the world. It consists of fissure-erupted lava plateaus and large conical volcanoes. Because of Iceland's position on the mid-ocean ridge, the island continuously in a state of tension as the east and western portions of the island are moved apart by sea-floor spreading: the oldest rocks occur in the eastern and the western part of Iceland. The rocks are dominated by basalt close to theolite in comparison; however, they exist large volumes of Alkalic rocks such as rhyolite [Maury, 1993]. In the Galapagos rift, the largest sheet flows are known to extend for 7 km from the eruptive center. Pillow lavas are marked by numerous bulbous protrusions up to a few meters in diameter and organized like accretion underwater phenomena. The distribution of volcanic features suggests that volcanism is episodic (about every 10,000 year phase). The Azores islands, annex of the Mid-Atlantic Ridge is an immense basaltic system yet active (the last volcanic event was in 1957).

Islands Related to Strike-Slip Faulting Zones in Plate Boundaries

Other kinds of islands related to plate boundaries correspond to the islands resulting of high tectonic activity in strike-slip faulting zone. In that case, the seismic activity is more important than volcanism. In such an environment, islands are generally more mountainous and bigger than islands from arc systems or oceanic ridges. Examples of these islands are widespread in the oceans: the Greater Antilles (Cuba, Puerto Rico, Hispaniola, Jamaica, and Caymen islands), and Trinidad in the Caribbean Sea; Fiji, New Caledonia, New Guinea, Solomon, and Seychelles in the Pacific ocean. They generally have a long geological period which remote frequently to the Cretaceous period. The seismic and tectonic activity is represented by frequent and high earthquakes, with catastrophic effects. These relatively extensive and highly complex islands are constituted by fault block mountains, and have a mixed metamorphic and sedimentary basement. They are rich in mineral resources: hydrocarbon in Trinidad, nickel in New Caledonia, and copper and manganese in Fiji.

Intraplate Volcanic Islands

Although about 90% of volcanic activity on the earth is concentrated within plate boundaries, other volcanoes exist at locations remote from these boundaries. This corresponds to the location of sea mounts and volcanic islands, which are themselves often found in island chains such as Hawaii, Marquise, Sala y Gomez or Society islands of the Pacific. Although sea mounts are widespread throughout all ocean basins, they are most abundant in the Pacific Ocean. A small proportion of volcanic features are high enough to break sea level and form oceanic islands. Only a few oceanic islands are known to have been historically active, such as in the Hawaiian islands.

The volcanic rocks of these intraplate islands consist of alkali basalts marked by less silica and higher alkaline content than theolites [Coulon, 1993]. Some of these sea mounts and volcano islands are long linear chains far from spreading axes which indicates they were tectonically controlled. Another major aspect of the linear island chains is that they are marked by a progressive increase in age of the volcanoes away from the ocean ridge. The magma that built the volcanoes is derived from a relatively fixed magma source in the upper mantle, which is called a "hot spot." Because the magma source lies beneath the crustal plate and because the plate is moving laterally, the active volcanoes are eventually separated from the hot spot, causing cessation of volcanism (Figure 2). The linear, island chain association is indicative of tectonic control on formation. Within chairs, the islands become progressively older with distance from the intra-plate "hot spots", locations at which magma wells up through the crust from the mantle. This is caused by movement of the tectonic plate away from the magma source, accompanied by a cessation of volcanism [Bullard, 1976]. If the Hawaiian seamount chain began to form more than 40 million years (40 Ma) ago [Kenneth, 1992], Marquises (7 Ma) and Society (5 Ma), are younger [Goslin and Maia, 1993]. The basaltic rocks contain few minerals of economic value, but the basic lava forms rich soils which attract human population.

These islands may rise to several thousand feet above the sea level close to the magmatic source, but crustal cooling and sinking lowers the profile. As island elevation and terrestrial run-off decreases coral reefs are able to form, and keep pace with the relative rise in sea level caused by crustal sinking. Ultimately, this can result, as for example in islands of French Polynesia, in the formation of a coral atoll [Caron *et al.*, 1992] with no external presentation of the original volcanic form, except for the approximately circular shape.

Other prominent features and oceanic islands standing 2 to 3 km shallower than the surrounding sea floor and free form earthquake activity are known as aseismic ridges. Different hypotheses have been proposed to account for the origin of these aseismic features:

1. Isolated fragment of continental crust as Rockall Plateau in the North Atlantic, the Mascarene Plateau (Maurice Island) in the western Indian Ocean, Crozet and Kerguelen Plateaus in the south. These isolated continental fragments have been called microcontinents.
2. Old linear volcanic features composed of oceanic crustal basalt as the Emperor volcanic chain in the North Pacific, the Ninetyeast Ridge in the Indian Ocean and the Walvis and Rio Grande rises in the South Atlantic.

Pre-continent Islands on Passive Margins and Sea Level Changes

On the passive continental margins, especially where coastal zones present steeped relief, exist many islands. In this case, the geological, petrographic and sedimentological characteristics of these islands are similar to the neighboring continent. They represent only the seaward prolongation of the continent: they are "pre-continent islands". These partly submerged relieves form "rias" and "fjords" which have generally folded metamorphic basement [Nicholson, 1974]. As the contrary of active margin shelves, the continental shelves of passive margins are very flat and wide (more than 200 miles west to the Great Britain). The coastal geomorphology of these shelf islands are deeply indented and characterized by alternance of cliffs and pebble or sand beaches [Klingebiel and Vernette, this volume], with tide flats in protected bays if tide amplitude is important. Sediments are essentially clastic due to the importance of back-land area and the active continental erosion.

Changes in sea level correspond to different processes which interact at various temporal and areal scales. On the continental shelf and coastal zone of passive margins, sea level changes are relatively more important than in active margins where volcanic and tectonic activities also occur: old metamorphic topographic highs may be transformed into islands due to sea level rise such as occurred during the Holocene period.

The Holocene sea level history in the last 12,000 years, based on radiocarbon dating, shows (Figure 3) a maximum peak of about 5 m above present mean sea level around

5,000 B.P. in Brazil [Martin *et al.*, 1987] and Tierra del Fuego, Argentine [Porter *et al.*,

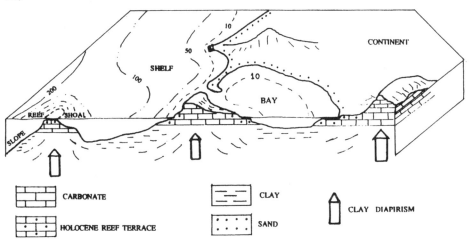

Fig. 3. Island issued by diapirism.

1984]. On the Atlantic coasts of France, the sea level rises rapidly between 10,000 to 7,000 years B.P., and becomes very slow during the last. In that country, no reliable evidence suggests the occurrence of a sea level higher than at present [Tiers, 1973]. The west side of Africa presents a similar trend. In the high latitude, many European and American continental shelf islands facing the Arctic Ocean show inverse trends due to the glacio-eustatism and isotatic response. For example, in Canada , sea level changes show that diverging trends may affect small islands of the Arctic sea several years after complete melting of the ice sheets [Pirazzoli, 1991].

Islands Created by Diapirism

Diapiric features consist of masses of low-density salt or mud mobilized after being deeply buried by sediments, generally as bulbous units and may form islands, especially on continental margins where sedimentation is very abundant. Examples of salt domes occur on the continental shelf off Texas [Jackson and Galloway, 1984] and northeastern Mexico and off Angola. Mud diapirs are usually found off large river deltas such as the Mississippi, Niger, or Magdalena, and result in compacting (vertical stress) of delta deposits. The consequent topographic highs are scattered specifically on the prodeltal of passive margin deltas. They may also result from tectonic stress; in that case, the fluid mud is mobilized along fault zones in accretionary sedimentary prism of plate boundaries [Griboulard *et al.*, 1991], and may outcrop such as in Barbados (Joe formation), south Trinidad or the southwestern part of the Colombian Caribbean Margin

[Vernette, 1985]. The source of the clay material is generally Miocene in age and the diaprism phenomena occurred from mid-Miocene to present [Duque-Caro, 1980].

On the shelf, some of these topographic highs emerge as islands. In the case of tectonic influence, the islands' location is related to the structural directions [Vernette *et al.*, 1992]. The sedimentation of these diapiric features is characterized by interbedded clay and sand clayed deposits. In tropical latitudes when the top of a diapir is located in the photic zone of a shelf, it may be colonized by coral reef (Figure 4), a source of carbonate sedimentation in a clastic sandy-clay environment.

Islands Constructed by Hydrodynamic Processes and Terrigenous Input

These islands result of sand accumulation on the coastal zone. They are built up by a complex combination of wave and tide action, and material input coming from rivers or coastal erosion. They are fringing evolving features. Their morphology depend essentially of the hydrodynamic processes and sediment budget of the coastal zone.

Barrier Islands

Barrier islands are long, straight features parallel to the shore and separated from the mainland coasts by lagoons, bays, or marshes. Barrier islands vary significantly in size. They may be only areas of beaches just above high water (long shore bar), or they may be major features up to 30 m in height with dunes and vegetation. These islands may consist of one or more ridges of dune sediment that mark successive shoreline position during progradation. Such islands are broken at intervals by tide inlets, and are kept

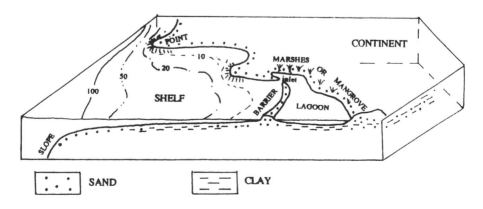

Fig. 4. Barrier island environment.

open by tidal effects, storm waves, or longshore transport [Gayet and Vernette, 1989; Fitzgerald, 1993].

Individual barrier islands range from a few kilometers to more than 100 km long and a few kilometers in width. These barriers are widespread features along most of the world. The longest barrier island in the world is Padre Island (Texas), which is about 200 km long and ranges from 1 to 8 km in width.

Barrier islands are supplied and molded essentially by marine coastal processes. They may be considered as the focal element of a much larger system that is termed the *barrier island system* [Oertel, 1985], which consist of six major coastal environments: (1) mainland; (2) back-barrier lagoon; (3) inlet; (4) barrier island; (5) barrier platform; and (6) shoreface. Other major environments of a barrier island are the beach, foredunes, washover fans, subaerial spits, tide flats and salt marshes [Leatherman, 1979]. The morphology of barrier islands is the result of a variety of marine and subaerial depositional and erosional processes.

The sedimentological features of barrier islands consist of three major clastic depositional environments:

1. Barrier beach complex and washover fans on landward.
2. Back-barrier region, or subtidal-intertidal lagoon.
3. Subtidal-intertidal delta and inlet channel complex [Reinson, 1979].

Sediment transport on the beach and shoreface is dominated by waves and wave-induced currents, although tidal currents may be important near inlets and channels. The wave swash runs up the foreshore and occasionally overtops the berm during high tides. The upper part of the barrier is an accumulation of loosed sediments influenced by aeolian and overwash processes. Dune ridges are contiguous with the berm and parallel to the shoreline. Barrier tidal flats are located on the lagoonal side of the barrier islands. Barrier spits occur at the ends of barrier islands and represent a process by which islands can migrate laterally.

Facies of barrier-beach and channel environments are mainly sand and gravel, whereas lagoonal deposits can be both organic-rich mud and sand (Figure 5). These sands come from the shelf or from the adjacent coastal lowlands.

Subsurface material that supports the barrier island is primarily related to the origin and evolution of the barrier island system [Leatherman, 1985]. Pre-Holocene topographic highs on submerged mainland surfaces provide platforms for future barrier islands. Some of the topographic highs are composed of lithified materials as in the west Florida coast where several barrier island systems have bedrock basement [Evans *et al.*, 1985].

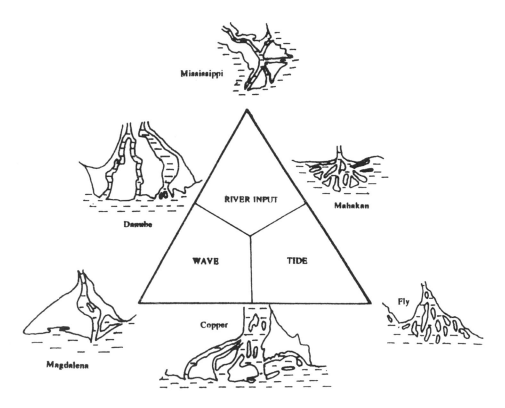

Fig. 5. Islands in estuarine and delta environment.

The present day barrier islands evolved during the Holocene period beginning about 10,000 years B.P. As the sea rose and the shoreline moved across the continental shelf, large masses of sand were moved with the migrating shore zone as beach deposits overlying a lagoonal carpet. Once sea level began to stabilize about 4,000 to 6,000 years ago, the barrier islands began to evolve into their present-day form [Swift, 1975]. Typical barrier environments similar to those present today took place between 2,000 and 1,500 years B.P. [Davis and Kuhn, 1985].

The bases of a large number of barrier islands in the world are at depths of 5 to 10 m, which is the depth at which sea-level rise slowed drastically about 5,000 years ago [Paskoff, 1987]. During the last 1,000 years, a continued slow rise of sea level has resulted in further transgressions of the barrier islands, primarily by overwash and inlet formation. The evolution of barrier islands is, therefore, closely related to sand supply, sea level variations, and the intensity of waves and currents over the inner continental shelf. Under conditions of continued sediment supply, stable sea level, and low-to-moderate subsidence rates, barrier islands prograde seaward. In contrast, a reduction in sediment supply, a rise in sea level, or a high rate of subsidence lead to the landward migration or destruction of barrier islands: they are very fragile coastal environments.

Estuary and Delta Islands

These islands, located at the limit between river and sea, are directly related to sedimentary construction of deltas and estuaries. The geomorphology of these features result from the interaction between river flow and hydrodynamics in the river mouth. The three main forcing parameters are river flow, wave energy, and tide range [Coleman and Wright, 1975]. If river outflow is preponderant, the morphology and sedimentary construction are of type delta (Figure 6); on the contrary, a high tide range allows for the construction of an estuary type.

The delta and estuary islands correspond to the sedimentary progradation of these two coastal environments. Deltaic systems are generally composed of three main environments: (1) delta plain, (2) delta front, and (3) pro delta. With the distributaries and levees, delta islands are the characteristic features of the delta plain. For a delta plain to form, the amount of sediment supplied by the river must be greater than the amount of sediment dispersed away from the river mouth by waves and tides [Allen *et al.*, 1979]. In the inner part of the river mouth, islands are elongated and parallel to the river. In the outer part (adjacent coastal zone), the sedimentary deposition is organized as spits and distributed on each side of the river's mouth. The outer spits are the younger ones. Nevertheless, on geological time scales, these islands are very young [Orton and Reading, 1993] and most of them are Holocene in age (<12,000 years).

These islands are mainly comprised of clastic, silty-sand and clay deposits, and are rich in organic mater: deltas and delta basins are among the better areas for hydrocarbon accumulation. The clastic material is issued from the erosion of the continent's upper-part regions. With no such consolidated deposit, these islands are very fragile. On the other hand, a serious diminution in the river input may, consequently, rapidly destroy the island due to the erosive action of the sea. Nevertheless, there are generally islands of important human settling because of the fertility of the material deposited (especially silt).

Deltaic islands are also vulnerable to sea level rise. In the absence of increased sedimentation or other compensating factors, a 3 m rise in sea level would flood for example 20% of Bangladesh and similar fraction of the Nile delta, Egypt [Broadus, 1993]. In the Mississippi delta, the historical trend of net land gain has been reversed and there is now a large net loss of land, resulting more from regional land subsidence than eustatic sea level rise.

Special Case of Coral Reef Construction

Coral biota lives in warm, clear and shallow waters of tropical latitudes. They may construct immense coral reefs and shoals, and they are the source of thick carbonate formations which are known from the Menozoic period to present. Geomorphologically, coral reefs can be classified into the following main types [Guilcher, 1984]:

1. Fringing reefs, which lie adjacent to the land. For the small volcanic oceanic islands, which have been built up by several successive eruptions over the major part of the Tertiary. Fringing reefs generally grow up on steep inaccessible coasts. For the bigger islands in this group, the rich volcanic soils may support intense farming. Fringing reefs provide artisanal fisheries and potential tourist attractions.

2. Barrier reefs which lie further offshore with a lagoon of varying depth separating them from the coast. The most spectacular example is the great barrier reef of Australia which is backed by shallow open sea several hundred kilometers in width. Biological resources of the reefs and lagoons provide high quality proteins due also to the breaches in the barrier which allow exchanges of water between lagoons and the ocean.

3. Atolls, which are subcircular reefs enclosing a lagoon. They have no island in their center but generally low-lying carbonate or patch reefs. Atolls such as Bikini, are

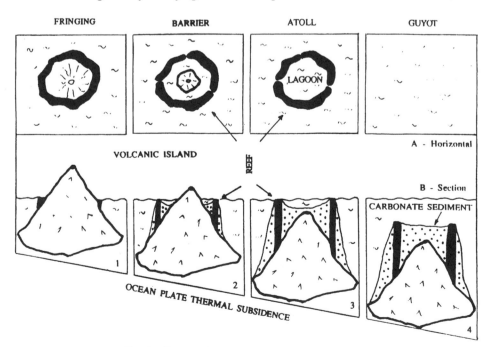

Fig. 6.   Coral reef evolution related to plate motion.

most common in the Pacific ocean. The size of atolls vary accordingly to the extent and depth of the bases but the portion above sea level is usually limited to a few square kilometers. So, they are very vulnerable, not only to Hurricanes, storms, and tsunami, but also for the lack of water, generally due to the high infiltration capacity of the carbonate basement.

In some cases, subsidence is replaced by uplift: atolls emerge and fresh water may fill the lagoons. Soils grow up and Karst landscapes appear. Central depressions and stepped sides suggest different stages of emergence.

Coral reefs, with their associated biota (calcareous algae, principally), are an important carbonate sediment source (conglomerate, sand, and mud). It is generally the only sediment supply; however; the maintenance of the coral reef growth is essential to avoid erosion of the beaches and coasts of these small islands [Bathurst, 1975].

Reef development is enhanced on the windward side of the islands, where wave action is more intense and waters have more open-ocean qualities. In the Caribbean Sea, growth of individual coral organisms is estimated between 1-20 cm/yr [Vicente et al., 1993], and are known to be up to 1.5 cm/yr for whole reef growth rates [Hendry, 1993]. But coral reefs have high vulnerability due to the multiplicity of stress variables (low salinity, high turbidity, storms, disease, overfishing, mass mortality in algal, and pollution of the water). Coral reef development is limited by colder surface currents or by upwelling; therefore, reefs usually occur on the western margin of the ocean basins.

Conclusion

This broad and simplified overview on small island geology shows the main differences for construction processes and geological features of sea small islands (Table 1). The main group (arc system, rift, tectonic, and intaplate islands) is related to plate motion and volcanic activity, and represented by volcanic basement and steep relief islands, with generally restricted shelves. In spite of the dangerous environments (high volcanism and seismicity), the basic lavas form rich soils which attract human population. The quiet precontinent islands established on metamorphic basement, allow development of fishing activities in the open sea, and sea-farming in protected bays, as conchyliculture in intertidal zones. Coral reefs which may develop on different kinds of substratum, but in clear and warm waters are a source of abundant carbonate sand and form islands that are very attractive for tourist activities.

Geological features also allow to establish a vulnerability assessment, related to physical characteristics. The most vulnerable islands are those produced by coastal hydrodynamic processes (barrier island systems, estuarine and delta islands). Their low altitude and no consolidated basement facilitate the erosive actions. The main destructive parameters are related to meteorological and marine processes: cyclone (hurricane and typhoon) and storms especially in tropical latitudes. A similar (hurricane and typhoon) and storms especially in tropical latitudes. A similar vulnerability exists for the low coral reef islands, but in that case, it is enhanced by the multiplicity of stress especially those related to sea water quality. Although less frequent than meteorological and marine constraints, earthquakes and volcanic eruptions are the major constraints of the islands related to plate motions. These highly fragile ecology and/or very vulnerable physical environments lead   the small islands to be easily dependent on foreign countries

TABLE 1

Island Types	Situation	Geological Environment and/or Cause	Petrographic Characteristics	Age	Fragility
Island arc systems	Plate boundaries	Subduction zone/ volcanism	Andesite	Ancient to actual	Moderate
Rift islands	Plate boundaries	Oceanic ridge/ volcanism	Basalt	Ancient to actual	Moderate
Intraplate volcanic islands	Intraplate	Hot spot/ volcanism	Basalt	Ancient to actual	Moderate
Precontinent islands	Passive margins	Sea level rise	Metamorphic	Ancient	Low
Diapiric islands	Passive margin or plate boundaries	Delta front or fault zone	Mud or salt	Ancient to actual	Moderate
Barrier island systems	Shelf	Coastal hydrodynamic processes	Sand	Holocene to actual	High
Estuarine and delta islands	Coastal zone	Hydrodynamic and river supply	Clastic sand and clay sedimentation	Holocene to actual	High
Coral reef islands	Shelf and offshore	Tropical latitudes	Carbonates	Ancient to actual	High

[Granger, this volume]. Nevertheless, the construction processes and geological features of small islands is part of the whole knowledge indispensable for the effective management and sustainable development, particularly to their coastal zone problems [Paskoff, 1985].

This overview emphasizes the complexity and the diversity of each island and lead to thank that generalizations for management have to be avoided [Maul *et al.*, this volume]: each small island has to be studied as a special case, and needs its own management program applied with its own value systems and cultures.

References

Allen, G.P., Laurier, D and J., Thouvenin, Etude sédimentologique du delta de la Mahakam, *Notes et Mémoires*, 15, Compagnie Française des Pétroles, Paris, 1979.

Baker, D.R. and D.H. Eggler, Fractionation paths of Atka (Aleutians) High-alumina basalts: constraints from phase relations, *Journal of Volcanology and Geothermal Research*, 18, 387-404, 1983.

Bardintzeff, J., Les éruptions explosives: produits rejetés et dynamismes, *Mém. Soc. géol. France*, No. 163, 155-166, 1993.

Bathurst, R., *Development in Sedimentology, Carbonate Sediments, and their Diagenesis*, 658 pp., Elsevier Scientific Publishing Company, Oxford, 1975.

Broadus, J.M., Possible impacts of, and adjustments to, sea level rise: the case of Bangladesh and Egypt, In *Climate and sea level change*, edited by Warrick, R.A., Barrow, E.M., and Wigley, T.M.L., Cambridge University Press, 263-275, 1993.

Bougault, H., Charlou, J-L., Fouquet, Y., Appriou, P., and P. Jean Baptiste, L'hydrothermalisme océanique, *Mém. Soc. géol. France*, No. 163, 99-111, 1993.

Bullard, F. M., *Volcanoes of the Earth*, University of Texas Press, Austin, 1976.

Butterlin, J., *Géologie structurale de la région des Caraïbes*, Masson édit., Paris, 255p., 1977.

Caron, J.M., Gauthier, A., Schaaf, A., Ulysse, J., and J. Wozniak, *Comprendre et enseigner la Planète Terre*, 271 pp., édit. Ophtys, Paris, 1992.

Coleman, J., and L. Wright, Modern river delta: Variabilities of processes and sand bodies, in *Deltas*, edited by M. Broussard, pp. 99-149, Houston Geol. Soc., 1975.

Coulon, C., Le Volcanisme de rifts Continentaux, *Mém. Soc. géol. France*, No. 163, 69-80, 1993.

Davis, R.A. and B.J. Kuhn, Origin and development of Anclot Key, wet-peninsular Florida, in: G.F. Oertel and S.P. Leatherman (editors), Barrier Islands, *Marine Geology*, 63 153-171, 1985.

Duque-Caro, H., Geotectonica y evolucion de la region Noroccidental Colombiana, *Bol. Geol. Ingeominas*, V. 23, 4-37, 1980.

Evans, M.W., Hine, A.C., Belknap, D.F. and R.A. Davis, Bedrock controls on barrier island development: West central Florida coast, in: G.F. Oertel and S.P. Leatherman (editors), Barrier Islands, *Marine Geology*, 63: 263-283, 1985.

Fitzgerald, D.M., Origin and Stability of Tidal Inlets in Massachusetts, in Formation and Evolution of Multiple Tidal Inlets, Coastal and Estuarine Studies, *American Geophysical Union*, V. 44, 1-61, 1993.

Gayet, J., and G. Vernette, Les Lagunes Côtières: Cours International d'Océanologie Côtière en région Caraïbe, *Bull. Inst. Geol. Bassin d'Aquitaine*, Bordeaux, No. 45, 107-121, 1989.

Goslin, J. and M. Maia, Les apports des travaux récents de géophysique à la comprehénsion du volcanisme intraplaque en domaine océeanique: évolution de l'hypothèse des points chauds, *Mém. Soc. géol. France*, No. 163, 113-128, 1993.

Granger, O., Geography of small tropical islands: implications for sustainable development in a changing world, Small Island Oceanography, *Coastal and Estuarine Studies, Volume 51*, American Geophysical Union, 1995.

Griboulard, R., Bobier, C., Faugères, J.C. and G. Vernette, Clay diapiric structures within the strike-slip margin of the southern leg of the Barbados prism, *Tectonophysics*, 192, 383-400, 1991.

Guilcher, A., *Coral Reef Morphology*, 248 pp., Wiley and Sons, England, 1984.

Hendry, M., Sea-level movements and shoreline change, Chapter 7, in G.A. Maul (ed.), *Climatic Change in the Intra-Americas Sea*, United Nations Environment Programme, Edward, Arnold Publishers, London, 115-161, 1993.

Jackson, M., and W. Galloway, Structural and depositional styles of Gulf coast tertiary continental margin: Application to hydrocarbon exploration, 225 pp., *AAPG* continuing education course note, series no. 25, 1984.

Juteau, T., Le volcanisme des dorsales océaniques, *Mém. Soc. géol. France*, No. 163, 81-98, 1993.

Kenneth, W., *Earth's Dynamic Systems*, MacMillan Publishing Company, New York, 647p., 1992.

Kennett, J. P., *Marine Geology*, 813 pp., Prentice Hall, Inc., Englewood Cliffs, 1982.

Klingebiel, A., and G. Vernette, Geology and Development Facilities of Small Islands Belonging to the Atlantic Margin of Africa and Europe, Small Island Oceanography, *Coastal and Estuarine Studies*, V. 51, American Geophysical Union, 1996.

Leatherman, S.P., *Barrier Island Handbook*, University of Maryland, College Park, MD, 1979.

Leathermen, S.P., Geomorphic and stratigraphic analysis of Fire Island, New York, In: G.F. Oertel and S.P. Leatherman (editors), Barrier Islands, *Marine Geology*, 63, 1-18, 1985.

Martin, L., Suguio, K., Flexor, J-M., Dominguez, J.M. and A.C. Bittencourt, Quaternary evolution of the central part of the Brazilian Coast: the role of sea level variation and of shoreline drift, UNESCO *Reports in Marine Science*, 43, 97-145, 1987.

Maul, G., Hendry, M., and P.A. Pirazzoli, Sea Level, Tides, Storm Surge, and Tsunamis, Small Island Oceanography, *Coastal and Estuarine Studies*, V.51, American Geophysical Union, 1995.

Maury, C., Les séries volcaniques, *Mém. Soc. géol. France*, No. 163, 39-55, 1993.

Nicholson, R., The Scandinavian Caledonides, in *The Ocean Basins and Margins*, edited by A. Nairn and F. Stehli, pp. 233-272, Plenum Press, New York, 1974.

Oertel, G.F., The barrier island system, In: G.F. Oertel and S.P. Leatherman (editors), Barrier Islands, *Marine Geology*, 63, 1-18, 1985.

Orton, G.J., and G. Reading, Variability of deltaic processes in terms of sediment supply, with particular emphasis on grain size, *Sedimentology*, 40, 475-512, 1993.

Paskoff, R., *Les littoraux-Impact des aménagements sur leur évolution*, Masson, Paris, 185 pp., 1985.

Paskoff, R., Les variations du niveau de la mer, *La Recherche*, No. 191, 1010-1020, 1987.

Pirazzoli, P.A., *World Atlas of Holocene Sea-Level Changes*, Elsevier Oceanography Series, 58, Elsevier, Amsterdam, 300 pp., 1991.

Porter, S.C., Stuiver, M., and C.J. Heusser, Holocene sea level changes along the Strait of Magellan and Beagle Channel, southernmost South America, *Quat, Res.*, 22, 59-67, 1984.

Reinson, G. E., Facies models 6: Barrier island systems, *Geosciences*, 6, 57-74, 1979.

Sugimura, A., and S. Uyeda, *Island Arcs: Japan and its Environments*, Elsevier Scientific Publishing Company, Amsterdam, 1973.

Swift, D. J. P., Barrier island genesis: Evidence from the central Atlantic shelf, eastern USA, *Sed. Geol.*, 14, 1-43, 1975.

Ters, M., Les variations du niveau marin depuis 10,000 ans, le long du littoral atlantique français, In: Le Quaternaire: Géodynamique, Stratigraphie et Environnement, *Suppl. Bull. Assoc. Fr. Et. Quat.*, 36, 114-135, 1973.

Vernette, G., La plate-forme continentale caraïbe de Colombie. Importance du diapirisme argileux sur la morphologie et la sédimentation, Ph.D. thesis, University of Burdeaux, 1,387 pp., 1985.

Vernette, G., Mauffret, A., Bobier, C., Briceno, L., and J. Gayet, Mud diapirism, fan sedimentation and strike-slip faulting, Caribbean Colombian Margin, *Tectonophysics*, 202, 335-349, 1992.

Vicente, V.P., Singh, N.C., and A.V. Botello, Ecological implications of potential climate change and sea-level rise, Chapter 11, in G.A. Maul (ed.), *Climatic Change in the Intra-Americas Sea*, United Nations Environment Programme, Edward, Arnold Publishers, London, 115-161, 1993.

Williams, H., and A. McBirney, *Volcanology*, Freeman, Cooper and Co., San Francisco, 1979.

12

The Geological Legacy of Small Islands at the Caribbean-Atlantic Boundary

Malcolm D. Hendry

Abstract

Contrasting geological structures form the gateway between the Caribbean Sea and Atlantic Ocean. To the north, islands of the Bahamas and Turks and Caicos are perched on extensive carbonate platforms whose surfaces have remained above the base of the euphotic zone through the Cenozoic. Southward, the generally volcanic Lesser Antilles have evolved in an island arc at the convergence zone of the Caribbean and American tectonic plates. A volcanically quiescent but seismically active branch of the arc diverges northeast of Martinique. Volcanism commenced in the Eocene, with islands building through the water column during the Cenozoic. The young non-volcanic island of Barbados, located above the accretionary fore-arc prism, first emerged above sea-level about 900,000 years ago.

Geological processes of millennia have controlled, and continue to influence, nearly all aspects of island economic life. While endowing the region with stunning beauty and environmental advantages, these processes have resulted in an uneven distribution of materials for construction; have left no significant metalliferous deposits; spawned no fossil fuels except in Barbados; and provided limitations to water supply in growing island populations. Widespread, severe risks associated with volcanic eruption, earthquakes, landslides, and longer term sea-level changes are also a consequence of these processes.

Small Islands: Marine Science and Sustainable Development
Coastal and Estuarine Studies, Volume 51, Pages 205–224
Copyright 1996 by the American Geophysical Union

The technological and human ability to measure, monitor, and manage this paradoxical legacy continues to develop. It is also appropriate to consider whether development can be sustainable when a nation's economic and social well-being can be influenced by the probabalistic events of extreme magnitude. As governments and institutions pursue their agenda for economic growth, greater emphasis on coordinated, cross-sectoral research, planning, and management is required. Inevitably, the fundamental realities of this geological legacy will be more broadly embraced and will help to build potential for defining and ultimately implementing a model for sustainable development of these islands.

Introduction

Efforts to define the special environmental and economic conditions of small islands have been ongoing [*e.g.*, Towle, 1985; Beller *et al.*, 1990], but like the concept of "sustainable development," they have yet to lead to clearly definable and implementable development models. Before the international research and development community move on to define another developmental concept, an opportunity exists to take existing theories some distance through the development and testing stage.

Small islands at the boundary of the Caribbean Sea and Atlantic Ocean have evolved from the interaction of geological and ocean/atmosphere processes which have remained active through millions of years. The products of these events, whether measured by the type and amount of resources and environmental attributes they have spawned, or the hazards associated with volcanism, earthquakes and landslides, provide the context within which sustainable island management may perhaps be understood and achieved.

The goal of this paper is to demonstrate geological processes responsible for this natural resource and environmental management heritage and to explore the rationale for an approach to small island management which emphasizes consideration for interdisciplinary activities and the need to integrate lessons from the geological past in planning for the future.

Island Geology - Process and Product

Physical Setting

Islands under consideration in this paper include the Bahamas and Turks and Caicos, and the Lesser Antilles, stretching from Grenada in the south to Anguilla in the north, where they are separated by the Anegada passage from the Virgin Islands and Greater Antilles. These two island groups illustrate strongly contrasting processes of geological development.

The Bahamas and Turks and Caicos Islands

The more than 700 islands of the Bahamas and Turks and Caicos (Figure 1) have evolved on a platform over 6 km deep, composed largely of shallow marine carbonate sediments locally interbedded with evaporites. Platform growth was initiated in the Jurassic period around smaller bank segments which were either relicts of a larger, remnant Cretaceous platform or of the rifted and/or eroded remnants of continental break-up [Schlager and Ginsburg, 1981].

While platform surfaces have remained in the euphotic zone or been exposed above sea level through the Cenozoic, Eberli and Ginsburg [1987] show they have formed by coalescence of smaller original bank areas. Two styles of sedimentation accommodated this merger: (a) aggradation in troughs; and (b) progradation of bank margins (Figure 2). Sediment was derived from shallower bank surfaces, fed by downslope gravity flow or settlement from suspended fine-grained muds themselves transported from bank and platform tops [Neumann and Land, 1975; Boardman and Neumann, 1984; Droxler *et al.*, 1988; Milliman *et al.*, 1993]. Off-bank sedimentation followed an east-west transport path, reflecting dominant meteorological and oceanographic conditions [Hine and Neumann, 1977; Hine *et al.*, 1981].

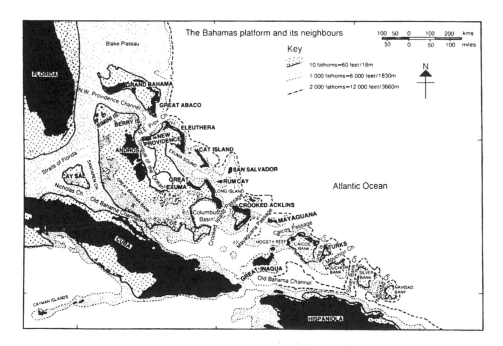

Fig. 1. The Bahamas platform [from Sealy, 1985].

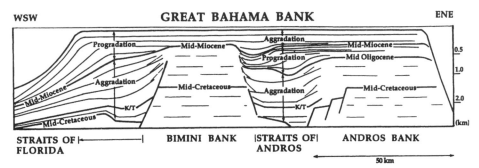

Fig. 2. Sequences of aggradation and progradation in the development of the Great Bahama Bank [from Eberli *et al.*, in press].

An important process controlling evolution of the young islands capping these platforms are the eustatic oscillations of sea-level which have accompanied global glacial and interglacial cycles over the last 3 million years. With a vertical range exceeding 100 m, these sea-level changes have repeatedly exposed and submerged platform surfaces in this island chain. Indeed, much of the vast submerged platform area, some 116,500 km^2, would have been exposed as dry land on numerous occasions through this time.

Consequently, island geology reflects the influence of shallow marine and supratidal to terrestrial sedimentary processes. The consequences of this land-sea interaction are strikingly demonstrated by geological maps of the islands. Figure 3 shows the surficial geology of San Salvador island, the numerous geological units of which reflect high and low-stand deposition of dunes, beach ridges, soil and solution features, intertidal to subtidal skeletal sands, ooid and peloid beds, and coral reefs [Hearty and Kindler, 1993].

Hence, it is the position of sea level and the character of ocean-atmosphere processes at this interface which have provided critical control on island history and will be arbiters of the island's future. These geological and oceanographic attributes underpin all approaches to resource management for these fragile island systems.

Islands of The Lesser Antilles

Volcanism in islands of the Lesser Antilles is a consequence of subduction of the American tectonic plate beneath the eastern edge of the Caribbean Plate (Figure 4). Magma generated by melting at depths of 80-160 km is the proximal source of igneous activity, rising to the surface in an arc some 850 km long.

In detail, however, the geological evolution of this arc is more complex than a model based on vent-derived extrusive volcanic rocks. Speed *et al.* [1993] describe the southern Lesser Antilles arc platform (SLAAP), which includes the area from Grenada to St. Lucia. The southern end of the SLAAP, including Grenada and the platform islands of the Grenadines, is an uplifted and east-tilted block of Eocene to Miocene pillow basalts of spreading origin, deep basin arc-derived volcanogenic turbidites, hemipelagites and pelagic and platform carbonates.   Composition of arc-derived sediment of late Eocene to early Miocene age suggests the existence of a nearby emergent, carbonate-fringed volcanic source at that time.   Shallowing, and possibly initial emergence of the SLAAP platform itself, may have occurred as early as the mid-Miocene [Speed *et al.*, 1993].

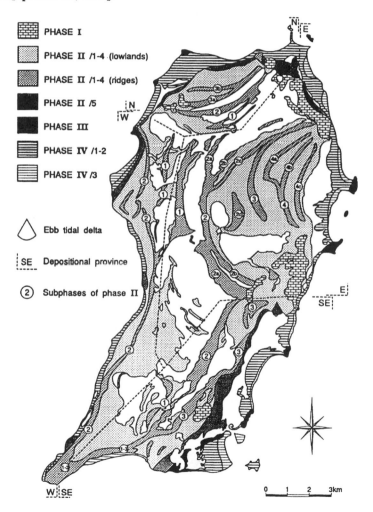

Fig. 3. Surficial geology of San Salvador Island, Bahamas. The complexity of the geological units is a function of repeated exposure and submergence of the surrounding platform under oscillating sea-level conditions in the Quaternary. The captioned phases recognize stages in the evolution of the island. From Hearty and Kindler [1993].

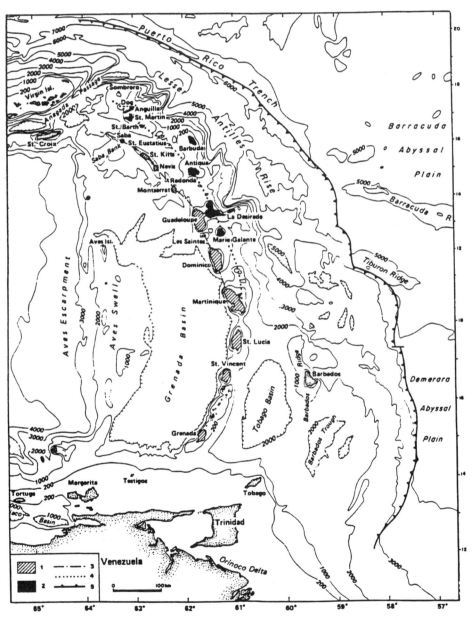

Fig. 4. From Maury *et al.* [1990]. General geology of the Lesser Antilles area. Islands discussed in this paper include all those of the arc from Grenada in the south as far north as Anguilla, east of the Anegada passage, plus Barbados. 1 = volcanically active branch of the arc; 2 = volcanically quiescent but seismically active branch of the arc; 3 = axis of inner arc; 4 = axis of outer arc; 5 = deformation front. Isobaths are in meters.

These rocks are intruded and overlain by Neogene magmatic and volcanic rocks with radiometric dates mainly younger than 12 Ma. With the exception of Kick 'em Jenny submarine volcano, magmatism may have ceased in the southern SLAAP by about 1-2 Ma.

Maury *et al.* [1990] describe divergence of the arc into two branches to the north of Martinique (Figure 4). The northeast branch includes an inactive line of low islands including Marie-Galante, Grande-Terre, La Desirade, Antigua, and Barbuda. Volcanic centers are not apparent in these islands which display a Miocene to Recent carbonate cover over middle Eocene to Oligocene volcanogenic rocks and sediments. Although volcanically inactive, these islands are still subject to seismic shocks, some of them in the higher Modified Mercalli impact ranges, with damaging consequences [Tomblin and Aspinall, 1975; McCann and Pennington, 1990].

The western branch consists of volcanically active islands including Dominica, Guadeloupe, Montserrat, St. Kitts, Nevis, Redonda, St. Eustatius, and Saba. There appears to have been a westward shift of volcanism from the presently inactive to the active volcanic branch in the Pliocene. The history of volcanic activity in the Lesser Antilles is summarized in Figure 5.

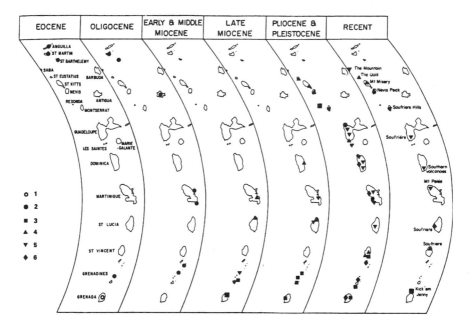

Fig. 5. Location of main Cenozoic volcanic series in the Lesser Antilles. 1 = volcanism of unknown affinity; 2 = orogenic type volcanism (precise affinities unknown); 3 = Mg-rich alkalic or subalkalic basalts and associated rocks; 4 = low-K volcanic series; 5 = medium-K volcanic series; 6 = high-K volcanic series. From Maury *et al.* [1990].

The Geological Legacy of Small Islands

Volcanism has persisted throughout recent times in the Windward and western Leeward Islands. The location, frequency, and nature of historic activity is shown in Figure 6. Continued volcanism, much of it very recent (Figure 5), has resulted in growth of these islands to elevations up to 1,700 m above sea-level (Table 1), a dominant counterpoint to processes of denudation which act to reduce land surfaces to the level of the oceans.

Of great interest is the submarine volcano Kick 'em Jenny, located 8 km north of Grenada, which is the most active Antillean volcano in historical times. Between 1974 and 1989 the volcano rose from a depth of 200 m to 140 m below sea level [Sigurdsson, 1989] and may emerge as the next Caribbean island before long.

Barbados is unique in the region, composed entirely of sediments deposited on a rising fragment of ridge above the accretionary prism of the fore-arc area [Speed, 1988]. The island edifice climbed through the water column, first emerging above the ocean surface about 900,000 years ago, when the earliest reef terraces were formed. Interaction of the

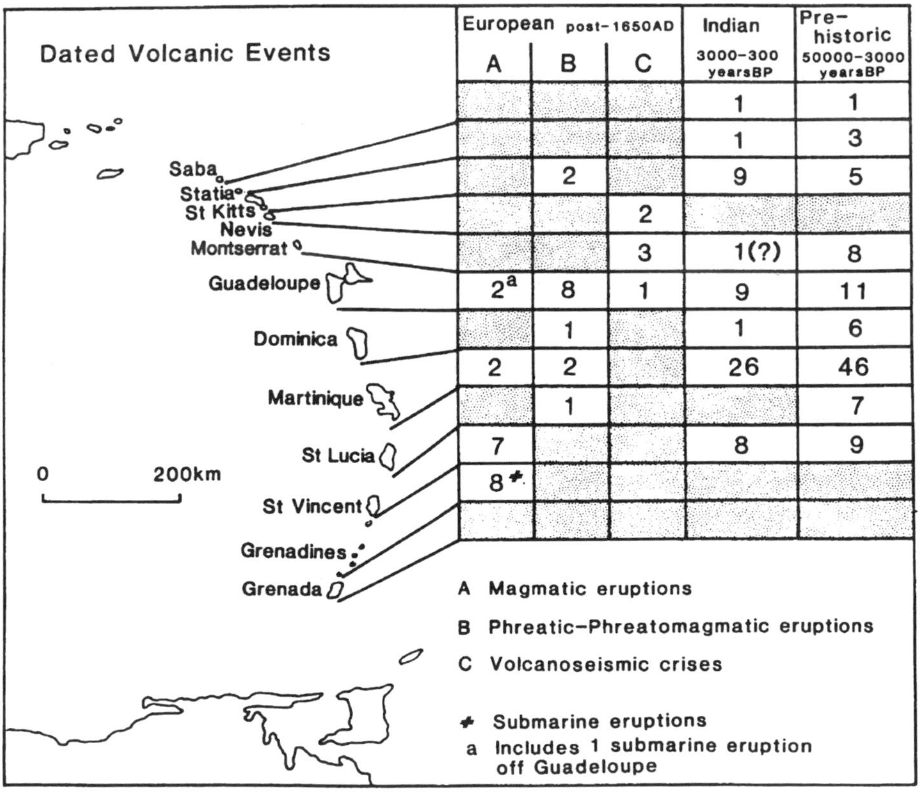

Dated Volcanic Events	European post-1650AD			Indian 3000-300 yearsBP	Pre-historic 50000-3000 yearsBP
	A	B	C		
				1	1
				1	3
Saba, Statia, St Kitts, Nevis		2		9	5
			2		
Montserrat			3	1(?)	8
Guadeloupe	2ª	8	1	9	11
Dominica		1		1	6
	2	2		26	46
Martinique		1			7
St Lucia	7			8	9
St Vincent	8✦				
Grenadines, Grenada					

A Magmatic eruptions

B Phreatic–Phreatomagmatic eruptions

C Volcanoseismic crises

✦ Submarine eruptions

a Includes 1 submarine eruption off Guadeloupe

0   200km

Fig. 6. Volcanic activity in the Lesser Antilles from pre-historic to historic times [From Roobol and Smith, 1989].

ascending island with oscillating sea level has created a distinctive step-like erosional cliff and depositional (reef) terrace topography [Bender *et al.*, 1979]. The main links with volcanism are the frequent ash bands interbedded in the chalk beds of the Eocene to Miocene Oceanic Series, an allochthonous unit transported onto a structural high from the fore-arc basin, and thin soils veneering elevated terraces, which may have greater mineralogical affinity with the volcanic ash emitted by eruptions on St. Vincent than with *in-situ* dissolution of exposed limestones in subaerial conditions [Muhs *et al.*, 1987].

Sea-level features which include constructional reef terraces are poorly to moderately developed on shelves of the active volcanic islands, although some terracing and reef development is apparent mainly on higher energy windward coasts [Adey and Burke, 1976; Macintyre, 1988; Wells, 1988]. The principal reason for limited reef growth is the weakly consolidated volcanoclastic sediments of the steep island flanks, where rapidly eroding sea cliffs, sand bars, and beaches do not provide adequate substrate for major reef development [Adey and Burke, 1976]. Influx of sediments contained in run-off from the islands, especially during the rainy season, is also a deterrent to reef growth.

This discussion has emphasized the relationship between geological and ocean processes in evolution of islands at the boundary of the Atlantic Ocean and Caribbean Sea. The ensuing sections discuss some key elements of the resource and environmental management legacy these processes have created.

Resource Management Implications and Issues

Physiography and Resource Management

Table 1 summarizes some major physical features of the islands. In the Bahamas, the volcanically inactive northeast branch of the island arc, and in the area of the Grenadine islands, where platform-like carbonate accumulation under changing sea-level conditions has produced broader shelf areas, the ratio of shelf to land area is very high. This contrasts with the low ratio in the volcanic Antilles, where underwater gradients are essentially a continuation of emergent island flanks.

Antecedent of island and shelf geology is the extent of the exclusive economic zone (EEZ). For politically discrete volcanic islands with near neighbors the oceanic EEZ boundaries are shared, diminishing the area of resource management control (Table 1). The extent of accessible shelf resources, including reefs and other biological communities, fisheries, and non-living resources including sand and aggregates, are also limited by shelf morphology in volcanic islands. The pelagic realm is closely adjacent to the shoreline of many of these countries: in the absence of shallow shelf environments this provides the basis of important pelagic fisheries (Mahon, Chapter 18). However, deeper benthic living and non-living resources are relatively inaccessible and not well understood.

Non-Metalliferous Resources

A key factor in economic development is the availability of sand, gravel, and lime for the construction industry. Drainage systems debouch over short, precipitous transport paths through fan-deltas to the sea, leaving minimal on-shore aggregate deposits in many Antillean islands. At best, localized, narrow shelves provide only limited sediment retention capacity [Hendry et al., 1993]. This is not the case in platform areas of the Bahamas, Turks and Caicos, Antigua-Barbuda, and the Grenadines of St. Vincent and Grenada, where more extensive sedimentary deposits are located, much of it carbonate material.

TABLE 1. Comparative physiographic data for some islands and island groups on the Caribbean/Atlantic boundary.

Country	Land Area (km^2)	Coast (km)	Shelf (km^2)	Ratio Shelf/Land Area	Maximum Elevation (m)	EEZ (km^2)
Antigua and Barbuda	440	153	3,570	8.1	402	110,103
Bahamas	13,934	3,542	116,500	8.3	60	759,402
Barbados	430	97	300	0.7	340	167,384
Dominica	751	148	716	0.9	1,730	15,092
Grenada and Grenadian Grenadines	344	121	1,600	4.6	833	27,440
Guadeloupe	1,709	306	1,650	0.9	1,467	26,200
Martinique	1,101	290	2,400	2.2	1,463	13,000
Montserrat	83	35*	106	1.3	915	21,100
St. Kitts and Nevis	262	---	850	3.2	1,156/985	11,319
St. Lucia	616	105*	520	0.8	950	16,121
St. Vincent and Grenadines	389	---	1,800	2.9	1,408	32,585

*Estimate.

As a consequence of availability and convenience, there has been large scale sand mining on the small pocket beaches of these islands. It is not surprising, therefore, that extensive beach erosion has resulted: natural rates of replenishment are clearly lower than rates of removal in most locations, even those with extensive offshore carbonate sources [Hendry, 1993]. Additionally, alteration to reef community structure induced by land-based sources of marine pollution and natural disturbances among reef-associated sediment-generating organisms [Tomascik and Sander, 1985; Hunte, 1987; Hunte and Younglao, 1988; Wittenburg and Hunte, 1992] raises the possibility of interference in sand production for the nearshore sediment budget in carbonate-based sedimentary regimes.

Alternatives to beach sand for construction are needed in the Lesser Antilles. Part of the aggregate problem is solved by importation of quartz sands of the Timerhi area in Guyana to several Caribbean islands. Crushed volcanic quarry stone is also available in most islands. Crushed rock does not possess the properties required for all applications, however, and does not produce quality sand grade material. This is true, for example, of beach renourishment--increasingly a requirement in tourism-dominated economies of the islands--for which required textural properties include good sorting and low angularity. For these purposes marine-derived sands are needed in the absence of local river sands and the identification of sources becomes a key element requiring basic geological and geophysical investigation.

The Bahamas, in particular, possess vast areas of marine sand on their broad shelf, providing a major source for the local construction industry over many years. Some local potential for sand also exists in the steeper island shelves of the volcanic Antilles, and on the Antigua-Barbuda platform and Grenadines platform. However, the wholesale, unregulated offshore mining of the shelves and platforms of Caribbean islands is not advocated without adequate preparatory studies because of the potential for damage to living marine resources including reefs, seagrasses, and fisheries.

With regard to lime, the only operational cement factory in the islands under discussion is located in Barbados, which exports to the other islands. The other primary sources are exported cement from Trinidad, Colombia, and Venezuela.

The methodology for broader scale assessment of sand and aggregate resources, including side-scan and geophysical surveys, has been applied in several regional studies of island shelves (Durand et al., Chapter 15) [Schwab and Rodriguez, 1992; Delcan, 1993]. Programs for assessing impacts of shelf mining and the requirements for mitigation or negation of harmful effects on adjacent biotic communities have also been undertaken [e.g., Hendry et al., 1993]. The careful identification and evaluation of the full range of sediment sources to provide least cost, minimal impact options should be the subject of further detailed examination.

Metalliferous Resources

There are no known offshore metalliferous deposits or placers in any of these islands, though the presence of underwater hydrothermal activity in a number of them, including

St. Lucia, Dominica, St. Kitts, and Montserrat raises the possibility that some local deposits may be found, though it is dubious whether they would be of economic value.

Unlike some of its close island neighbors in the Greater Antilles and Virgin Islands, the Lesser Antilles and Bahama islands have no significant onshore metalliferous deposits [Kesler *et al.*, 1990], and no mining associated with such deposits is known.

Energy from Fossil Fuels

The geologic history of these islands has not been kind enough to endow them with oil, gas, or coal. The exception is Barbados, where oil and gas are sourced from Eocene quartz-rich sandstones, pelagic and hemipelagic rocks at depths of over 7 km [Speed *et al.*, 1991]. Oil production is relatively small at about 1,500 barrels per day, supplying about 10% of domestic requirements, used to supplement oil imports.

Hence, there is a dependence on oil imports throughout these islands, which creates a drain on foreign exchange, but the geology and hydrology of several countries provides other opportunities for alternative energy sources.

Alternative Energy: Geothermal and Hydroelectric

In the volcanic Antilles, steep watercourses and dependable interior rain over high mountains has provided opportunity for installation of numerous small-scale hydroelectric generating plants. Dominica has the best established hydropower system, contributing approximately 31% of effective (dry season) capacity by 1989 figures [CCA/IRF, 1991a]. While in many cases economically feasible, there are some concerns over net benefits in relation to environmental consequences in some countries. It is interesting to note that damage to watersheds, causing heavy river siltation, appear to have eliminated significant options for hydropower in St. Lucia [CCA/IRF, 1991b].

Many of the volcanic islands have potential for geothermal energy, though this has yet to be realized. Numerous studies and drilling programs have been conducted over the years. They point out the many problems, including limited cost-advantage over other energy sources, maintenance costs, dependence on foreign technology and equipment, and environmental costs [Los Alamos National Laboratory, 1984]. The most recent study concludes that best potential for geothermal energy is in Dominica, Nevis, and Montserrat, where production could exceed domestic demand [UNTCD, 1992]. At present, benefits are mainly found in the spas and spectacular fumaroles visited by tourists in places like the Soufriere of St. Lucia, St. Vincent, and Dominica.

Many other forms of alternative energy are in use to varying degrees in the islands, including wind, solar, biomass, and charcoal. A discussion of these sources is not germane to the main thrust of this paper, however.

Geology and Water Resources

Geology plays a vital role in water resources, from physiographic control of rainfall patterns to the availability of surface or subsurface supplies. The problems of water supply in low-lying, limestone islands where all resources are subsurface are discussed by Cant (Chapter 20). Antigua represents another difficult case of low-island water resource problems, where due to low and at times unpredictable rainfall, and high demand, desalinization plants have been installed [APUA, 1989].

In the volcanic islands the bulk of water supply is from the surface due to limited percolation, steep terrain and high run-off, and steady rainfall on higher mountain areas. Their is potential for development of groundwater sources, which will become more important with increasing impacts on watersheds from agriculture and increasing demand from expanding populations. For example, in Grenada, with a projected increase in well capacity, as much as 21% of supply may be produced from groundwater sources by the late 1990's [Waal, 1987]. In St. Kitts, drilling has identified groundwater as having potential to assist with surface water in satisfying foreseeable demand [KPA, 1988].

Geology and Ocean Processes

Small island size determines that the effects of a direct encounter with a tropical storm or hurricane is devastating, because the entire land area may be engulfed within the cyclonic radius. Shelf configuration is also influential: while the narrowness of many shelves tends to restrict the extent of set-up due to storm surge [Mercado et al., 1993], the reverse is true of storm waves which develop across unrestricted fetch in the Atlantic Ocean and approach shorelines with minimal loss of energy because deep water is found relatively close to land.

The general susceptibility of low islands of the region to rising sea level has been discussed elsewhere [Hendry, 1993] and initial modeling of socio-economic consequences of different sea-level scenarios has also been undertaken [Engelen et al., 1993]. One of the major problems is to determine the cause of variation between sea-level signal in the islands, differences which cannot be easily attributed to tectonic motion. This is partly a consequence of inadequacies in using extrapolated linear ground displacement rates and partly of limited coincidence between sea-surface stations and sites of measured tectonic motion [Hendry, 1993]. Moreover, because few tide gauges are deployed, extrapolation of regional or assumed sea-level rise data to other island areas becomes an inadequate but unavoidable technical procedure for modeling and design purposes [Hendry, 1994]. As Maul et al. (Chapter 6) point out, their is pressing need for improved measurement and monitoring capability of sea level at the island-specific scale, where this critical interface is a prime determinant of many long-term physical planning and development control questions.

For example, determination of coastal development set-back limits in the region has, perhaps curiously, tended to rely on technically inadequate statutory set-backs of a few tens of feet. These traditional set-back criteria will become increasingly obsolete if pressure to construct immediately adjacent to Caribbean shorelines continues. They will be even more problematic if the potential increase in storm frequency due to climate change should be realized [Wigley and Santer, 1993] and if there is a greenhouse-induced pulse of accelerated sea-level rise in the region.

Geological Hazards

Landslides

Steep slopes, deep weathering profiles, and periodic intense rainfall predispose higher islands of the Lesser Antilles to landslide effects. These problems are aggravated through deforestation and road cutting. The costs of landslides are financial and mortal: for example, 25 lives were lost in landslides in Dominica between 1924 and 1985, while expenditure on debris removal and road repair totaled EC $1,263,750 (U.S. $468,055) for the period 1983-1987 [De Graff, 1989]. For small islands these costs represent a perpetual burden on the national exchequer. In Dominica, clean-up and repair costs alone represented almost 0.3% of average annual government expenditure over this four-year period. This does not include costs associated with crop loss, relocation, rehabilitation, and downstream environmental consequences such as siltation of streams and coasts.

Volcanic and Seismic Hazards

Despite the many natural blessings bestowed on the Lesser Antillean islands, their location and geological configuration beget other menacing characteristics that are of critical concern in the national economic context. These are the ever-present volcanic and seismic risks.

Earthquakes are common, and have foci at varying depths in the benioff zone (Figure 7). The most damaging earthquake of recent times occurred on October 8, 1974, when damage estimated at Modified Mercalli intensity VIII and VII effected Antigua and Barbuda, respectively [Tomblin and Aspinall, 1975].

While it is difficult to predict earthquakes, significant benefits are to be derived from monitoring and measurement. These benefits include the establishment of standards for construction and for identification of patterns of land use to minimize effects of future events [Shepherd, 1989]. Identification of earthquake activity is itself strongly indicative of a prelude to volcanic eruption, but these tremors may only be felt close to the vent [Shepherd et al., 1988]. Shallow earthquake swarms near volcanoes, together

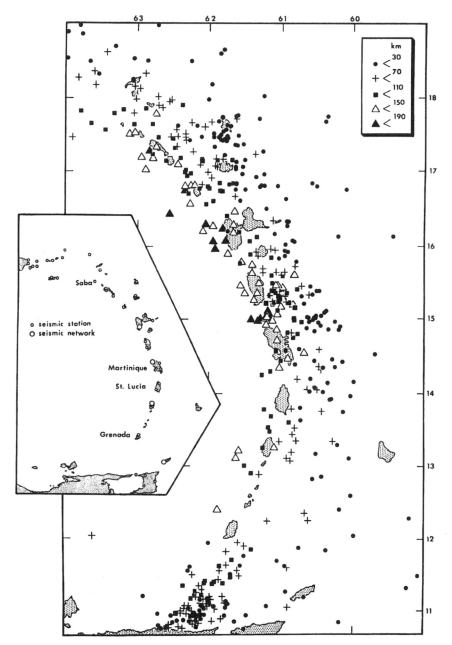

Fig. 7. A map of earthquake epicenters deeper than 30 km in the Lesser Antilles from July 1978 to April 1984 south of 15°N and from July 1980 to April 1984 north of 15°N. The symbols denote depth intervals of 40 km. Inset shows the location of seismic stations and networks. From Wadge and Shepherd [1984].

with increased rates of steam and gas emission, can be detected using straightforward techniques [Shepherd, 1989]. Hence, maintenance of monitoring activity is central to preparedness for events which can cause catastrophic loss of life and economic dislocation.

Eruptions in the West Indies can be devastating. The St. Vincent eruption in 1812 killed 56 persons directly, and in 1902 an estimated 1,565 persons died, mainly due to ash flows. The Mt. Pelee eruption on Martinique in 1902 killed over 30,000 people. However, even if detected, it is not possible to predict whether the eruption will be major or minor [Shepherd, 1989]. Hence, decisions to evacuate are crucial. They may be timely and life-saving, as with eruptions in Martinique in 1929-1932 and St. Vincent in 1979 [Shepherd et al., 1988], but even so there is significant disruption and cost.

For example, it was estimated that the evacuation of part of Basse-Terre (Guadeloupe) in 1976, when the eruption died out without major phreato-magmatic activity, cost between U.S. $200-$400 million. This represents the entire gross domestic product (GDP) for many eastern Caribbean island economies for whom such costs would be unbearable. Sadly, it appears that part of the contribution to debate on evacuation of Guadeloupe was an acrimonious public row between respected scientists, further fueled by opinions of self-appointed experts, underpinned by the all-important policy of the French central government that there should be no risk to life whatsoever [Shepherd et al., 1988].

Major improvements in the decision-making process since that time, put into practice on St. Vincent in 1979, are allied to continued advances in monitoring capability. Full-time volcanic and seismic monitoring capacity is maintained throughout the eastern Caribbean by the Seismic Research Unit of the University of the West Indies (UWI) and the Institute de Physique de Globe, of the University of Paris, in the French West Indies (see Figure 7). The systems are telemetered, with real-time relay from station to base, and with close cooperation between the UWI and French units. A critical missing link in this chain is a tsunami warning system, especially in light of the risk from Kick 'em Jenny submarine volcano [Smith and Shepherd, 1993] (Maul et al., Chapter 6).

The Caribbean community (CARICOM) supports a full-time disaster relief unit based in Barbados [CDERA, 1992], which also confronts response requirements to other regional hazards including those presented by hurricane and landslide. Regional governments maintain a national disaster response capability headed by disaster management coordinators and support committees with representation from key public sector agencies that possess response capacity.

Despite these efforts, lessons derived from fundamental geological understanding of geodynamic processes in the region do not always reflect in long-term strategic planning initiatives. This is demonstrated by the tendency of populations to relocate directly onto areas which have proven vulnerable to volcanic impact in the past. While the French government chose to relocate the capital of Martinique from St. Pierre to Fort de France after the 1902 eruption, there are still many population centers in islands of the volcanic Antilles sited in locations where major historical volcanic activity occurred [Roobol and Smith, 1989].

This may be a question of resource allocation in economies that have to grapple with day to day realities of unemployment, and provision of basic utilities, health, and education services, but the strategic approach (or lack of it) places an awesome burden on scientists and administrators who must of necessity monitor these hazards and advise on reactions based on assessed levels of risk. Long-term planning is not the only solution to this ceaseless problem, but it could ameliorate much of the danger and help to reduce the enormous costs attendant on the inevitable recurrence of these catastrophic events.

Conclusion: Geological Processes and Sustainable Development

Inhabitants of islands on the border of the Caribbean Sea and Atlantic Ocean have inherited a paradoxical geological legacy. On one hand is the splendor of volcanic peaks, rich soils supporting lush forests, waterfalls, golden beaches, and the other arresting environmental attributes which now form the basis of economic activity in many countries, through tourism and agriculture.

In contrast, the islands were not endowed with fossil fuels (except Barbados); they have no major metalliferous deposits; raw materials for construction are irregularly distributed; and water resources have to be carefully managed for growing populations. Moreover, as a consequence of their geological character they are extremely vulnerable to major hazards including landslides, earthquakes, volcanic eruptions, tsunami, and the longer-term effects of sea-level changes, in addition to meteorological and oceanographic threats presented by large storms. Further, there is substantial evidence that negative environmental impacts have resulted from inappropriate utilization of natural resources.

This mixed endowment provides opportunities and problems to which populations and governments have had to conform and in which the underlying threat of catastrophic natural events provides a solemn counterpoint to the fight for national economic development. Is development sustainable when a nation's economic and social well-being can be threatened by probabilistic events of extreme magnitude?

Emerging from this discussion is the inescapable connection between geological processes and economic development, the understanding for which, and management of, are cross-sectoral and interdisciplinary. As regional governments and institutions pursue their agendas, a greater proportion of resources must be allocated to initiatives from basic research through planning and management which embrace and identify with these fundamental realities.

References

Adey, W. H., and R. Burke, Holocene bioherms (algal and bank barrier reefs) of the eastern Caribbean, *Geol. Soc. Am. Bull.*, *87*, 95-109, 1976.

APUA, *Review, 1*, Antigua Public Utilities Authority Newsletter, 1989.

Beller, W. S., G. D'Ayala, and P. Hine (Eds.), *Sustainable Development and Environmental Management of Small Islands*, 419 pp., Parthenon Publ., Paris, 1990.

Bender, M., R. Fairbanks, F. Taylor, R. Matthews, J. Goddard, and W. Broecker, Uranium-series dating of Pleistocene reef-tracts of Barbados, West Indies, *Geol. Soc. Am. Bull.*, *90*, 557-594, 1979.

Boardman, M. R., and A. C. Neumann, Sources of periplatform carbonates: Northwest Providence Channel, Bahamas, *J. Sed. Pet.*, *54*, 1110-1123, 1984.

CCA/IRF, Dominica: Country environment profile, 239 pp., Caribbean Conservation Assoc. and Island Resources Foundation, 1991a.

CCA/IRF, St. Lucia: Country environment profile, 332 pp., Caribbean Conservation Assoc. and Island Resources Foundation, 1991b.

CDERA, *Caribbean Disaster News, 1*, Caribbean Disaster Emergency Response Agency, Bridgetown, Barbados, 1992.

De Graff, J. V., Assessing landslide hazard for regional development planning in the eastern Caribbean, in *Proc., Expert Meeting, Hazard Mapping in the Caribbean*, edited by D. Barker, 136 pp., United Nations Disaster Relief Organization, 1989.

Delcan, Coastal geophysical survey, Delcan International Ltd., Government of Barbados, 1993.

Droxler, A. W., J. W. Morse, and W. A. Kornicker, Controls on carbonate mineral accumulation in Bahama basins and adjacent Atlantic Ocean sediments, *J. Sed. Pet.*, *58*, 120-130, 1988.

Eberli, G. P., and R. N. Ginsburg, Segmentation and coalescence of platforms, Tertiary, northwest Great Bahama Bank, *Geology*, *15*, 75-79, 1987.

Eberli, G. P., G. Kendall, P. Moore, G. L. Whittle, and R. Cannon, Testing a seismic interpretation of the Great Bahama Bank with a computer simulation, *Am. Assoc. Petrol. Geol. Bull.*, in press.

Engelen, G., R. White, and I. Uljee, Exploratory modeling of socio-economic impacts of climatic change, in *Climatic Change in the Intra-American Sea*, edited by G. Maul, 389 pp., United Nations Environment Program, Edward Arnold Publ., London, 1993.

Hearty, P. J., and P. Kindler, New perspectives on Bahamian geology: San Salvador Island, Bahamas, *J. Coast. Res.*, *9*, 577-594, 1993.

Hendry, M. D., Sea level movements and shoreline changes, in *Climatic Change in the Intra-American Sea*, edited by G. Maul, 389 pp., United Nations Environment Program, Edward Arnold Publ., London, 1993.

Hendry, M. D., Future sea-level change, in *Water Levels for Barbados*, edited by D. Smith, Delcan International Ltd., Government of Barbados, 1994.

Hendry, M. D., W. Hunte, B. Humphrey, C. Parker, and R. Bateson, Assessment of the impact of beach renourishment on the marine environment at Jalousie Plantation Estate, Bellairs Research Institute, Barbados, 78 pp., 1993.

Hine, A. C., and A. C. Neumann, Shallow carbonate-bank-margin growth and structure, Little Bahama Bank, Bahamas, *Am. Assoc. Petrol. Geol. Bull.*, *61*, 376-406, 1977.

Hine, A. C., R. J. Wilber, J. M. Bane, A. C. Neumann, and K. R. Lorenson, Offbank transport of carbonate sands along open, leeward bank margins, northern Bahamas, *Mar. Geol.*, *42*, 327-348, 1981.

Hunte, W., A survey of coral reefs near Grand Anse, Grenada, Bellairs Research Institute, for Org. of American States, 68 pp., 1987.

Hunte, W., and D. Younglao, Recruitment and population recovery in the black sea urchin *Diadema antillarum* in Barbados, *Mar. Ecol. Prog. Ser., 45*, 109-119, 1988.

Kesler, S. E., E. Levy, and F. C. Martin, Metallogenic evolution of the Caribbean region, in *The Caribbean Region*, edited by G. Dengo and J. E. Case, pp. 459-482, The Geology of North America, Vol. H., Geological Society of America, 1990.

KPA, St. Kitts-Nevis exploratory drilling project no. 866-11044, final hydrogeology report, Ker, Priestman and Associates, for Canadian International Development Agency, Ottowa, 1988.

Los Alamos National Laboratory, Evaluation of the St. Lucia geothermal resource: Summary report, prepared for government of St. Lucia, 1984.

Macintyre, I. G., Modern coral reefs of western Atlantic: New geological perspective, *Am. Soc. Petrol. Geol. Bull., 1*, 1360-1369, 1988.

Maury, R. C., G. K. Westbrook, P. E. Baker, Ph. Bouysse, and D. Westercamp, Geology of the Lesser Antilles, in *The Caribbean Region*, edited by G. Dengo and J. E. Case, pp. 141-166, The Geology of North America, Vol. H., Geological Society of America, 1990.

McCann, W. R., and W. D. Pennington, Seismicity, large earthquakes and the margin of the Caribbean Plate, in *The Caribbean Region*, edited by G. Dengo and J. E. Case, pp. 291-306, The Geology of North America, Vol. H., Geological Society of America, 1990.

Mercado, A., J. D. Thompson, and J. C. Evans, Requirements for modeling of future storm surge and ocean circulation, in *Climatic Change in the Intra-American Sea*, edited by G. Maul, 389 pp., United Nations Environment Program, Edward Arnold Publ., London, 1993.

Milliman, J. D., D. Freile, R. P. Steinen, and R. J. Wilbur, Great Bahama aragonitic muds: Mostly inorganically precipitated, mostly exported, *J. Sed. Pet., 63*, 589-595, 1993.

Muhs, D. R., R. C. Crittenden, J. H. Rosholt, C. A. Bush, and K. C. Stewart, Genesis of marine terrace soils, Barbados, West Indies: Evidence from mineralogy and geochemistry, *Earth Surf. Processes and Landforms, 12*, 605-618, 1987.

Neumann, A. C., and L. S. Land, Lime mud deposition by calcareous algae in the Bight of Abaco: A budget, *J. Sed. Pet., 45*, 763-786, 1975.

Roobol, M. J., and A. L. Smith, Volcanic and associated hazards in the Lesser Antilles, in *Proc. in Vulcanology, 1, Volcanic Hazards*, edited by J. H. Latter, pp. 57-85, Springer Verlag, 1989.

Schlager, W., and R. N. Ginsburg, Bahama carbonate platforms - the deep and the past, *Mar. Geol., 44*, 1-24, 1981.

Schwab, W. C., and R. W. Rodriguez, Progress of studies on the impact of Hurricane Hugo on the coastal resources of Puerto Rico, *U.S. Geol. Surv., Open-File Rept. 92-717*, 95 pp., 1992.

Sealy, N. E., *Bahamian Landscapes*, Collins Educational Publ., 93 pp., 1985.

Shepherd, J. B., Earthquake and volcanic hazard assessment and monitoring in the Commonwealth Caribbean: Current status and needs for the future, in *Proc., Expert Meeting, Hazard Mapping in the Caribbean*, edited by D. Barker, 136 pp., United Nations Disaster Relief Organization, 1989.

Shepherd, J. B., K. C. Rowley, L. L. Lynch, D. M. Beckles, and W. Suite, Summary proceedings of Lesser Antilles volcanic assessment seminar (LAVAS), 31 pp., University of the West Indies, St. Augustine, Trinidad, 1988.

Sigurdsson, H., Submarine investigations in the crater of Kick 'em Jenny volcano, *Bull. Sci. Event Alert Network, 4*, 45-49, 1989.

Smith, M. S., and J. B. Shepherd, Preliminary investigations of the tsunami hazard of Kick 'em Jenny submarine volcano, *Natural Hazards, 7*, 257-277, 1993.

Speed, R. C., Geological history of Barbados: A preliminary synthesis, in *Trans. 11th Carib. Geol. Conf.*, Ministry of Finance, Barbados, 1988.

Speed, R. C., L. H. Barker, and P. L. B. Payne, Geologic and hydrocarbon evolution of Barbados, *J. Petrol. Geol., 14*, 323-342, 1991.

Speed, R. C., P. L. Smith, K. S. Perch-Nielsen, J. B. Saunders, and A. B. Sanfilippo, Southern Lesser Antilles arc platform, pre-Late Miocene stratigraphy, structure and tectonic evolution, *Geol. Soc. Am. Spec. Paper 277*, 98 pp., 1993.

Tomascik, J., and F. Sander, Effects of eutrophication on reef-building corals, I: Growth rate of the reef-building coral *Montastrea annularis, Mar. Biol., 87*, 143-155, 1985.

Tomblin, J. F., and W. P. Aspinall, Reconnaissance report of the Antigua, West Indies earthquake of October 8, 1974, *Bull. Seismol. Soc. Am., 65*, 1553-1573, 1975.

Towle, E., The island microcosm, in *Coastal Resources Management, Development Case Studies*, edited by J. R. Clark, 749 pp., National Parks Service, Dept. of Interior, Washington, D.C., 1985.

UNTCD, Exploration for geothermal resources in the eastern Caribbean, United Nations Department of Technical Cooperation for Development, Bridgetown, Barbados, 1992.

Waal, de, L., Grenada water supply sector development plan, United Nations Department of Technical Cooperation for Development, Bridgetown, Barbados, 1987.

Wadge, G., and J. B. Shepherd, Segmentation of the Lesser Antilles subduction zone, *Earth Planet. Sci. Lett., 71*, 297-304, 1984.

Wells, S. M., Atlantic and eastern Pacific, *Coral Reefs of the World, 1*, 373 pp., 1988.

Wigley, T. M. L., and B. D. Santer, Future climate of the Gulf/Caribbean Basin from the global circulation models, in *Climatic Change in the Intra-American Sea*, edited by G. Maul, 389 pp., United Nations Environment Program, Edward Arnold Publ., London, 1993.

Wittenberg, M., and W. Hunte, Effects of eutrophication and sedimentation on juvenile corals, I: Abundance, mortality, and community structure, *Mar. Biol., 112*, 131-138, 1992.

13

Geology and Development Facilities of Small Islands Belonging to the Atlantic Margin of Africa and Europe

André Klingebiel and Georges Vernette

Introduction

From a geological and oceanographic point of view, the eastern side of the Atlantic Ocean is characterized by a passive margin structure, related to the opening of this ocean since the lower Mesozoic period and to the mechanism of seafloor spreading responsible for continental drift. Among the three major continental plates (African, Iberic and European), only the European Plate contains numerous small coastal islands; this difference seems to be due to both a difference in climatic erosional processes, *i.e.*, effects of Pleistocenic glaciations on the European Plate, and chiefly to stronger and younger tectonic and isostatic events that occurred during Mesozoic and Cenozoic periods in the northern hemisphere. The main coastal island systems from south to north are Bissão and Los Islands next to the Guinean coast and all the numerous coastal islands of Europe from the Bay of Biscaye to the Norwegian Sea.

However, the three plates are cut by transverse faults through which magmatic extrusions build up volcanic relief and oceanic islands. These main fractures and volcanic systems from south to north are Cameroon, Fernando-Po, Príncipe, São Tomé, Senegal, Cape Verde Islands, Atlas Mountains (Morocco), Canary Islands, the Azores, northern Ireland, the Faeroe Islands (Denmark), and Iceland. Thus, various kinds of islands may be distinguished within the eastern margin of the Atlantic Ocean:

1. Volcanic islands, comprised of basaltic plateau and volcanic cones.
2. Pre-continental islands, comprised of old metamorphic basement and folded sedimentary substratum.
3. Coastal barrier islands.

Small Islands: Marine Science and Sustainable Development
Coastal and Estuarine Studies, Volume 51, Pages 225–237
Copyright 1996 by the American Geophysical Union

No coral reef islands occur in this oceanic system.

Knowledge of the geological structure of the insular areas is basic to understanding the main characteristics of the land and shelf morphology, the location and nature of potential renewable and/or nonrenewable resources, elaborate economic development schemes, and coastal management programs.

Oceanographic and atmospheric data are also basic within a global model of fluxes occurring in these insular systems and contributing to their environmental budget. Rainfall and fresh water resources, solar energy, wind energy, hydraulic and sea energy, and the determination of exceptional physical conditions are all factors that may limit biosphere and hydrosphere usage.

Volcanic Islands

From a geological point of view, two types of volcanic structures may be distinguished: flat basaltic plateau and conic eruptive accumulations.

Basaltic Plateau Structure

Basaltic plateau structure is built up by a series of flat, subhorizontal lava flows with an average thickness of about 10 m, becoming gradually thinner towards peripheric zones, while the more viscous feldspar basalt flows, on the other hand, terminate abruptly with a blunter type of wedging out. Columnar structures may occur but are generally weak and irregularly developed. The individual flows are clearly separated by welded scoriaceous horizons and interbasaltic tuffs and tuffaceous clayey sediments of great importance for paleontology, hydrogeology, and mineral resources.

As to the origin of the basaltic islands, it is likely that they were built by fissure eruptions. Locally and at various periods, there were land areas in the form of islands, perhaps in the form of rows of islands as in the Hawaiian Islands which nowadays stand up as the highest parts of a submarine volcanic ridge also built up above a system of fractures. Within the eastern Atlantic margin, two groups of islands present a basaltic plateau structure, with a large submerged shelf area.

Madeira (Portugal)

The main island, Porto Santo, is an elongated (58 km long and 22 km wide) basaltic plateau with very high cliffs on its northern coast that reach 1861 m of altitude at Pico Ruivo. Its name is a remembrance of its initial forestal cover, now totally destroyed due to agriculture. Two small islands, Desertas and Salvagems, have very few inhabitants (less than 3,000) and are eastern emerged prolongations of this basaltic system. Most of the population (264,000 inhabitants) lives on Porto Santo (average population density: 220/km^2) around the town of Funchal. Main resources are agriculture, fisheries, and trade. The lack of available living space on this very steep island results in constant emigration.

Faeroe Islands (Denmark)

This group of 18 small islands extends from north to south over a distance of about 118 km and from east to west over a distance of about 75 km. The total land area is 1400 km^2; the population is about 44,000 inhabitants (average population density: 32/km^2). Due to strong winds and storms, the only vegetation is grass. However, the climate is temperate. Main resources are fisheries and cattle. One coal-bearing sequence interbedded in the basalt series has been exploited for local purposes only.

The Faeroe Islands belong to the North Atlantic basaltic province which extends from northern Ireland (Antrim) and western Scotland to Iceland and Greenland (Figure 1). They form a high-lying zone on the Wyville-Thomson Ridge, an elongated shallow water feature across the North Atlantic mainly built from fissure eruptions. From sea charts one can observe that the southeast-northwest trend of the Wyville-Thomson Ridge is reflected in the trends of the Faeroe Islands fjord system, as in the trends of the drowned fjords off southeast Iceland. Judging from gravity measurements and bottom samples, the Wyville-Thomson Ridge is composed of basalt, as is the Faeroe Bank and the Rockall and Porcupine Banks west of Scotland. If these now separated areas of basaltic plateau, which together cover about 250,000 km^2, were once part of one land mass, this land mass must have had an area of 2-3 million km^2.

These islands have an elongated form with a pronounced northwest-southeast trend. The average height is about 300 m and the highest point is 882 m. The land slopes gently down towards the southern and eastern coasts; the western and northern coasts are steep and sometimes extend vertically up to a height of 750 m (e.g., Myling cliff at the northwest point of Strömö, the highest abrupt cliff in the world). Below sea level, there is an abrupt fall for the first 50 m after which the seafloor slopes regularly and gently down to a depth of 200 m [Rasmussen and Noe-Nygaard, 1970].

Although tide amplitude is relatively low (1 to 2, 5 m), tide currents around and in channels between the islands are very strong (1 to 12 knots) and changing, with back-currents and time differences.

Conic Eruptive Structure

The other groups of volcanic islands are chiefly eruptive, belonging to strombolian or volcanic-type structures, with steep conic relief built up by trachytic and andesitic viscous lavas, products of explosive eruptions and tuff deposits.

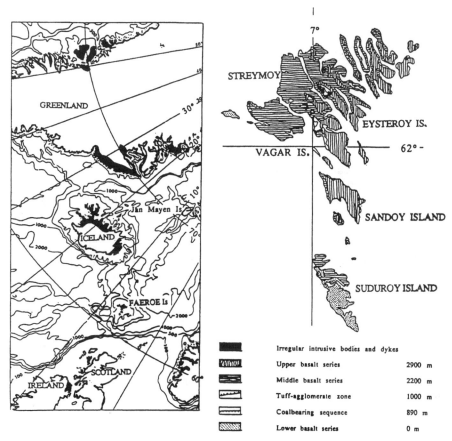

Fig. 1. Location of Faeroe Islands (Denmark) in the North Atlantic basalt province; geological map and schematic geological section through the Faeroe basalt series [after Rasmussen and Nygaard, 1970].

Canary Islands (Spain)

This group of seven islands (and some uninhabited inlets and skerries) extends between 13°30'W and 18°30'W and 27°30'N and 29°20'N; it has a land area of 7273 km² and a total population of 850,000 inhabitants (average population density: 116/km²). Volcanic relief is very steep with high cliffs; the highest volcano is Pico de Teide (3711 m) on Tenerife Island. The shelf is very narrow; its largest extension is less than 4 km, and the seafloor steeps rapidly at depths exceeding 3500 m, except on the eastern side of the archipelago, where in the Canary Channel, 100 km wide between Fuerteventura and Cape Juby on the Africa coast, the depths do not exceed 1400 m. The volcanic activity is now over, as in the eastern part of the system composed of Cenozoic basaltic extrusions in the Anti-Atlas and Atlas mountains.

Azores Archipelago (Portugal)

This archipelago is an eastward, 600 km long annex of the Mid-Atlantic Ridge more than 1,000 km from the African continental margin. Three groups of islands lie between 36°50'N and 40°N latitude and 25°W and 31°20'W longitude. The total land area is near 2500 km^2 and the population is approximately 320,000 inhabitants (average population density: 128/km^2). The Azores belong to a huge basaltic mid-oceanic system and, although the ocean floor is very irregular, it steeps gently to depths generally less than 2,000 m.

The central part of this volcanic system is yet active in the western extremity of Faial Island, along a series of little volcanic cones constituting the Peninsula de Capelo. The major height of this island is 1043 m around the 2 km large central crater named 'Caldeira'. An historic eruption was described in 1672 that built up the Cabeço do Fogo (571 m). The last volcanic event was a strombolian-type eruption beginning in September 1957 at the western point of Faial Island near Cape Capelinhos lighthouse. This eruption, which ended in October 1958, created a new volcanic inlet, 86 m high and mainly composed of scoriaceous and pyroclastic materials [Castello Branco et al., 1959].

A general west-northwest and east-southeast trend of island morphology and fractures is visible in the central archipelago and submarine relief of the surrounding oceanic floor. This is, peculiarly, the case of the elongated form of Pico Island where a volcanic cone culminates at 2284 m, and of the São Jorge Island, which is an elongated basaltic plateau with very steep cliffs.

The region of the Azores is well known as one of the major anticyclonic centers of the tropical zone due to the presence of the cold Canary Current. The climate of this region is warm and rainy. The large range of altitudes allows for an intensive cultivation activity and production of both temperate and tropical climate fruits, vegetables, and grains, as well as a large production of cattle. These products provide good support for an active agro-industry, as well as for textile production (linen). Exploitation of clays produced by tuff and volcanic ash alteration allows for good ceramic production. These islands are also of high strategic interest (U.S. military base on Santa Maria Island).

Cape Verde Islands

This archipelago is the sole insular state of the western Atlantic region located 500 km west of the Senegal coast between latitudes 14°40'N and 17°20'N and longitudes 22°30'W and 25°30'W (Figure 2). Its global area is about 4700 km^2, and the population is approximately 150,000 (average population density: 32/km^2). The relief of these volcanic islands, as well as that of the surrounding seafloor, is very steep. The areas able to be exploited for cultivation are very narrow, and the agriculture resources are limited by aridity of the climate. Therefore, a segment of the population is submitted to emigration.

The Cape Verde Islands probably began to form prior to the end of the Jurassic period by submarine volcanic processes interbedded in fossiliferous argilaceous limestones and cherts, which are deep water deposits [de Assumpçao *et al.*, 1968]. These beds crop out in Maio Island and dip east and west away from the center of the volcanic structure. They were, apparently, warped by an uplift of the island and are covered by horizontal Tertiary and Quaternary sediments with many volcanic interbeds [Machado *et al.*, 1969]. Between middle Cretaceous and upper Miocene times, a major tectonic disturbance caused upwarping of the islands which were eroded at sea level. The volcanic rocks resulted from eruptions mostly from Miocene and Quaternary times. The last eruption occurred on Fogo Island in 1951 [Ribeiro, 1954].

Fernando-Po and Annobon (Spain); São Tomé and Príncipe (Portugal)

This row of volcanic islands of submarine volcanic relief (guyots) is situated in the Gulf of Guinea, along a northeast-southwest direction, constituting the Medio-Guinean

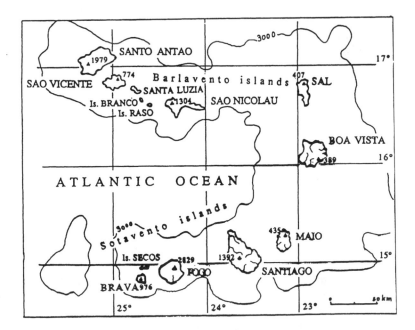

Fig. 2. Cape Verde Islands location and surrounding bathymetric map (altitudes and ocean floor depth in meters).

Ridge, a prolongation of a major fracture of central Africa from Mount Cameroon to Tibesti. São Tomé has an elevation of 2024 m and an area of 824 km^2 and 55,000 inhabitants (average population density: 64/km^2). Príncipe has a land area of 110 km^2 and culminates at 948 m (5,000 inhabitants; average population density: 45/km^2). Annobon is an islet of 15 km^2 culminating at 685 m with 1,440 inhabitants (average population density: 95/km^2). Fernando-Po has an area of 2017 km^2 and culminates at 3106 m; the population is 40,000 inhabitants (average population density: 20/km^2).

One can see that the relief of these islands is very steep; observation of sea charts reveals the same kind of relief for the surrounding seafloor. Under 1000 m of altitude, the land is covered by a dense tropical vegetation due to equatorial temperatures and abundance of rainfall (1200 to 5200 mm/yr). At higher altitudes, the lower temperature leads to a temperate but very wet climate. The substratum is composed mainly of basaltic and phonolithic lavas, as in São Tomé and Annobon, with andesitic and trachytic dikes. Some Miocene and Holocene paleo-beaches are observed at various levels around these islands [Cornen et al., 1977], indicating some isostatic instability of this volcanic ridge.

Today, seismic activity is observed at Mount Cameroon and Santa Isabel in Fernando-Po. The earliest volcanic activity known on São Tomé was of Paleocene age. The main phases of eruption and building up of the Medio-Guinean Ridge occurred during the Miocene (18 to 24 My) and Pliocene (2,5 to 4 My) periods. Volcanic activity on Fernando-Po and Mount Cameroon is from the recent Pleistocene (0-7 My) period.

Economic Geology of Volcanic Islands

Volcano-basaltic systems are poor in minerals and rocks of any particular importance:

1. Basalt and andesite: boulders and slabs are used as building materials. They also have good characteristics for crushed stone usage.
2. Tuff: quarried for its powdered form for fine granulate usage and as a substitute for sand. Tuff is known for its high porosity and water content.
3. Clays: various types occur in considerable amounts as products of tuff alteration; one can find clay minerals corresponding to all degrees of silicate evolution, from Fuller's earth (smectite = montmorillonite) to kaolinite and fire clays, these latter being the result of normal terms of weathering processes of the equatorial climate.
4. Lacustrine deposits: interlayered in the basaltic subhorizontal flows, lacustrine deposits can be of some economic value for local purposes: one can find diatomaceous earth, sometimes associated with peat bogs. In the Faeroe Islands, two coal fields were actively exploited in the past.
5. Soils: generated by weathering and alteration of the alkaline volcanic rocks and tuffs, soils have a good ion exchange capacity and a high fertility; their agricultural value depends on relief slope, rainfall abundance, and seasonal distribution.

Pre-Continental Islands

These islands are relicts of unsubmerged reliefs of the continental margin. The steeper the continent relief, the more numerous the islands. This is well demonstrated on the west coast of Ireland, Scotland and Norway where Pleistocenic rivers and glaciers dug deep valleys now drowned by the Holocene transgression, forming 'rias', 'forthes', and 'fjords'. The relief, the geological structure and nature of these islands, is very similar to those of the next continental area of which they are only a seaward prolongation.

With the exception the little Los Islands next to the Guinea coast, the margin of Africa does not have pre-continental islands. The same observation can be made for the Iberic Plate. On the contrary, there are numerous pre-continental islands along the Atlantic coast of Europe, from the southern limit of the European Plate at the south end of the Brittany French coast in the Bay of Biscaye, to the Norwegian and Finland coasts. The continental shelf around these islands is very flat; it steepens very slowly down to the 200 m isobath which is generally more than 100 km off the Atlantic coast of France, and reaches 225 miles west of the United Kingdom and Ireland [Owen, 1974]. It is mostly covered by relict sediments produced by continental erosion and deposited during the Pleistocene low sea level periods; these deposits contain fluvial and, perhaps, morainic characteristics north of 50°N latitude where they extended to the Scandinavian islands during the last glacial period.

The oceanographic and climatic conditions are similar on the continental coastal zones and on the coastal islands, but with an accentuation of sea influence on islands due to the higher ratio of "land area/length of coastline." On the western side of Europe, the vicinity of the ocean is the cause of a warming modification of climate originating from the Gulf Stream. The coastal forms are deeply indented, as they are the result of the differential fluvial, glacial, and then coastal erosion into rocks of varying hardness. They are characterized by alternance of cliffs on the ocean-facing zones of the islands and by pebbles or sand beaches and tidal flat formations in protected bays. The high frequency of storms and the high average amplitude of waves provoke active coastal erosion that needs an overcalibration of port protection and other coastal defenses.

The large range of tidal amplitude in this northern part of the Atlantic Ocean is the source of strong tidal currents, distributing all kinds of sediments along the coastal zone and along the inner continental shelf. The diverse nature of the seafloor is the source of a large diversity of bioproductions, allowing numerous fishing activities in the open sea, sea-farming in protected bays and, in the wide intertidal zones, intensive activities of conchyliculture. However, insularity may be a limiting factor concerning the development of economic activities. Most of the coastal islands of Europe experienced an intense emigration flow during the last century and now support only a low density of population. Some of them are now classified as faunistic and faunistic "natural reserves," and are sometimes dedicated to sea-bird and fauna preservation. Economic development is generally based on touristic activities and may be limited by reference to legal constraints about environmental quality preservation, particularly fresh water resources and possibilities of waste storage. These two problems surged from projects of coastal zone valorization by industrial or tourism development programs which induced major fresh water consumption and produced major needs about polluted effluents and waste treatments. The example of some of the French pre-continental islands shall be used as demonstration.

Two kinds of pre-continental islands may be distinguished on the basis of structure and nature of their substratum.

Coastal Islands with a Folded and Metamorphic Basement

The western coast of the European Plate is cut by the Armorican [Owen, 1974] and the Caledonian [Nicholson, 1974] chains of mountains whose basements are extensive and varied in age, lithology, and structure. The example of the French island Belle-Ile en Mer is demonstrative of environmental management. The insular area is about 84 km^2 and the permanent population under is 4,900 inhabitants (average population density: 58/km^2). Main resources are agriculture, on a plateau culminating at 65 m of altitude, and fisheries. In summer, the population grows to 15,000 and even 18,000 in August. Fresh water comes from various springs and is stored in three artificial lakes with a global capacity of about 850,000 m^3. Domestic waste collection produces 3600 T/year (daily average = 10 T) with wide seasonal variations. These products are crushed and buried in a special site managed to prevent ground water, river, and coastal water pollution by means of an intermediate lagoonage treatment. It is only a provisional process, however, because the storage capacity of the site shall be saturated within 10 years.

Coastal Islands with Sedimentary Unfolded Rock Substratums

These islands border sedimentary basins such as the Aquitaine Basin (France: Ré and Oléron Islands) and the North Germany Basin (Heligoland Island). This latter island, composed of triassic sandstone, was intensively eroded and has lost most of its area (four-fifths) since the 9th century [Pawlowski, 1905]. Ré and Oléron Islands are good examples of economic management problems encountered in such pre-continental insular areas which were recently connected to the next continental area by a road bridge. This connection brought solutions to some administration, security, and trade problems, as well as water supply in the summer and waste exportation to the continent. It also facilitated the touristic fluxes and consequent seasonal fluxes of the population.

Oléron Island has a very flat and low (culminating at 12 m) area of 175 km^2. Its permanent population is about 14,400 inhabitants (average population density: 82/km^2). Coastal erosion of the substratum, composed of limestone in subhorizontal beds of upper Jurassic and upper Cretaceous age, produce cliffs several meters high, locally covered by coastal sand dunes (culminating at 34 m) prograding landward, parallel with coastal sand drift. The silicoclastic sand of the beaches and coastal dunes is inherited from relict sands covering the near shelf. Main resources of the island are vineyards, wheat, early vegetables, and cattle breeding. Oyster culture, widely developed in the salt marshes and protected embayments of the east coast, is a dominant activity that utilizes the salt marshes and intertidal areas. Fresh water supply depends on the surficial ground water table in limestones. Winter rainfall is yet sufficient to renew the high summer consumption due to the important seasonal touristic influx. However, the thin fresh water table is greatly influenced by the underlying seawater if one looks at surficial ground water resistivity values measured in the wells used for the water supply of numerous villages [Bendhia, 1973].

Ré Island has an area of about 80 km^2 and a permanent population of 9,000 inhabitants (average population density: 110/km^2). This east-west elongated (28 km long, and 0,1 to 5 km wide) and flat island is partly composed of a Jurassic limestone substratum and mainly composed of salt marshes and coastal dunes bordering the two sides north and south of the island on which occur an intense coastal sand drift and a coastal erosion due to eastward incidence of waves and strong tidal currents [Germaneau, 1985]. Main resources of the island are vineyards, early vegetables, wheat, salt marshes, and oysters. For the past few years, the island has been connected to the near by continent. This connection, providing easy accessibility to the sand beaches, produced a rapid increase in touristic investments, and numerous houses were built on the fields bordering the coast. Unfortunately, coastal erosion has been the cause of various individual material deposits: rocks and pebbles coming from intertidal outcrops, sands coming from small coastal dunes, concrete blocks, etc., without any significant result. The need of a global coastal management program for the island became evident [Germaneau, 1985; Renard, 1985]. Basic problems such as fresh-water supply and waste elimination are now solved by an exchange with the continent through the bridge. Ré is no more an island.

Barrier Islands

These islands are always very close to the continent they are fringing. They are coastal sand accumulations built up by a complex combination of wave action and littoral sand drift; they are generally enhanced by sand dunes on their sea face. Their morphology is permanently evolving and is strictly controlled as well by coastal hydrology (wave patterns and tide currents) and sediment budget of the coastal system (fluxes of sediments from the inner shelf and from the fluvial input).

Along the coast of west Africa, small barrier island systems are located in the delta front of the Sanaga River in Cameroon and out of the Guinea Bissão estuary in the Bissagos Archipelago. Some coastal sand banks are observed along the Mauritania coast (d'Arguin Bank), but the seaward winds are not able to build coastal dunes.

South of Portugal, a row of barrier islands is an evolving separation between the ocean and the Ria Formosa lagoon [Granja *et al.*, 1984]. The coastal dunes are low (3 to 4 m high only) and are not able to prevent the islands from wave and tidal current erosion. Since 1942, Baretta Island has been prograding seaward at a distance of about 300 m; Culatra Island migrated eastward at a distance of about 600 m; Armona Island migrated westward at a distance of about 500 m, inducing modifications to the channels in the lagoon and between the lagoon and the ocean. Quantifications of sand movements led to a global scheme of sediment budget (Figure 3). The main source of sand seems to be the eroding western coast of this system (196,000 m^3/year) and not the fluvial input into the lagoon; the budget of sand transport in the channels between the islands by tidal currents has a resulting balance oriented from sea to lagoon.

North of the Netherlands, the Wadden Islands are located on the northern portion of a great tidal flat area with parallel the sand dikes of Groningen and Friesland. In 1932, the Zuidersee was closed off from the tidal flat area by a great dike. The Wadden Islands lie in a dynamic area where there is an important sand drift from west to east with a high rate of erosion and progradation on various parts of the islands [Jelgersma, 1992]. Fortunately, the coastal dune strip is 1 to 5 km wide and 5 to 30 m high, storing a volume of sand able to slow coastline evolution. These dunes are also able to prevent these low lands from a sea-level surge.

Natural resources of sandy barrier islands are only sand beaches and recreation activities on the seaward face, pine forestry, drinking water supply in coastal sand dunes, and culture and salt marshes in the back lowlands of the inner face. To facilitate land use of these barrier islands, some of them were connected to the near by continent by an unsubmersible dam; that is the case of Römö Island on the west coast of Denmark.

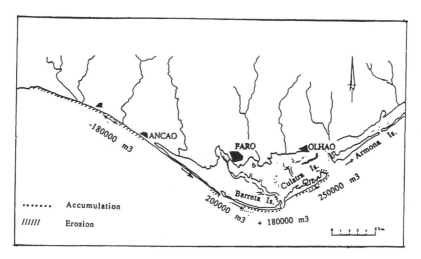

Fig 3.   Ria Formosa barrier islands (south Portugal) and coastal sand drift processes [after Granja *et al.*, 1984].

Conclusion

The aim of this description of the geological and physical characteristics of the Atlantic eastern margin islands is to help to understand and to compare their potentialities of development. The volcanic islands are related to major transverse faults of the oceanic floor. They are built up by basaltic and fluid or andesitic and viscous lavas. Eruptive and seismic activity is constant around and under these islands. With the exception of the Faeroe and Azores Archipelagos, where the ocean floor is regionally enhanced by basalt up lift and forms wide shelves, the seafloor around the volcanic islands steeps down rapidly to oceanic depths.

Of the volcanic islands of this Atlantic region, only the Cape Verde Archipelago is an independent state. Others groups of islands belong to the near by continental states, with some of them having a particular status of relative autonomy. The coastal islands all belong to the near continental states; some of them lost their insularity after they were connected to the continent by bridges. A tentative comparison of physical characteristics and demography of these various islands leads to a few observations (Table 1):

1. Volcanic islands have a lower relief index (area/culminating altitude = km^2/m) than pre-continental and overall barrier islands.
2. The relation of wideness of the shelf with the relief index is not absolute and needs a more precise morphological approach, taking into account particular values of each island.
3. Climate has no direct influence on average population density.

TABLE 1. Characteristics of Eastern Atlantic Islands

Island	Substrate	Relief	Shelf	Climate	Total Land Area (km^2)	Average Population Density (km^2)	Coast
Annobon	Volcanic	0,02	Restricted	Equator	15	96	9.5
Azores	Volcanic	1,09	Wide	Temperate	2500	128	3.8
Belle Ile	Metamorphic	1,30	Wide	Temperate	84	58	6.5
Canary	Volcanic	1,92	Restricted	Temperate	7273	116	2.5
Cape Verde	Volcanic	1,71	Restricted	Arid	4700	32	16.2
Faeroe	Basaltic	1,63	Wide	Temperate	1400	32	57.2
Fernando-Po	Volcanic	0,29	Restricted	Equator	2017	20	4.0
Madeira	Basaltic	0,64	Restricted	Temperate	1200	220	0.7
Oléron	Limestone	5,11	Wide	Temperate	175	82	3.2
Príncipe	Volcanic	0,11	Restricted	Equator	110	45	7.5
Ré	Limestone	6,66	Wide	Temperate	80	112	3.6
Ria Formosa	Barrier	1,00	Wide	Temperate	--	--	--
São Tomé	Volcanic	0,40	Restricted	Equator	824	64	1.90
Wadden	Barrier	15,00	Wide	Temperate	--	--	--

To appreciate the relative importance of potential coastal zone use, in relation to global land use by the insular population, in each of the groups of islands, a "*coastal-population-index*" has been calculated. This is an average value of the coastal length (meters) that each inhabitant of the island can use. This index is a ratio of the theoretical length of the shoreline of the island or archipel and the population. Inverse values can express the population pressure on the coastal zone. This index seems to be of some interest in appreciating the potentialities of future economic developments and facilities for a coastal management based upon human resources, and geological and oceanographic characteristics.

References

Assumpçao, C. T. de, F. Machado, and A. Sarralhiero, New investigations on the geology and volcanism of the Cape Verde islands, Rep. 23rd Session Intern. Geol. Congress, Czechoslovakia, Proc., Section 2, 9-16, 1968.

Bendhia, H., Contribution à l'étude géologique et hydrogéologique de l'Ile d'Oléron, Thèse 3ème cycle No. 1095, Univ. Bordeaux, 102 pp., 24 fig., 1973.

Castello Branco, A. de, G. Zbyszewski, F. Moitinho de Almeida, and O. Da Veiga Ferreira, Le volcanisme de l'Ile de Faial et l'éruption du volcan de Capelinhos, Servicios Geologicos de Portugal, Memoria No. 4, Lisboa, 1959.

Cornen, J., P. Giresse, and G. Kouyoumontzakis, La fin de la transgression holocène sur le littoral d'Afrique équatoriale et australe, Buèll. ASEQUA No. 50, p. 62-64 et 68-83, 1977.

Dewey, J. F., The geology of the southern Caledonides, in *The Ocean Basins and Margins*, edited by Nairn and Stehli, pp. 205-231, vol. 2, Plenum Press, New-York, London, 1974.

Germaneau, J., L'île de Ré, occupation et exploitation des milieux naturels: Dangers et remèdes, Cahiers de la mémoire, ADRT édit. Poitiers, 13-26, 1985.

Granja, H., J. M. Froidefond, and T. Pera, Morpho-sedimentary evolution process of the Ria Formosa coastal lagoon (Portugal), *Bull. Inst. Geol. Bassin d'Aquitaine, 36*, 37-50, 1984.

Jelgersma, S., Vulnerability of the coastal lowlands of the Netherlands to a future sea-level rise, in *Impacts of Sea-Level Rise on European Coastal Lowlands*, edited by M. J. Tooley and S. Jelgersma, pp. 94-123, Inst. British Geographers Spec. Publ., Univ. of Bristol, U.K., 1992.

Machado, F., J. Azeredo Leme, and J. Monjardino, O complexo sienito-carbonatico da ilha Brava, Cabo Verde, Garcia de Orta, Lisboa, *15*(1), 93-98, 1969.

Nicholson, R., The Scandinavian Caledonides, in *The Ocean Basins and Margins*, edited by Nairn and Stehli, pp. 161-203, vol. 2, Plenum Press, New-York, London, 1974.

Owen, T. R., The geology of the western approaches, in *The Ocean Basins and Margins*, edited by Nairn and Stehli, pp. 233-272, vol. 2, Plenum Press, New-York, London, 1974.

Pawlowski, A., L'île d'oléron à travers les âges d'après la géologie, Pa cartographie et l'histoire. Les transformations du littoral français. Bulletin de Géographie historique et descriptive. pp. 217-236. Imprimerie Nationale, Paris.

Rasmussen, J., and A. Noe-Nygaard, Geology of the Faeroe Islands, Geol. Surv. of Denmark, I, Series No. 25, Copenhagen, 142 pp., 1970.

Renard, V., Le problème foncier et l'avenir de l'île de Ré: Les cahiers de la mémoire, ADRT édit. Poitiers, 51-58, 1985.

Ribeiro, O., A ilha do Fogo e as suas erupçaoes, *J. Inv. Ultramar*, Lisboa, 1954.

14

Beach Erosion and Mitigation: The Case of Varadero Beach, Cuba

Guillermo García Montero and José L. Juanes Martí

Abstract

Varadero Beach, one of the most important tourist resorts of Cuba, has been experiencing an erosive trend of its shoreline for the past 25 years. Shoreline retreat has been estimated at 1.2 m/yr with an average sand loss of 50,000 m^3/yr. The main results of a research program for beach erosion studies are presented; this includes the results of a mitigation program applying artificial beach nourishment that has been developed since 1987, with a total of 700,000 m^3 of sand nourished to the beach. The main causes of beach erosion and the reasons for specific mitigation actions are explained. The application of these results to other small island countries could be inferred.

Introduction

In many small island developing countries, tourism represents an important economic activity, and its development is based on the recognition of environmental beauty as a valuable natural resource. In other words, tourism activities are today very sensitive to the quality of the environment. But, when talking about small islands in general, "the environment" could be understood as the marine and coastal zones from which they obtain their basic support. Small islands countries are, in fact, "coastal-marine" countries. Moreover, the importance of the coastal zone for small island countries should be emphasized because beaches are perhaps one of the most important resources available to them. Beaches are, therefore, an important factor of their sustainable development. As with any

Small Islands: Marine Science and Sustainable Development
Coastal and Estuarine Studies, Volume 51, Pages 238–249
Copyright 1996 by the American Geophysical Union

coastal system, beaches are the result of the combination of many natural factors influencing their behavior in such a way that in stable systems the seasonal variations are compensated for and, in a long-term period, no sand losses are produced which means no erosion occurs. However, small islands have several common characteristics (Table 1) that affect the natural equilibrium of the coastal zone in general and of the beaches in particular. The alteration of these factors (even just one of them) has been the origin of beach erosion at national, regional, and international scales and, in fact, one could say that beach erosion is at present observed on the coasts of most countries all over the world, affecting in various degrees the development of the tourism industry. Many cases could be mentioned to show the effect of this serious problem for several geographic zones. The reader could be well informed by means of the work of Paskoff [1992] for the coasts of North Africa; Shuisky [1986] and de Heer [1986] for several countries; Kirk [1987] for New Zealand, *etc.* Unfortunately, lack of data about erosion processes for small island countries, in general, and for those of the Caribbean region, in particular, do not allow us to show statistics as the ones offered for the continental zones. Exceptions to this are the data available for a few eastern Caribbean islands [Cambers, 1985] and for Cuba [Juanes, 1986a].

For an important part of the community (scientists, environmental managers, and policy-decision makers), the general character of the shoreline retreat has been associated, to some extent, with climate changes and, in particular, to continuous sea

TABLE 1. Small Island Common Characteristics*

---

1.   Susceptibility to natural environmental events (hurricanes, cyclones, *etc.*)

2.   Little overall climate variability, generally, but potential for climate upsets.

3.   Tendency towards ecological instability when isolation is breached.

4.   Almost immediate repercussions on the coastal zone and marine environment of terrestrial events.

5.   Extensive land-sea interface which increases the fragility of coastal ecosystems and the demand for coastal zone management.

6.   No interior hinterland or central terrestrial core areas that are essentially distant from the sea, such that coastal resource planning and management is essentially synonymous with national resource planning and management.

7.   Small land mass to ocean space makes islands especially vulnerable to global environmental phenomena such as sea level rise.

---

*source: Doc.   No.   CCS/UNECLAC/ACM/RTM1/3:   Regional   Technical   Meeting   for   the Atlantic/Caribbean/Mediterranean, Port Spain, Trinidad and Tobago, July 12-16, 1993.

level rise. It is already accepted, however, that shoreline retreat is, particularly for beaches, the immediate response of the coast to many human-induced activities, including not only those resulting from erroneous environmental planning of the coastal space, but also those arising from mitigation actions improperly implemented.

In the particular case of Cuba, coastal erosion has been evident with sea penetration in the mangrove lower zones of the southern coast and with loss of sand at beaches that extend along the northern coast, particularly at Varadero Beach, the main tourist resort of the country. At Varadero Beach, the falling of trees, the outcrop of rocky surfaces, and the destruction of tourist facilities due to the force of sea waves are the best evidence of the advance of the erosion processes that have occurred during the last 25 years. As a consequence of this, a research program was designed and implemented in 1978 to establish the general scheme of sediment dynamics in the region, find causes and rates of erosion, define the ecologically sustainable measures to mitigate the effects of the erosion, and restore, as far as possible, the natural conditions to the beach. The positive results obtained from this work are presented.

Analysis

The coastal zone is a complex system where interaction between land and sea is a permanent condition for its adequate functioning. Any external or abnormal factors influencing this interaction could lead to a disturbance of this system and, as a consequence, to the generation of conditions for coastal erosion (Figure 1).

Varadero Beach, one of the most important coastal resorts of Cuba, located 130 km east of Havana on the north shore of the 22-km-long Peninsula de Hicacos (Figure 2), is a good example of the negative influence that external factors could exert on a beach system. The Peninsula de Hicacos is characterized by long and broad sandy beaches and well developed dune fields all along the shore. The shelf (north shore) is characterized by its narrow extension to the southwest near the Canal de Paso Malo and, on the contrary, by its broad extension to the northeast.

The most common causes of erosion that have been postulated for Varadero Beach are:

1. Mining of beach sand.
2. Interference of bulkheads and jetties.
3. Buildings, houses, and other structures built over the dune.

Mining of beach sand on Varadero Beach was a common practice until 1978, with an estimated 2,000,000 m^3 of sand removed. As is clearly obvious, this practice originated from the construction of buildings and hotels, caused the depletion of natural sand reserves, not only in the submerged zone, but also in the dune fields. This resulted in changing the topography of several areas which, in turn, affected the natural sediment exchange between beach and dune. It has to be borne in mind that this exchange is probably the most important natural process for beach stability.

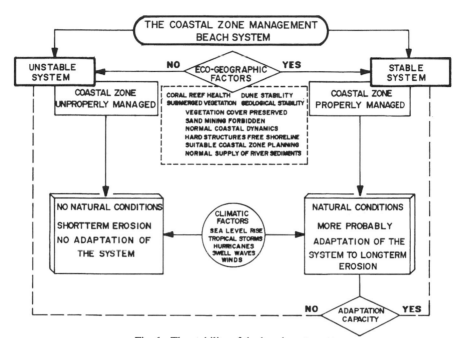

Fig. 1. The stability of the beach system.

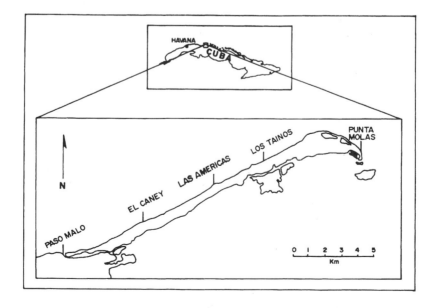

Fig. 2. Location map of Varadero Beach, Cuba.

Besides this, the Canal de Paso Malo, built in 1956 to provide fishing boats in the lagoon with a shorter route to the open water of the Straits of Florida, became an important factor disturbing the normal shore dynamics. The Canal entrance on the north is bracketed by two jetties, both projecting more than 100 m toward the sea, and starting sand deposition on the updrift side of the longshore current and erosion on the downdrift side. However, sand deposition on the updrift side of the longshore current (east side of the Canal) is also influenced by erosion due to the "seawall" design of the jetty channel and the effect of the wave fronts hitting nearly perpendicular against this side.

The major cause of beach erosion on Varadero Beach is the presence of more than 150 buildings, houses, and other hard structures that were constructed near the coastline or even over the dune, on a 5 km long space on the beach. Due to environmentally erroneous planning of the coastal space in former times, the effects created by these touristic facilities and private houses are as follow:

1. Reflection of sea wave energy scours the beach and nearby areas.
2. Interference with local coastal processes affect the shore current system; hence, sediment transportation is carried out in the wrong direction.

Although very destructive, the erosive mechanisms of these effects are simple: sea wave energy is reflected on hard structures located on the coastline; a larger quantity of sediment (other than normal) is moving away from the affected shore current system; and sediment transportation offshore is more intense, reaching deeper zones where beach sand is out of the long term natural system equilibrium (Figure 3). It is easy to understand that under these conditions the beach, generally speaking, fails to be an efficient "coastal structure" for the natural defense of the coast. Finally, the effect of sea level rise on beach erosion, although certainly present, has to be estimated in the future, although it does not reach the magnitude of other human-induced factors.

It should be pointed out that mitigation actions against beach erosion must be properly implemented. Any action improperly implemented is a source of trouble for the coastal system. For this reason, beach erosion solutions have to be, in the first place, those which can be oriented to eliminating the factors causing the erosion. Therefore, it is not recommended to introduce any hard anti-erosion structure (e.g., seawalls) without a depth study of the coastal zone and its dynamics system. On the contrary, artificial beach nourishment is an aesthetic method which meets environmental considerations, and it is also appropriate since it cures the root cause of beach erosion which is an insufficient sand supply [Paskoff, 1992].

It could be summarized that one of the best solutions for beach erosion mitigation could be accomplished by eliminating, as far as possible, those factors causing the problems and increasing the sand supply to the beach system. However, the situation is even more complex because coastal zone management requires not only science and engineering skills, but also legal instruments (e.g., Shores Act) that regulate planning and utilization of the coastal zone.

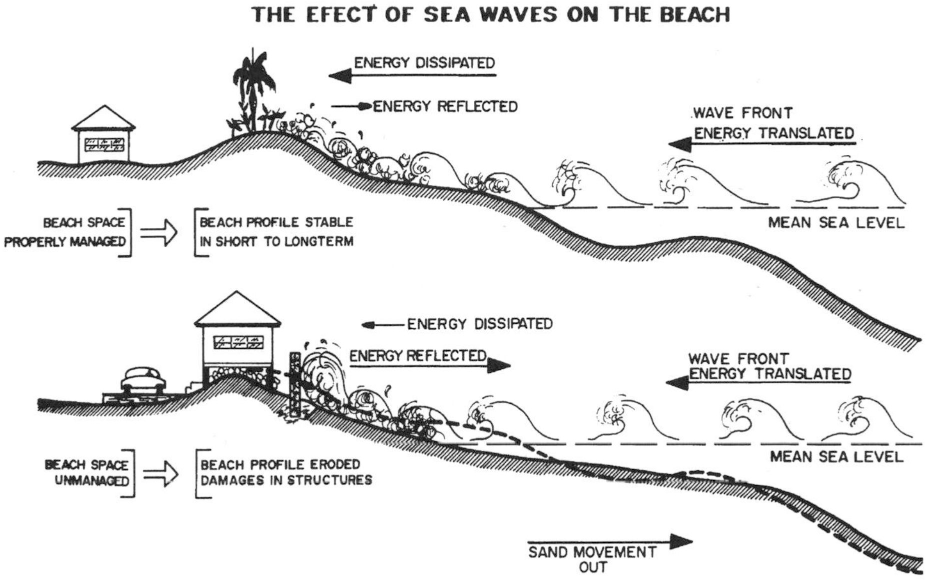

THE EFECT OF SEA WAVES ON THE BEACH

Fig. 3. Sea wave effects on the beach.

Erosion mitigation at Varadero Beach is the consequence of a combination of actions directed towards the knowledge of coastal processes, identification of causes, elimination or control by legal regulations of all or some of these causes, and the artificial extra supply of sand to the beach.

Discussion

The basic scientific purposes of this research program were directed towards the knowledge of:

1. Geomorphological characterization of the shelf.
2. Seasonal and spatial morphological variations of beach profiles.
3. Tendency of sediment transportation.
4. Composition and distribution of sand.

Seismic profiles were useful to describe the way in which sand at the shelf is distributed in the different basins [Foyo Herrera, 1982]. It was determined that these basins are limited by rocky ridges that served as plinths to the actual reef formations and the accumulation of complete sand cover in the basins. The three main basins described are located on the northern shore along the peninsula and are also characteristic due to their narrow extension to the southwest and to their wide extension to the northeast. Observations made on board the research mini-submarine *Argus,* up to a depth of 600 m, revealed the existence of depressions perpendicular to the rocky ridges like small submarine canyons, through which the sand could go out to the shelf.

Topographic measurements of beach profiles made monthly since 1978 allow us to identify seasonal and long-term erosion [Juanes *et al.,* 1985]. In winter, cold fronts in the Gulf of Mexico generate high waves which travel south as swell waves towards Cuba's northern coast, causing expansive beach erosion. In summer when the beach accretes, it reaches its maximum, as regularly represented in profiles of October and November (Figure 4).

Shoreline measurements show an erosion trend of 1.2 m/yr, while the erosive trend in terms of sand loss has been demonstrated by seasonal calculations taken during 14 years of monitoring 11 profiles along the 22 km of beach. Calculations show a general erosive tendency, with an estimated 50,000 m³/yr loss of sand. Figure 5 shows this erosive trend in three chosen locations [Juanes and García, in press].

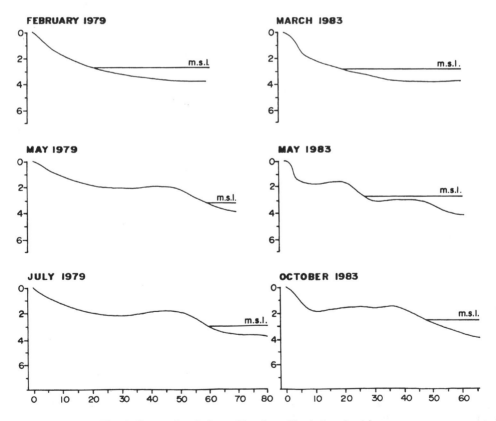

Fig. 4. Seasonal variations of beach profiles in Las Américas.

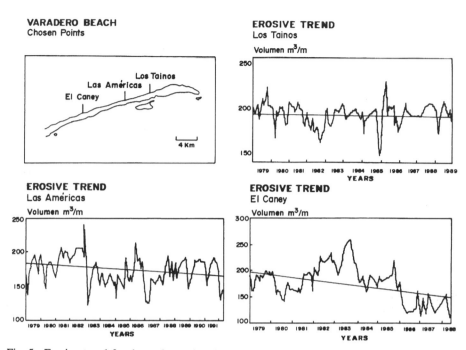

Fig. 5. Erosion trend for three chosen locations at Varadero Beach (Los Tainos, Las Américas, and El Caney).

The study of sediment transportation trends, based on the wind's characteristics, was made through the use of the Knaps method. In this method, a one-dimensional index of energy transferred by the wind to the sea surface was established, while the directional analysis of this index allowed us to obtain its components parallel and perpendicular to the coast. The results obtained with the application of this method [Juanes *et al.*, 1986b] suggest that the greatest tendency of sediment transportation at the peninsula is established from east to west during most of the year, even though there is a transport in contrary direction during winter months, which corresponds to the scheme of shore dynamics determined from the study of morphological variations of beach profiles.

Studies of the composition and distribution of sand in the region were based on granulometric analyses and microscopic observations of more than 1,000 samples obtained from the shelf and beach during different expeditions. The samples thus obtained allowed us to show that 90% of the sand content was formed by thin fragments of benthonic organisms, among which 40% are the algae *Halimeda*, 25% are foraminifera, and 20% are mollusks [Juanes *et al.*, 1986a].

Results obtained from Varadero Beach were fundamental in describing its main characteristics, understanding the natural functioning processes of the beach, and selecting the proposal of artificial beach nourishment as the most appropriate solution for mitigation because of its well recognized ecological and aesthetic advantages over other techniques [Centre for Civil Engineering Research, 1990]. Hence, the basic parameters for our artificial beach nourishment project (such as the borrow area, necessary sand volume, sand characteristics, *etc.*) were easily determined, thanks to the results of this research program. An example of a typical project on this matter could be found in Juanes *et al.* [1993].

The selection of the borrow area is a particularly important step in the process of designing the project. The seismic characterization of the shelf allowed us to locate a suitable sand basin, with over two millions cubic meters of sand in reserve at 20 km from the beach. This basin was described [Instituto Cubano de Hidrografía, 1988] as a sand accumulation deposit at 10 m average depth with sand characteristics very similar to that of the beach. It was also determined that this deposit did not belong to the natural sand sources of Varadero Beach; therefore, this basin is out of the dynamic system of the beach.

Another important step is to determine the correlation between the native sand and that to be introduced. This was done by means of microscopic analysis of a set of samples taken from both locations (Table 2) and by determination of the Overfill Factor [Coastal Engineering Research Center, 1984] which allowed us to quantify how much the volume of sand has to be increased to compensate for sand loss due to granulometric differences. Since 1987 to 1992, five campaigns for artificial nourishment of the beach were done over a shoreline of about 4 km from Canal de Paso Malo to El Caney, with a total of 700,000 m^3 of sand nourished. The results of the two first years of replenishment work can be found in Schwartz *et al.* [1991].

Studies of sand volume on the beach one year after the mitigation work showed that sand loss was not greater than 20%. According to international experience, this result is satisfactory [Leonard *et al.*, 1990]. Figure 6 is only one example on how the aesthetic and recreation conditions of the beach have been recovered and also on the protection given to many buildings and houses. It has to be pointed out that in 1985 around 20 of these installations were partially or totally destroyed due to the effects of big swell waves associated with Hurricane Juan in October 1985, but in March 1993 similar effects associated with the "storm of the century" did not cause any damage, demonstrating not only the effectiveness of the method employed for mitigation, but also how the beach recovered its capacity as the more efficient natural defense of the coast.

TABLE 2. Composition of Beach and Borrow Area Sand

	Beach (Native)	Borrow Area (Introduced)
Diameter (mm)	0.10-0.25	0.25-0.50
Composition		
Calcareous algae	40%	41%
Foraminifera	25%	23%
Mollusks	20%	28%

Fig. 6. Before and after the artificial beach nourishment at El Caney: (a) August 1988; (b) November 1988; and (c) October 1991.

It could be added that mitigation actions should also include those directed at regulating or forbidding any human activity that could be a source of trouble for the coastal zone, and the beach in particular. On Varadero Beach, different steps have been taken to reduce the human influence as far as possible. As a consequence, a plan was established for the general and adequate management of all activities related with the environment, including the following:

1. Sand mining is strictly forbidden.
2. Construction work is prohibited in a precise zone inside the beach and dunes.
3. Any building or house over the beach or dune that was damaged by sea action could not automatically be rebuilt; preferably, it should be demolished.
4. All development action or work in the peninsula has to be coordinated with, and approved by, management authorities.

The application of these regulations has been very useful in the process of elaborating upon the project of the Shores Act of Cuba, which is, at present, under study.

Conclusions

Varadero Beach is a good example not only of trouble due mainly to human-induced activities, but also of success in mitigation actions to reduce or eliminate these troubles. Experiences and results obtained thus far are also applied to the solution of beach erosion in other small island countries.

One has to understand that beaches are, simply speaking, the consequence of the correlation between two major elements: a source of sand to replace that lost by natural erosion and a dynamic mechanism to facilitate natural beach replenishment and profile maintenance. Man must avoid hurting nature in order to preserve its resources and to use them properly. It should be borne in mind that "the sea can be either friendly or hostile. It is calm and beautiful one day, furious and terrifying the next. On days calm enough to make surveys and do construction work one must not forget that before long unleashed violence will follow." [Bascom, 1980].

Acknowledgments. Preparation of this paper was possible thanks to the help of the Department of Geology, Institute of Oceanology, Academy of Sciences of Cuba. The collaboration of Mr. Carlos García, Mr. Raúl Martell and Mrs. Argelia Fernández is gratefully acknowledged.

References

Bascom, W., *Waves and Beach: The Dynamics of the Ocean Surface*, Anchor Press, Garden City, N.Y., 366 pp., 1980.

Cambers, G., An overview of coastal zone management in six east Caribbean Islands. UNESCO/ROSTLAC/COMAR, 2a, reimpr, Montevideo, 54 pp., 1985.

Centre for Civil Engineering Research, Codes and Specifications/Delft Hydraulics Laboratory, Manual on artificial beach nourishment, Report, No. 130, 195 pp. (Recommendations 1), 1990.

Coastal Engineering Research Center-USA, Shore Protection Manual, 1984.

Foyo Herrera, J., Distribución de los sedimentos arenosos en la plataforma norte de la Península de Hicacos, Rep. Invest. Inst. Oceanol. Acad. Cien., Cuba, no. 2, 23 pp., 1982.

Heer, R. J. de, Beach nourishment by dredging, International Institute for Hydraulic and Environmental Engineering, 116 pp., 1986.

Instituto Cubano de Hidrografía, Informe científico sobre los trabajos geofísicos-hidrográficos para la localización de fuentes de abasto de arena para la rehabilitación de la duna de la playa de Varadero, 28 pp., 1988.

Juanes, J. L., and C. García, Análisis de la tendencia erosiva de la playa de Varadero 1978-1992, in press.

Juanes, J. L., E. Ramírez, and V. S. Medvediev, Dinámica de los sedimentos en la Península de Hicacos, Cuba, I: Variaciones morfológicas del perfil de playa, *Ciencias de la Tierra y del Espacio, 10*, 69-84, 1985.

Juanes, J. L., Investigaciones de morfología litoral y dinámica de arena en la Península de Hicacos, Informe. Arch. Cient. Inst. Oceanol. Acad. Cien., Cuba, 32 pp., 1986a.

Juanes, J. L., E. Ramírez, M. Caballero, V. S. Medvediev, and M. G. Yurkevich, Dinámica de los sedimentos en la Península de Hicacos, Cuba, II: Efecto de las olas de viento en la zona costera, *Ciencias de la Tierra y del Espacio, 11*, 93-101, 1986b.

Juanes, J. L., C. García, R. Martell, and R. Rodríguez, Proyecto de regeneración de la playa en el sector costero Peñas de Bernardino-Punta de Chapelin, Varadero, *Inst. de Oceanol. Acad. Cien. Cuba, 19* pp., 1993.

Kirk, R. M., Managing the coast, in *Southern Approaches: Geography in New Zealand*, edited by P. G. Holland and W. B. Johnston, pp. 239-259, Christchurch, Chapter 12, 1987.

Leonard, L. A., K. L. Dixon, and D. H. Pilkey, A comparison of beach replenishment on the U.S. Atlantic, Pacific and Gulf coasts, in *Artificial Beaches*, edited by M. L. Schwartz and E. C. F. Bird, *J. Coast. Res.*, Special Issue, *6*, 127-140, 1990.

Paskoff, R., Sandy beaches evolution and management in North Africa, *UNESCO Tech. Pap. Mar. Sci., 64*, 187-195, 1992.

Schwartz, M. L., J. L. Juanes, J. Foyo Herrera, and G. García Montero, Artificial nourishment at Varadero Beach, Cuba, in Proceedings, Coastal Sediments '91, Specialty Conference/WR Div./ASCE, Seatle, WA, June 25-27, pp. 2081-2088, 1991.

Shuisky, Yu. D., Problems of drift balance investigation in the coastal zone (in Russian), Moscow, Nauka, 238 pp., 1986.

15

Surficial Geology on the Insular Shelf of Martinique (French West Indies)

Françoise Durand, Claude Augris, and Patrice Castaing

Abstract

The small islands, like Martinique, are characterized by the fragility of the environment respect to the considered surface area smallness. Their economic expansion is connected to the knowledge of their surrounding ocean, particularly with relation to the development of their marine resources. On the insular shelf, sedimentary accumulations can be picked out by prospecting methods (seismic, corings, *etc.*), leading to the assessment of marine sand exploitation sites.

Sedimentological investigations on the insular shelf of Martinique by IFREMER and the University of Bordeaux 1 used complementary techniques such as side-scan sonar and echo sounder, sediment grab samples, and camera-towed photography to investigate the seafloor. The results show two major sedimentary facies: the volcanic facies, situated north and northwest of Martinique, transported as detrital inputs or by the wind, and the organogenic facies, located on the eastern and southeastern sides of the island, produced by the destruction of autochthonous coral and algal formations. These two gradients, governing the patterns of the littoral and neritic deposits, mix in variable proportions and characterize different sedimentary areas.

Introduction

The Archipelago of the Lesser Antilles stretches 850 km from north to south, from the Anegada Passage that separates it from the Greater Antilles, up to the coastline of

Small Islands: Marine Science and Sustainable Development
Coastal and Estuarine Studies, Volume 51, Pages 250–265

Venezuela. This curl, stretching across the Atlantic, cuts off the Caribbean Sea to the west. Located in the heart of the latter (Figure 1), Martinique, with a surface area of 1080 km^2 [Westercamp *et al.*, 1989], extends longitudinally over a distance of 65 km.

Martinique, an asymmetric island due to a double volcanic history, features a pronounced relief to the north and northwest, as the result of recent volcanogenesis. To the south and southeast central plains appear, where hillocks ranging from 100 to 300 m in altitude follow one another. This particularity is also true along the coasts [Pons *et al.*, 1978]. The Atlantic seaboard (lined with coral reefs), as well as the northern and southern coastlines (opposite Dominica and St. Lucia), present slight gradients, whereas along the Caribbean coast the incline of the seafloor is very substantial (Figure 2). This asymmetric submarine topography controls the distribution of sedimentary facies.

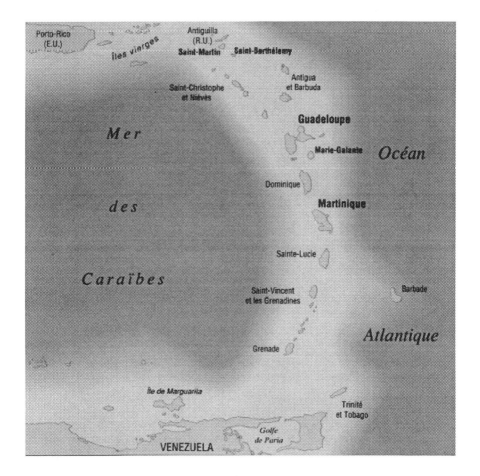

Fig. 1. Lesser Antilles Archipelago [Westercamp and Tazieff, 1980].

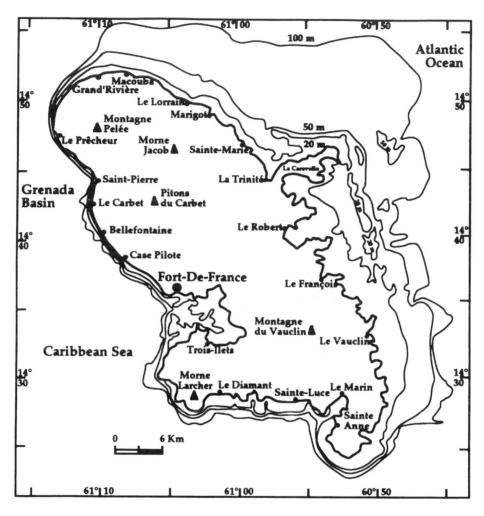

Fig. 2. Simplified bathymetric map of the insular shelf of Martinique (20 m, 50 m, and 100 m isobaths).

Within the framework of two marine geology surveys (IGMAR 1 and IGMAR 2), the mapping of the sedimentary units of the Martinique coastal area was performed jointly by IFREMER and the University of Bordeaux. This mapping was based on detailed reconnaissance surveys using complementary techniques such as side-scan sonar and echo sounder analysis, sediment grab samples, and camera-towed photography along the seafloor. Taking into account all the past surveys made on the insular shelf of Martinique since 1976 (GUYANTE, CARACOLANTE 1 and 2, CORDET 1 and 2), a substantial data bank has been established. There are thus 300 geophysical profiles totaling 3200 km, 496 sediment grab samples, and 15 submarine photography profiles.

Methodology

Side-scan sonar is an instrument used for mapping geological units of the seafloor in depths from 10 to 200 m. It provides an acoustic "image" in a gray range, the shade of which varies depending upon submarine morphology and nature. The information contained in the sonographs corresponds either to a change in shade (boundary between two facies) or to morphological information (characteristic of current ripple fields, consolidated formations, etc.). They require calibration of the different shades by sediment grab samples and submarine photography profiles to obtain the equivalent in geological terms (Figure 3).

Based on plotting the routes, information contained in the sonographs was first reported on a 1:15,000 scale for Fort-de-France Bay and on a 1:10,000 scale for the eastern, northeastern, and northwestern seaboards. Syntheses were then performed on various scales; the maps presented here are on a 1:300,000 scale for the Atlantic coast (Figure 4a) and on a 1:100,000 scale for the Caribbean coast (Figure 4b).

Grab samples were analyzed by traditional methods used in sedimentology such as grain-size distribution. Indices characterizing the sediments (medium grain, standard deviation, and skewness) were calculated.

Discussion

Processing of the sonographs, along with the echo sounder tapes, grain-size analysis of the samples, and study of the submarine photographs, led to identification of the surficial geology. This study enabled the development of maps of the sedimentary facies for the majority of the shelf (Figures 4a and 4b), supplementing the results already obtained [Pons and Julius, 1972; Pons et al., 1978; Pons and Julius, 1984; Froidefond et al., 1985; Pons, 1988; Henocq et al., 1990].

Sedimentation results from a duality between volcanic sediments and biogenic sediments, associated in variable proportions:

1.   Montagne Pelée and the Pitons du Carbet range are responsible for the presence of volcanoclastic elements in the surficial sediments, transported by run-off water and the wind.

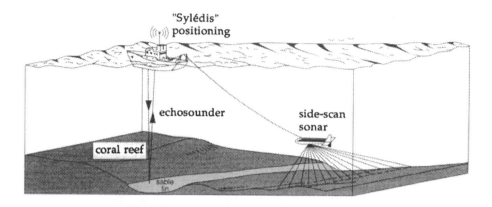

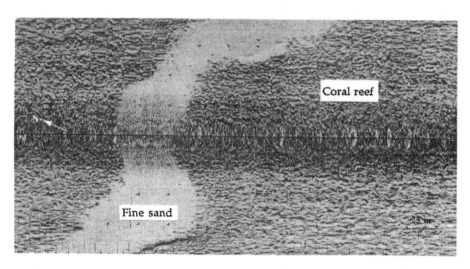

Fig. 3. Principle of the side-scan sonar and example of a sonograph [Augris *et al.*, 1992].

2.   Martinique, due to its location in a tropical environment, promotes the development
     of substantial submarine life, contributing bioclastic sediments from reef and algal
     construction.

The eastern shelf, bathed by Atlantic waters, represents approximately 75% of the
surface of the insular shelf of Martinique (Figure 2). Its outer limit corresponds to the -
100m isobath. Off la Trinité (northeast), the insular shelf reaches its maximum
extension.

Durand et al.   255

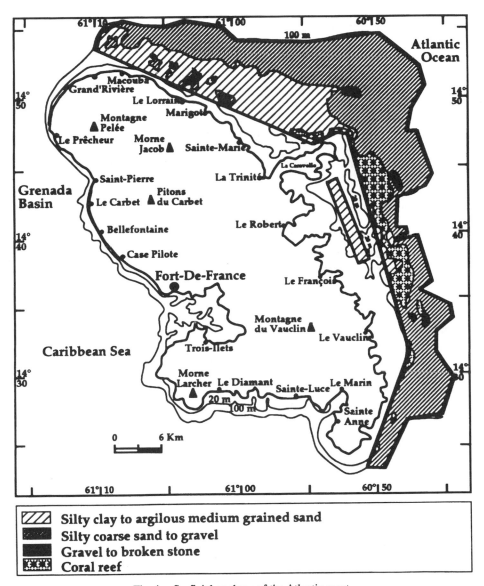

Fig. 4a. Surficial geology of the Atlantic coast.

To the southeast and east, the insular shelf, punctuated by barrier reefs, plays a double role by supplying biogenic elements and by considerably reducing the exchange between internal and external waters, thus trapping the terrigenous and littoral sediments (Figures 4a, 5, and 6). The surficial geology between Presqu'île de la Caravelle and the southern tip of Martinique is characterized by sediments rich in carbonate which are the result of two antagonistic effects of the swell [Pons *et al.*, 1978]: triggered by a permanent regime of trade winds, it ensures the development and destruction of the reef constructions. Some deposits are almost completely made up of the biogenic phase (Figure 7a). Near the coasts the sedimentation is finer, essentially composed of clay and silt, with a dominating presence of carbonate.

In the northeast sector, the sediments of volcanic and terrigenous origin are predominant, entailing a substantial quantity of fine material onto the shelf (Figure 4a). Towards the open sea, deposits are at first composed of volcanic sand (Figure 7b) but, with the presence of coral formations, the biogenic fraction assumes greater importance and becomes dominant (Figures 7c and 8).

Consolidated formations were evidenced in the region close to Marigot (in the northeast). The width and height of these banks reach several meters; the length is variable, ranging up to a few hundred meters. Their orientation is between N70 and N150. During the last cruise, a core has been performed, showing up and down three levels: a well-preserved coral reef formation above a fossil coral reef in the process of disintegration and a volcanic level. They are consolidated lavas topped by ancient coral reefs (Figure 9).

The Caribbean coast, with the exception of Fort-de-France Bay, presents a very limited insular shelf. Isobaths appear relatively close, sometimes very lobed (notably in the northern half), but do not show the existence of any submarine canyons. From the north to the south they open out slightly upwards, indicating a discreet widening of the insular shelf (Figure 2). As for the northeast, the specificity of the northwestern deposits is essentially due to the combined presence of close volcanic reliefs and a high hydrological regime. The resulting sedimentation is composed of a substantial volcanic phase to the detriment of a slightly represented biogenic phase (Figure 4b). The deposits are characterized by a few reefs surrounded for the most part by volcanic mud and sands.

Fort-de-France Bay is shallow: the 100 m deep seafloor, present outside the bay, climbs quickly towards the -20 m isobath (Figure 2). In the south and west it shows large coral heads, the silting-up of which decreases out towards the open sea. The east and north, under the influence of industrialization and fluvial input [Castaing *et al.*, 1986], are characterized by substantial fine sedimentation, creating many mud patches (Figures 4b, 7d, and 7e).

In addition to its erosive action, the swell, created mainly by east to northeast sector trade winds, shapes symmetric ripples which, to the south of the Presqu'île de la Caravelle, appear parallel to the coast whereas in the northern part, they have an obliquity of approximately 40° with the shoreline. The ripple lengths of the sedimentary figures range from 1 to 4 m (Figures 5 and 10).

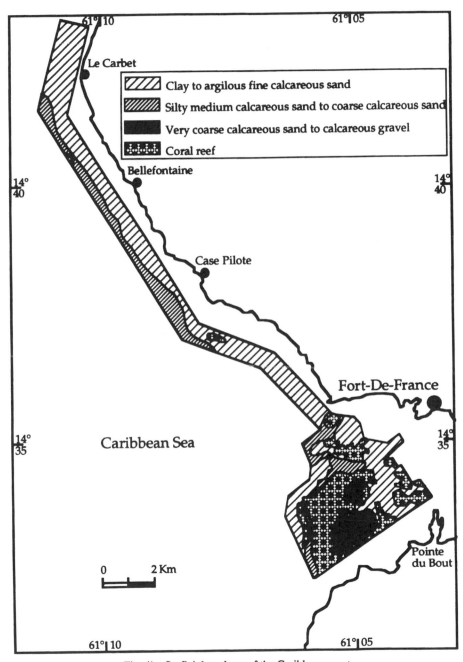

Fig. 4b. Surficial geology of the Caribbean coast.

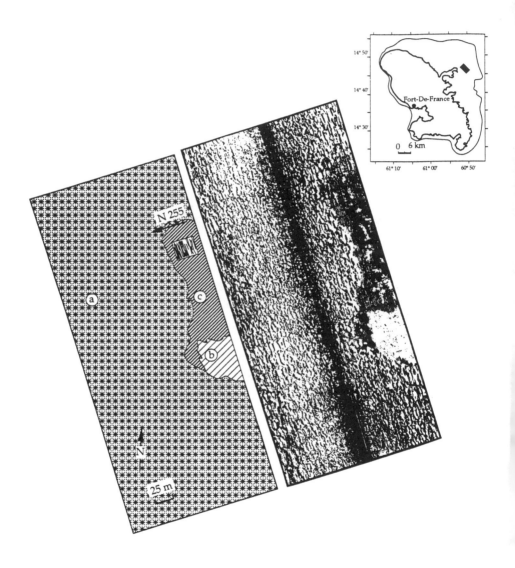

Fig. 5. In a depression, in the heart of a coral reef flat (a), an accumulation made up of two types of sediment (b and c) can be observed. In (b), there is silty clay or argilous fine sand juxtaposed on coarse sand (c). In the latter, sedimentary figures, set up in shallow water (25 to 30 metres), are shaped either by the swell, by a N 255 unidirectional current, or by their combined action. These slightly wavy figures are not very divided and their ripple length does not exceed 2 metres.

Durand et al. 259

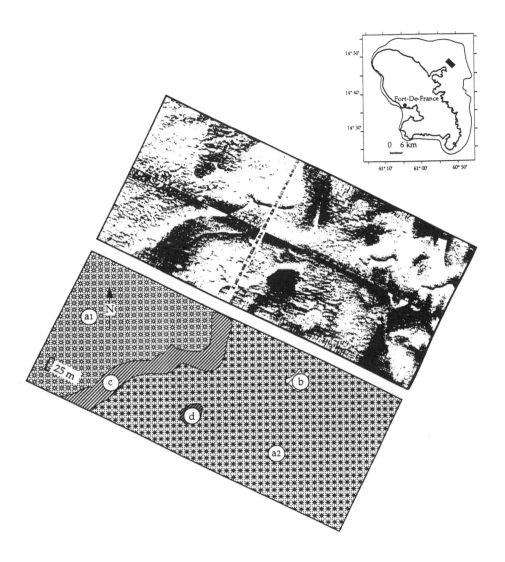

Fig. 6. To the Northeast of the Presqu'île de la Caravelle, a transition area between sediments of volcanic origin and sediments of biogenic origin, coral reef flats of different outcropping modes are juxtaposed (a1 and a2). An area filled with coarse sand (c) separates these two edifices. These reefs offer micro-depressions where sediments of various types pile up, ranging from a very coarse pole (d: gravel) to a finer pole (b: silty clay to argilous fine sand).

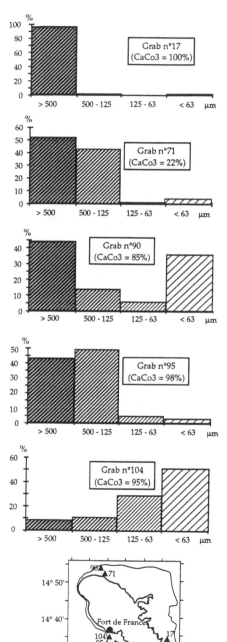

Figure 7a. Exclusively carbonated, clean, light beige, sediment, sampled at 39 metres in depth, containing pieces of coral (30%) and algal incrusts (70%). Grain-size analysis indicates a medium grain of 3.8 mm (gravel), a standard deviation of 1.04 (not very classified) and a skewness of 2.04 (well classified towards the fines).

Figure 7b. Essentially sandy (96%), slightly carbonated (22%) sediment, containing substantial amounts of volcanic lithoclasts, some debris of algae and few shells. Analysis of the sand shows a medium grain of 0.7 mm (coarse to very coarse sand), a standard deviation of 1.43 (slightly classified) and a skewness of -0.51 (symmetric curve). Depth : 47 metres.

Figure 7c. Greyish-beige, essentially carbonated (85%) sediment, showing a rather substantial amount of mud (36%), with a substantial amount of algal and coral fragments to the detriment of the shelly debris. The medium grain is 1.2 mm (very coarse sand), with a standard deviation of 2.11 (very slightly classified) and a skewness of 0.50 (symmetric curve). Depth : 71 metres.

Figure 7d. Sampled on a shoal in the bay of Fort-de-France (depth : 8 metres), this grey sediment is characterized by worn carbonated constituents (98%), with equal proportions of shelly and coral fragments. Grain-size analysis shows a medium grain of 0.4 mm (medium to coarse sand), a standard deviation of 1.18 (slightly classified) and a skewness of -0.16 (symmetric curve).

Figure 7e. This sediment is differentiated by a substantial quantity of mud (51%), providing a greenish-black colour. The coral fragments (30%) and shelly debris (70%) constitute the carbonated phase (95%). The medium grain is 0.2 mm (fine sand), with a standard deviation of 1.91 (slightly classified) and a skewness of -1.61 (well classified towards the coarse grains). Depth : 25 metres.

Fig. 7. The different kind of sediment of the insular shelf of Martinique.

Location of the sediment grabs

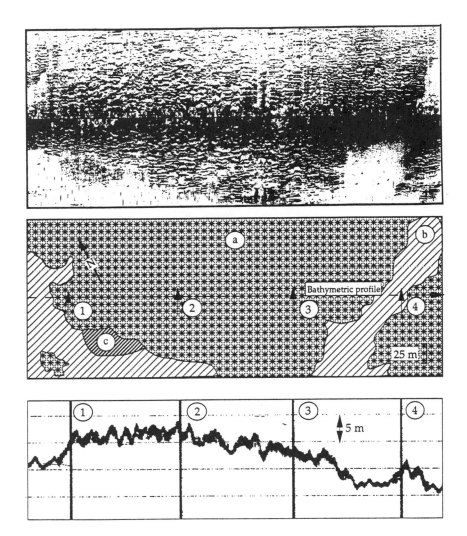

Fig. 8. Sonograph positioned on the N-East coast, compared to an extract of the corresponding echosounder tape. It shows a 10-metre high coral reef (a) surrounded by fine sediments (silty clay to silty, very fine sand) piled up in the depressions (b). A deposit of coarse sand (c) can be distinguished flattened against this reef (c).

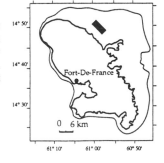

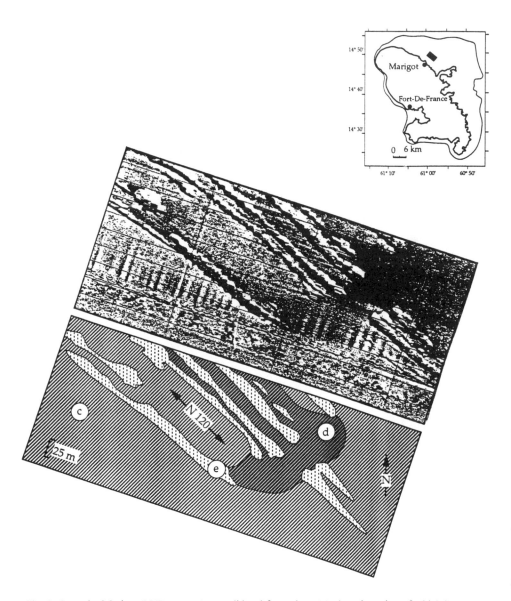

Fig. 9. Opposite Marigot (N-East coast) consolidated formations (e), the orientation of which is N 120, were evidenced: they are either consolidated lavas or vestiges of ancient barrier reefs. They are characterized by a width of some twenty metres, a length of several hundred metres and a height of approximately five metres. They appear in coarse sand (c) but gravel (d), resulting from their stripping, is also visible.

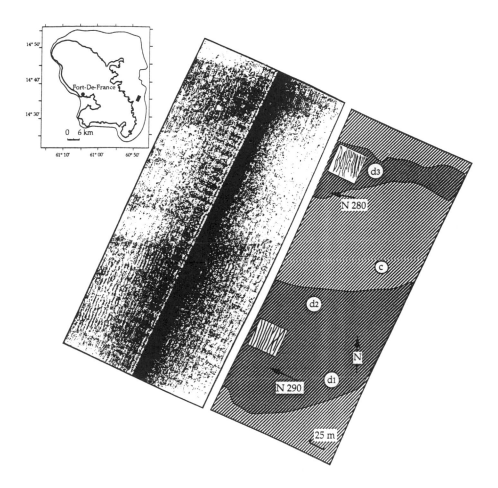

Fig. 10. Over a short distance (20 metres), two current ripple fields with various characteristics are visible. They both appear in a very coarse sediment (d1: gravel and/or broken stone), at a depth of thirty metres. In d2, the ripples are not very divided and their ripple length is about 3 to 4 metres. In d3, the ripple length is under 2 metres and they seem more divided, with acute angle dichotomies. This difference in ripple length may be related to a finer grain size for the d3 field. The elongation direction of the N10/N20 ridges indicates that these ripples were generated by a N280/N290 unidirectional current. This direction does not correspond to the propagation direction of the swell but certainly to a direction resulting from the combination of the swell and the local morphology.

Conclusion

The opposition of Atlantic and Caribbean margins, constituting the insular shelf of Martinique, is translated by the presence of different sedimentological and morphological environments. The eastern insular shelf is wide, with an internal area including many modern reef constructions and an external area where the average gradients are not very high (0.5%), except on the slope ruptures (13% to 18%) [Pons, 1988]. This type of deposit is mainly comprised of carbonated, autochthonous sediments resulting from the destruction of modern or fossil formations. Few volcanic elements participate in this sedimentation.

Unlike the eastern shelf, the western shelf is narrow, displaying high declivities. The essential part of the sedimentation is ensured by two types of volcanoclastic elements: epiclastic elements (from the leaching of reliefs such as Montagne Pelée and the Pitons du Carbet) and pyroclastic elements (transported by stream-lavas or by air) [Pons, 1988].

A more in-depth study of the sedimentary figures, as well as the examination of the current data collected during the two recent surveys, will enable studies already performed to be completed [Brucks, 1971; Stalcup and Metcalf, 1972; Castaing et al., 1986; Pujos et al., 1992] and provide better assessment of the sediment transport in the coastal and shoreline areas.

Acknowledgments. I wish to acknowledge the Sabli`ere Guadeloupeenne for supporting my research.

References

Augris, C., R. Assor, P. Clabaut, A. Grotte, and H. Ondréas, Carte des formations superficielles du plateau insulaire de la Guadeloupe, avec ses îles environnantes et Saint-Martin, 10 cartes au 1/25 000, IFREMER, 1992.

Brucks, J. T., Current of the Caribbean and adjacent regions as deduced from drift bottle studies, *J. Mar. Res., 21*(2), 455-465, 1971.

Castaing, P., A. De Resseguier, C. Julius, M. Parra, J.-C. Pons, M. Pujos, and O. Weber, Qualité des eaux et des sédiments dans la baie de Fort-de-France (Martinique), *Rapport CORDET, 84,* L 0896, Bordeaux, 93 pp., 1986.

Froidefond, J.-M., L. Berthois, R. Griboulard, C. Julius, and J.-C. Pons, Terrasses submergées d'origine récifale, variations du niveau marin et activité tectonique sur le plateau sud et est de la Martinique, *Géodynamique des Caraïbes,* Symposium, Paris, February 5-8, 1985, 143-154, 1985.

Henocq, C., M. Pujos, J.-C. Pons, and S. G. Odin, Sédimentologie du plateau insulaire nord-est de la Martinique: Impact du volcanisme en région récifale; comparaison avec la baie de Fort-de-France, *Sci. Géol. Bull., 43*(1), 15-28, 1990.

Pons, J.-C., Genèse et répartition des produits détritiques dans un contexte volcano-sédimentaire tropical: Exemple de la Martinique et de ces bassins adjacents, *Bull. Inst. Géol. Bassin d'Aquitaine* (Bordeaux), *43*, 5-151, 143 fig., 72 tabl., 1988.

Pons, J.-C., and C. Julius, Contribution à l'étude d'une mangrove de la région du Vauclin (Martinique), *Bull. Inst. Géol. Bassin d'Aquitaine* (Bordeaux), *12*, 181-186, 1972.

Pons, J.-C., and C. Julius, Potentialités en argile d'une île volcano-sédimentaire sous climat tropical: Exemple de la Martinique (French Lesser Antilles), *Bull. Inst. Géol. Bassin d'Aquitaine* (Bordeaux), *35*, 135-151, 6 fig., 1984.

Pons, J.-C., C. Julius, A. Klingebiel, and M. Pujos, Morphologie et sédiments de la plate-forme continentale de la Martinique, *Géologie en Mijnbouw*, *57*(2), 287-292, 1978.

Pujos, M., J.-L. Gonzalez, and J.-Cl. Pons, Circulation des eaux sur les plateaux insulaires de Martinique et Guadeloupe. Evolution des littoraux de Guyane et de la zone caraïbe méridionale pendant le Quaternaire, ORSTOM Ed., Symposium PIGC 274/ORSTOM (Guyane), November 9-14, 1990, 415-435, 1992.

Stalcup, M. C., and W. G. Metcalf, Current measurements in the passages of the Lesser Antilles, *J. Phys. Res.*, *77*, 1032-1049, 1972.

Westercamp, D., and H. Tazieff, Guide géologique des Petites Antilles, Masson et Cie Ed., Paris, 134 pp., 1980.

Westercamp, D., P. Andrieff, P. Bouysse, S. Cottez, and R. Battistini, Carte géologique de la Martinique au 1/50 000, BRGM, 1989.

16

Littoral Ecological Stability and Economic Development in Small Island States: The Need for an Equilibrium

Vance P. Vicente

Abstract

Caribbean islands often share the following confining physical and biological coastal scenario: (1) a limited coastline; (2) a restricted shelf; (3) a permanent thermocline; (4) low nutrient waters; and (5) a sparsity of upwelling zones. Nonetheless, the littoral systems surrounding the Caribbean islands are: (1) nuclei of biodiversity; (2) centers of origin; (3) responsible for coastal organic production; and (4) critical to small island state economies. This chapter emphasizes the values of submerged marine systems which, among other functions, create beaches, sustain fisheries, wildlife and protected resources, and protect shorelines from natural disasters and from impacts associated with sea level rise. However, the submerged littoral systems responsible for the above functions are driven by light reaching the bottom. The vulnerability of benthic-driven tropical coastal systems to non-point sources of pollution and turbidity are also discussed. Hopefully, this chapter will alert small island state officials to the economic value of the three marine tropical phenomena which create and modify coastal resources: coral reefs, mangroves, and turtle grass beds. The vulnerability of these systems represents the potentially irreversible loss of a not yet fully exploited nor understood resource which may be the ultimate source of income for small islands when other economic attempts fail.

Small Islands: Marine Science and Sustainable Development
Coastal and Estuarine Studies, Volume 51, Pages 266–283
This paper not subject to U.S. copyright.
Published in 1996 by the American Geophysical Union

Introduction

For centuries, civilizations around the world have exploited the littoral environment for human settlement, navigation and transportation of goods, recreation and tourism, consumptive purposes, and as waste disposal sites. Historically, the littoral zone has been the coastal arena where the traditional clash between coastal development and marine ecosystems takes place.

In coastal management jargon, "littoral" refers to a wide spectrum of near shore systems. In an ecological context, however, the littoral zone generally includes communities living on the bottom (benthos). They are commonly classified into: littoral (sometimes referred to as eulittoral) or intertidal; supra-littoral (those occurring above the tide line); and inner-sublittoral (bottom communities permanently submerged and which extend to about 200-300 ft of depth). The outer-sublittoral (benthic communities between 200 and 750 ft), bathyal (750-1500 ft), and deeper systems (e.g., abyssal, hadal) are found below the inner-sublittoral zone and develop under dark, aphotic conditions [see Hedgepeth, 1957, for details].

The littoral systems defined above occur along an inshore-offshore depth gradient, are subjected to different physical and biological variables, are different in age and structure, develop differently, and react in distinctive ways to natural and anthropogenic disturbances associated with economic development.

Public awareness of the importance of even the most elemental biological components of an inner-sublittoral system was aroused about 60 years ago. Between 1930 and 1933 practically all eelgrass beds (Zostera marina) were wiped out in the north Atlantic region by the "wasting disease." The aftermath was as follows:

1. Collapse of the bay scallop (Argopecten irradians) fisheries [Fonseca et al., 1992; Stevenson and Confer, 1978].
2. Landing values of cod and plaice plummeted [Milne and Milne, 1951].
3. The first historical species extinction of a marine invertebrate (Lottia alveus) [Carlton et al., 1991].
4. Decline of wildlife resources [Moffit and Cottam, 1941; Thayer et al., 1975].

Since the wasting disease event, 60 years of scientific research on littoral systems, including tropical island studies, have repeatedly demonstrated that to some degree inner-sublittoral systems have most or all of the following attributes:

1. They are the most productive natural ecosystems in the world.
2. Serve as nursery, breeding, and feeding grounds for inshore and offshore fishery resources, wildlife, and endangered species.
3. Stabilize the shoreline.

4. Protect the shoreline from storms by acting as natural barriers.
5. Mitigate the coastal impact of sea level rise by acting as self-accreting, shore-protecting systems.
6. Sustain the economy of small islands by providing food resources, recreational sites, and touristic attractions.
7. Serve as nuclei of marine biodiversity.

In this paper emphasis is given to three systems: seagrass beds, mangroves, and coral reefs. Their response to various sources of disturbances and their implications to small islands are discussed.

Analysis

Tropical Islands

Economic progress in small island states is largely based on tourism, agriculture, light and heavy industries, and fisheries. The extent to which any of the above endeavors are adopted by an island state depends largely on the location, size, topography, and geology of the island. However, cultural, political, and socioeconomic factors often determine the type of economic activities within a given island.

Whatever the cultural values, political status, or socioeconomic problems that a given island may have, littoral ecological stability should strongly be considered during the planning and developmental stages of an island. The underlying basis for this statement is that Caribbean islands often share the following confining physical and biological coastal scenarios: (1) limited coastline; (2) restricted shelf; (3) permanent thermocline; (4) oligotrophic waters; and (5) sparsity of upwelling zones.

A permanent thermocline prevents nutrient rich deep water from reaching the photic zone. Therefore, the transformation of deep water nutrients into biomass becomes inhibited. The oligotrophic condition (*e.g.*, NO_3 = 0.1-0.3 ug at/l) limits the primary production of oceanic waters to about 100 gC m^{-2} yr^{-1}. The sparsity of upwelling regions prevents vertical nutrient mixing between deep water and surface water. This results in low primary energy production and holds true for most tropical oceanic islands. It may not hold true for continental islands adjacent to upwelling areas. How then, are marine fisheries and wildlife resources nourished in offshore, tropical oceanic islands?

The vast marine biological diversity, as well as the multispecies fisheries characteristic of tropical islands, are largely supported by three peculiar littoral tropical phenomena: coral reefs, turtlegrass beds, and mangroves (*e.g.*, the mangrove root community). Other inner-sublittoral systems such as hard grounds, coral communities, gorgonian flats, sponge grounds, and calcareous green algal beds are also ecologically meaningful. In other words, the biological diversity and fisheries resources of tropical island states are sustained largely by benthic primary production.

Coral reefs and seagrass beds are inner-sublittoral systems while mangroves (*e.g.*, *Rhizophora mangle*), although classified as forested wetlands [Cowardin *et al.*, 1992], may also provide substrate for the development of littoral and sublittoral communities (*e.g.*, mangrove root communities). In contrast to the surrounding oceanic waters, these systems are some of the most productive in the world.

Seagrass Beds

Of all littoral systems, tropical Atlantic seagrass beds have probably the highest primary productivity rates of all natural systems in the world. For example, in Puerto Rico, primary production (6,898 gC m^{-2} yr^{-1}) and biomass (2,260 gC/m^{-2}) of seagrasses are very high, even under non-optimal conditions. Under pristine conditions, seagrass primary production surrounding small islands may well exceed 10,000 gC m^{-2} yr^{-1}. Turtlegrass (*Thalassia testudinum*) beds are 54 times more productive than tropical phytoplankton and are almost twice as productive than the most productive pelagic waters off Peru (3,650 gC m^{-2} yr^{-1}), and 14 times more productive than terrestrial temperate grasslands [for details see Vicente, 1992b]. Furthermore, contrary to temperate systems, net primary production of tropical meadows continues all year long. Seagrass beds do not have to be extensive to be important from a trophic standpoint. For example, it has been shown that shallow estuarine areas with sparse cover (17%) can account for about 70% of the total production.

Any of the six Caribbean seagrass species (Table 1) may form either isolated patches or vast expanses (called meadows or seagrass beds), depending on the water quality of the environment, the nature of the substrate, and the geomorphology of the coast. Vast expanses of seagrass beds are usually found on island shelves when these are: (1) extensive enough to allow excessive wave energy dissipation; (2) shallow; (3) consist of sandy or muddy bottoms (or other non-consolidated substrates); (4) contain waters that are clear and oligotrophic; and (5) are somewhat protected from high wave action [Vicente, 1992c].

The influence of seagrass beds on coastal processes and their importance to mankind may be summarized as follows: seagrasses significantly modify the physical, chemical, and geological properties of coastal areas; they provide nutrients, primary energy and

TABLE 1. Species of seagrasses within the Wider Caribbean region. The vernacular names given may vary depending on the geographical location. Only the general seagrass distribution is given in the table [see de Hartog, 1970, for detailed distributional information]. Only *Halophila johnsonii*, a type of sea vine has a very restricted distribution and may become classified as a protected resource.

Nomenclature	Vernacular	Geographic Location
Thalassia testudinum	Turtle grass	Southeast coast of North America, West Indies, South America
Halophila decipiens	Sea vine	Tethyan distribution
Halophila engelmannii	Sea vine	Southeast coast of North America, Gulf of Mexico, Puerto Rico
Halophila baillonis	Sea vine	Antilles, Brazil, South America
Halophila johnsonii	Sea vine	Southeast Florida
Syringodium filiforme	Manatee grass	West Indian region
Halodule wrightii	Shoal grass	East Pacific, West Indian region
Ruppia maritima	Wigeon grass	Worldwide

habitats which sustain coastal fisheries resources; they create foraging grounds for endangered species; and they enhance biological diversity and productivity. Other than providing nutrients, primary energy, and habitats for fisheries species populations, tropical marine meadows play other important functions as well. The leaves of turtle grass, shoal grass, manatee grass, and sea vines enhance water quality by precipitating suspended matter, preventing resuspension of anoxic sediments, and transforming nutrients into biomass.

The roots and rhizomes of seagrasses hold the sediment together and prevent coastal erosion. The above roles are usually referred to as the "trapping-binding function" of seagrass beds. West Indian seagrasses are also a staple diet for two federally listed endangered and threatened species: the West Indian manatee, *Trichechus manatus*, and the green turtle, *Chelonia mydas*, which is the only reptilian seagrass consumer [Vicente, 1992d]. The hawksbill turtle, *Eretmochelys imbricata*, may also be found foraging on sponges (*e.g.*, *Chondrilla nucula*) within seagrass meadows surrounding tropical islands [Vicente and Carballeira, 1992].

Coral Reefs

Coral reefs have been able to support small island fisheries resources, protect the shoreline from erosion, create and/or nourish sandy beaches, and represent an invaluable, perhaps the most valuable, coastal resource of many island states in the Caribbean. Although coral reefs generally have a low biomass (*e.g.*, 280 gC m^{-2}), they are highly productive (between 2,900-4,200 gC m^{-2} yr^{-1}) and support a wide diversity of taxa, many of which have commercial or recreational value [see Bohnsack, 1992; Rogers, 1977; Goenaga and Boulon, 1991; Appledoorn *et al.*, 1992].

Most present coral reefs developed between 6,000 and 9,000 ybp, when most of the small island state shelves became flooded by sea level rise following the Wisconsian Glaciation. A slower eustatic sea level rise, shallow water, high water transparency, and proper substrate conditions following the mid-Holocene allowed the development of coral reefs within the shelf systems of most Caribbean islands.

Many reefs developed on drowned eolianitic structures [Kaye, 1959] deposited parallel to shore during the Wisconsian glacial period [Goenaga, 1988] and have been able to develop and grow vertically in harmony with mid-Holocene changes in the oceanic (relative sea level) and atmospheric conditions of the time.

During the last decades, however, natural disturbances (not to speak of anthropogenic disturbances) have had a significant negative impact (on an ecological time scale) on the physical and biological integrity of the system. These disturbances include massive die-offs of sea urchin herbivores, hurricanes, coral bleaching, changes in sea surface temperature, and competitive displacement of corals by algae and sponges [Vicente, 1993].

There should be an increasing awareness and protection of coral reefs surrounding small islands within the Caribbean region because of the following scientific facts:

1. The tropical surface water of the Caribbean Sea is not necessarily the ideal water mass for coral reef development [Vicente, 1992a]. This water mass receives over 20% of the fresh water discharged annually by the world's rivers. For example, the dispersal of the Amazon River's discharge alone affects surface salinity, phytoplankton concentration, and phytoplankton species composition throughout the western tropical Atlantic [Müller-Karger *et al.*, 1988]. These factors do not favor maximum reef development since coral reefs grow best in oligotrophic waters with low primary productivities and high water transparency.
2. Coral reefs within the West Indian region are also frequently (and intensely) weakened by hurricanes since a large portion of the Caribbean reef tract lies within the hurricane belt. At times, however, density independent events (such as storms and hurricanes) may enhance local reef diversity when they disturb the system at an intermediate level [Connell, 1978].

3. Caribbean reefs are also more prone to bioerosion than reefs elsewhere [Highsmith, 1980] and biological diversity is lower than in the Indo-Pacific (there are about 80% more genera and species of corals in the Pacific than in the Caribbean). In the Indo-Pacific and in the Red Sea corals extend to great depths. For example, in the Gulf of Eilat the depth of the photic zone is 70 m whereas in Puerto Rico 1% of the light level in many cases is 20 m offshore. Nearshore Secchi depth values are often between 1 m and 2 m.

4. The massive demise of the sea urchin *D. antillarum* which spread throughout the Caribbean from January 1983-1984 and the major coral bleaching event of 1987 have decreased coral cover within many Caribbean reefs.

5. Furthermore, Caribbean corals are frequently subjected to various forms of diseases (*e.g.*, Black band and white band disease) and stresses as reviewed by Peters [1984]. Based on the above, we may state that the ecological integrity of reefs, particularly those within the Caribbean, is at present threatened.

The above circumstances are very relevant when one considers the ongoing debate on whether modern reefs will be able to cope with predicted climate change (*e.g.*, an eustatic sea level rise of 6 mm/yr over the next century). Buddemeier and Smith [1988] and Smith and Buddemeier [1992] state that the projected climatic scenario is well within the range of reef accretion rates (a rate of 10 mm/yr is the consensus value for maximum sustained reef vertical accretion rates). This cannot hold true for many Caribbean coral reefs characterized by low coral cover. The fate of a given coral reef in situations where bioerosion rates are greater than accretion rates for extended periods of time is the loss of its physical and ecological integrity and, therefore, of its functional values.

There is additional evidence which suggests that West Indian reefs are in a fragile state. This evidence was submitted during a symposium held at the University of Miami during June 1993. For example, after 20 years of monitoring coral reefs at Curacao and Bonaire, Bak and Nieuwland [1993] found a significant decrease in coral cover and colony number in the upper part of the reefs. A long-term annual monitoring of coral reef communities on the north coast of Jamaica has revealed dramatic changes during the last 20 years by natural and human stress. For example, coral cover at replicate sites and depths has declined from 27-77% in the 1970's to less than 5% in 1990 [Hughes, 1993]. Six coral reef stations monitored between Miami and Key West lost coral species (a loss of 13%-29% of their species richness) and five of the six areas lost live coral cover [Porter and Meier, 1992]. In Puerto Rico, Vicente [1993] reported local coral population extinctions and dramatic decreases in coral cover during the last 10 years within reefs of the southwest coast of the island under non-polluted conditions.

Management measures oriented towards conservation, restoration, and enhancement of Caribbean coral reefs (if implemented promptly) may result in the preservation of their functional values and in the maintenance of their ecological integrity. Otherwise, the resistance of coral reefs to natural or enhanced climatic changes or to any additional external sources of stress (whether natural or anthropogenic) will continue to decline. Meanwhile, Caribbean reefs are vulnerable and should be considered as a threatened ecosystem.

Mangroves

Among the 80 species of mangroves, five are found in the Wider Caribbean region, but only four are found within island boundaries: *Avicennia germinans* (L.) Stearn, *Conocarpus erectus* (L.), *Laguncularia racemosa* (L.), and *Rhizophora mangle* (L.). *Pelliciera rhizophorae* is restricted to the coast of Colombia [Calderon, 1984]. Within the region, there are about 3,230,000 hectares of coastal shoreline dominated by mangrove vegetation which represents about 15% of the world inventory of mangroves [Snedaker, 1993].

Like coral reefs and seagrass beds, the functional values of mangrove forests are often critical to fisheries, wildlife resources, and biodiversity. Furthermore, these systems are also highly productive with 730 gC m^{-2} yr^{-1} being a representative value for net above ground production [Mann, 1973; Benner and Hodson, 1985]. In some small island situations, mangroves can become exceedingly productive. For example, mangroves at Joyuda lagoon (on the southwest coast of Puerto Rico) produce over 20 tons of organic matter per hectare per year [Levine, 1981].

Mangrove production has important trophic implications. For example, the bulk of the leachable fraction of red mangrove leaves are transformed into microbial biomass with a very high (30%) efficiency [Benner and Hodson, 1985]. Mangrove systems also nourish open water systems (the concept of "outwelling") and provide critical habitats for reef fish (*e.g.*, groupers and snappers), invertebrates (oysters and lobsters), and other commercial and recreational (snooks, tarpoons) species [Thayer *et al.*, 1987].

Another important role of mangrove forests is the "trapping and binding function" of their extensive prop root systems (prop roots of red mangroves penetrate over 10 ft below the sediment). Furthermore, fringing mangroves have been reported to contribute to coastal accretion as much as 25-200 m/yr! This becomes a particularly important function if one considers present and projected sea level rise scenarios and coastal erosion problems [see Maul, 1993].

Much like seagrasses, mangroves provide habitat, food and shelter for fish, shellfish, wildlife, and endangered species. At least two endangered species of birds (the brown pelican, *Pelecanus occidentalis,* and the yellow shoulder blackbird, *Agelaius xanthomus*) and many species of herons and egrets roost or nest on the mangrove canopy fringing tropical islands.

The roles of mangroves in stabilizing the shoreline, controlling flood conditions, transforming nutrients into biomass, serving as sediment filters, and detoxifying contaminants in tropical island situations are well established in the literature. The reader is encouraged to review Lugo and Brown [1988] and Lopez *et al.* [1988] for a general analysis of tropical island mangroves and wetland contributions to coastal processes.

The impacts of economic activities on mangrove forests have historically been disastrous. However, they have been protected in many small island states during the last 20 years. This has resulted in the natural restoration of mangrove forests in some cases. For example, in Puerto Rico a total of 16,476 cuerdas (1 cuerda = 3930 m^2) were left (after vast destruction between the 1950's and 1970's) in 1974. At present, however, there are 22,971 cuerdas, an increment of 39.5% as reported by Torres-Rodriguez [1993].

Mangroves are more eurytopic, more resilient and restorable than seagrass beds and coral reefs. Furthermore, the importance of mangroves have been well publicized. For example, there are more than 6,000 publications on mangroves alone [Rutzler and Feller, 1988] and inventories continue to be updated and published. Although the general public is well aware of their functional values, mangrove forests continue to be threatened by non-sustainable logging practices, mariculture, and coastal development.

Other Littoral Systems

As long as good water quality is maintained and the physical conditions remain stable, intertidal communities (when recruitment is not limited) are resilient and robust. There are certain characteristics shared by most intertidal communities whether the system consists of a rocky intertidal coastline, sandy beach, or mud flat. The littoral zone is characterized ecologically as an unpredictable system where physical factors largely determine the community structure, productivity, and diversity of the system. According to the stability time hypothesis [Sanders, 1968], these systems may be defined as physically controlled systems where resilient colonizing (r-selected) populations (some of which are referred to as biofouling) prevail. However, the role of coactive forces (*e.g.*, predation, competition) will often create the mosaic pattern found on rocky intertidal systems.

The rocky intertidal environment found on Caribbean tropical islands, although not favorable for seagrass development, generally support a diverse macrophytic assemblage of red (rhodophyta), brown (phaeophyta) and green (chlorophyta) algae. However, macrophytic algae on tropical rocky coasts do not influence biological diversity, fisheries resources, and primary energy production as seagrasses do. Contrary to seagrasses, macroalgae do not develop an extensive rhizome-root mat, do not have high standing crop values, and depend on external sources of nutrients. Generally, these systems are not self sustained and are ephemeral.

However, in temperate latitudes [Mann, 1972] the rocky littoral environment is one of the best habitats for macrophytic algal growth. For example, within the inner-sublittoral zone, macroalgae can produce from 1,000-2,000 gC m^{-2} yr^{-1} and about 500-1,000 gC m^{-2} yr^{-1} intertidally [Mann and Chapman, 1975]. Furthermore, littoral rocky habitats in the temperate zone support significant fisheries resources (*e.g.*, abalone, kelp, green urchins).

Tropical intertidal communities have their own intrinsic value, but normally do not perform the functions of seagrass beds and coral reefs which are so valuable to man. In addition, fisheries resources associated with rocky coastal systems are few. Perhaps the only significant fishery resource associated with the tropical rocky intertidal region is the West Indian top shell (*Cittarium pica*) known in the West Indies as "vulgao" or "whelk." However, tropical macroalgae can also be productive. For example, in Curacao, fleshy and filamentous algae can produce 3.3 gC m^{-2} day^{-1} or 1,204.5 g m^{-2} yr^{-1} [Wanders, 1976].

Discussion

The type of inner-sublittoral systems surrounding a small tropical island is determined largely by geological features (*e.g.*, extension of the island shelf, promontories), coastal barriers, wave energy, water transparency, substrate type, salinity and nutrients, and historical events. On the other hand, when there are no or few physical constraints on the community structure, biological diversity and productivity within a system will largely depend on coactive [Hutchinson, 1953] or deterministic processes such as biological recruitment rates, predation, competition, intensity and frequency of disturbing processes, and local nutrient dynamics [Connell, 1961, 1978; Paine, 1966; Paine and Vadas, 1969].

The stability, productivity, and biological diversity of tropical littoral systems depends on a multitude of variables which operate on different scales of time and space. For example, the high biological diversity found within tropical seagrass beds, mangrove root communities, coral reefs, or within other littoral system types cannot be simply explained by productivity alone. To explain the diversity phenomena of littoral tropical systems in general one has to consider evolutionary factors, historical trends, deterministic processes, as well as stochastic and anthropogenic factors. Several hypotheses based on non-equilibrium principles (*e.g.*, the intermediate disturbance hypothesis, equal chance hypothesis, gradual change hypothesis) have been proposed to explain biological diversity. The situation becomes more complex if one considers species equilibrium based hypotheses (*e.g.*, the niche diversification hypothesis, circular network hypothesis, and the compensatory mortality hypothesis). Some intermediate levels of disturbance (whether physical or biological) appear to be important in maintaining local diversity in some systems [Connell, 1978].

For this discussion it is important to become aware of the complexities of ecological processes, but it may not be necessary to go into further hypothetical explanations on biodiversity. What is important for this discussion is that tropical islands have unique littoral systems which are biologically diverse, extremely productive, and that are very valuable to small island economies. At the same time, many of these systems are on the brink of disaster, which represents the potentially irreversible loss of a not yet fully exploited nor understood resource which may be the ultimate source of income for small islands when other economic attempts fail.

Particularly in small islands, man has often exploited and modified the littoral resources beyond their self-sustaining capability. Frequently, the structural and functional changes which occur within these systems are irreversible on the scale of time of human interest. Attempts to restore the functional values of these systems often fail.

The impact of man on littoral systems can be divided into two broad categories: (1) point-source pollution problems (e.g., outfalls); and (2) nonpoint source pollution (e.g., sediment and nutrient runoffs). Historically, coastal environmental regulations have given more emphasis to point-source pollution, since toxic waste and sewage have caused serious human health problems in the past (e.g., the Minamata disease, Love Canal). Point-source pollution represents a threat to the integrity of littoral systems but these problems have been addressed with some success since the 1970's.

Nonpoint source pollution has received more recent attention [Burroughs, 1993]. It appears to affect littoral systems on much larger temporal and spatial scales. I have prepared a diagram to help explain how nonpoint pollution sources disturb sublittoral systems in order to develop specific recommendations to mitigate their effects on the environment (Figure 1).

In this scheme, nonpoint sources are divided into upland and shore origins. The type of activities which create nonpoint source disturbances are included in the third row of Figure 1: deforestation and farming activities (two examples of upland sources), landfills, dredging, and industrial transport (three examples of shore sources). Deforestation, or other types of land clearing activities, increase sediment runoff which adds sediment particles to the water column, particularly during the rainy season. Clay particles often remain suspended for extended periods of time in the water column and have often been reported to physically impact littoral systems by blocking light transmission and physically injuring benthic primary producers such as scleractinian corals and seagrass beds. Landfilling operations (also commonly associated with coastal development) also add sediment to the water column, particularly when caliche or soil is

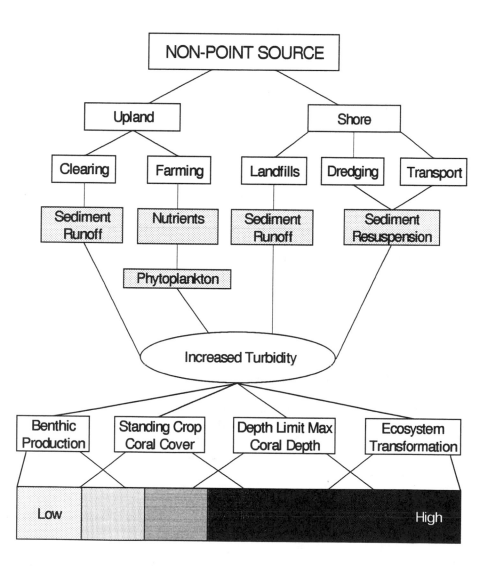

Fig. 1. Effect of non-point source pollution on sublittoral systems.

used as filling material. On the other hand, dredging industrial or recreational transport resuspends (but does not add) sediment into the marine environment.

Farming and culturing activities often result in overfertilization of nearby waters. In the normally well lit surface waters of the Caribbean, nutrients associated with these activities (nitrates, nitrites, ortho-phosphates) can readily be transformed into phytoplankton production which also can block light from reaching the bottom. For example, if phytoplankton reaches 2 g/m^3, the compensation depth (the depth of the photic zone) is changed from 100 m to 3.5 m of depth.

All of the above factors, directly (*e.g.*, sediment inputs) or indirectly (*e.g.*, phytoplankton production) promote light scattering and light absorption which results in turbid, murky-looking water. The effect of turbidity on submerged tropical systems can best be explained from a turbidity gradient point of view, hereby referred to as the "turbidity gradient hypothesis." If turbidity increases slowly over short periods of time, the effect will be reflected at a physiological level. For example, there may be a drop in *primary productivity* at an organismic level (individual coral heads, individual colonies of sponges and gorgonians, turions of seagrasses, thalli of macroalgae). The response of benthic systems to economic activities which result in a higher level of turbidity sustained over longer (but not permanent) periods of time could be measured in terms of deviations (declines) in the *standing crop values* of seagrasses and macroalgae or in changes in the percent live *coral cover*.

High turbidity levels will result in a decrease in the depth limit of seagrass beds [Vicente and Rivera, 1982] and in a reduction of the *depth limit* of corals (bottom row, Figure 1). This implies the loss of deeper seagrass beds and coral reef habitats. Under extreme *sustained* turbid conditions, benthic systems deteriorate and shallow water coastal areas will no longer be supported by benthic production. At this point an ecosystem transformation occurs where the system changes from a benthic-based production to a plankton-based production. What this means is that the ecological integrity of reefs and seagrass beds within the affected area is lost: the death of these systems. Usually when an ecosystem reaches this stage the changes are irreversible within an ecological time scale. Furthermore, the above situation enhances biofouling populations which bioerode the reef framework and the seagrass turions are replaced by benthic filter feeders such as ascidians, sponges, ectoprocts, and hydrozoans.

Light is critical for maintaining the ecological integrity of seagrass beds and coral reefs. A light transmittance of at least 20% is required to maintain a solid stand of seagrass. Furthermore, qualitative aspects of light may also become important. For example, Buesa [1975] demonstrated that the most useful wavelengths for turtle grass (*Thalassia testudinum*) were green (525 nm) and red (620 nm). In coral reefs, water quality is of primary importance for maintaining all the functional coral forms. For example, 60% of the light reaching the surface is required for the development of branching corals, 40% for massive corals, and from 4-1% for platy coral growth.

General Economic Analysis: Coral Reefs and Seagrass Beds

The following example attempts to value indirectly some of the functional values of a coral reef. Supposing that a given island state requires the construction of a sea wall in order to protect a 1 km stretch of beach. The construction of the sea wall would cost approximately U.S. $1-3.3 million since sea walls cost between $300-$1,000 per foot [Vicente, 1989b]. This is an approximate value of creating *one* reef function, shoreline protection. In order to value a coral reef as a whole, one would still need to account for at least six more coral reef attributes:

1. Their resistance to wave action.
2. Their self accreting, self maintaining capability.
3. Their sand producing function.
4. Their fisheries resources directly associated with the reef.
5. Their ecotouristic, aesthetic values of the system.
6. Their scientific and educational values.

The natural self accreting, self maintaining capability of a reef is almost impossible to replicate. Most Caribbean small island states are within the hurricane belt, and become exposed to north Atlantic winter swells. This implies an increased maintenance cost to artificial, protective structures.

The sand producing function of coral reefs should also be adequately valued. A live coral reef with close to 100% live cover (coral and calcareous algae) produce 10 Kg (= 22 lbs.) $CaCO_3$ m^{-2} yr^{-1}, much of which transform to white calcareous sand and fine sediments through bioerosion. To value this properly one may want to remember that the cost of beach replenishment is between U.S. $1 million to U.S. $6 million per mile, and that the restoration of Miami Beach alone was $64 million dollars [Vicente, 1989b]. On the other hand, coral reefs nourish beaches without cost.

The fisheries resources of Caribbean islands is largely dependent on benthic reef habitats and represent an industry of millions of dollars per year. For example, the annual yield of lobsters from the island shelves and banks of the Lesser Antilles (not including U.S. Virgin Islands) are about 1,000 mt. The retail value of this yield in restaurants would be about U.S. $40 million [FAO, 1993].

The ecotouristic value of reefs is recently receiving more and more socio-economic attention. For example, while conventional tourism grows worldwide at the rate of 4%, ecotourism grows at a rate of 18%-25%! It has been estimated that SCUBA diving alone generates close to U.S. $1 billion for the Caribbean area [Gonzalez, 1993]. Furthermore, the scientific and educational value of reefs is unquestionable. Learning about past climatic conditions from the analysis of reef sediments, the discovery of new secondary metabolites, and the discovery of new species are priceless.

Seagrass beds share many of the functional values of coral reefs and, therefore, must be valued similarly. From a biological standpoint, seagrass beds (*e.g.*, turtle grass beds) are often valued for their role as nursery, feeding and reproductive grounds for many commercial taxa (*e.g.*, shrimps, conchs, lobsters). Conchs (*e.g.*, *Strombus gigas*) are integral components of turtle grass beds in the Caribbean. The production of conch on the island shelves of the Lesser Antilles is responsible for a U.S. $40 million industry [FAO, 1993].

The economic potential of other seagrass herbivore species in the Caribbean (*e.g.*, the white urchin *Tripneustes ventricosus*) has not been fully explored and may represent another source of income for small island states. For example, the temperate green urchin *Strongylocentrotus droebachiensis* was not economically important until 1987. Between 1987 and 1992, however, 30 million pounds were landed in Maine and now this species represents a U.S. $15 million per year industry to the state [Griffin, 1993]. Similarly, the Caribbean white urchin could be exploited responsibly and efficiently by small island states (see Daniel, Chapter 24). For example, assuming a conversion efficiency of 15%, one hectare of turtle grass can produce about 12 tons of white urchins per hectare per year. Sea urchins are highly valued for their gonads.

The trapping-binding function of seagrasses and the sand generating flora (*e.g.*, *Halimeda* spp.) associated with this type of system are invaluable in preventing coastline erosion, creating natural beaches, and nourishing existing beaches. Many white sandy beaches in the Caribbean are formed by fragments of the calcareous green alga *Halimeda*. Seagrass beds (and coral reefs) become increasingly valuable considering the shoreline erosion of the Caribbean associated with sea level rise. For example, WMO/UNEP [1990] have produced the first set of cost estimates for coastline protection considering a sea-level rise of 1 m in 100 years in the Caribbean. For all island territories the cost of new construction alone would be U.S. $11.1 billion at an average of .20% of the gross domestic product [Hendry, 1993].

Marine ecologists often agree that development in small island states has occurred in disharmony with littoral systems. Since we have failed to implement effective preventive measures, we must resort to restoration, enhancement, and preservation of these systems in order to maintain their economic resources. Enhancement of coastal water quality (*e.g.*, water transparency) is indispensable if we want to restore and maintain small island coastal benthic systems.

References

Appledoorn, R., J. Beets, J. Bohnsack, S. Bolden, D. Matos, S. Meyers, A. Rosario, Y. Sadovy, and W. Tobias, Shallow water reef fish stock assessment for the U.S. Caribbean. NOAA Tech. Memo. NMFS-SEFSC-304, 70 pp., 1992.

Bak, R. P. M., and G. Nieuwland, Twenty years of change in coral communities over deep reef slopes along leeward coasts in the Netherland Antilles, Proceedings, Global Aspects of Coral Reefs: Health, Hazards, and History, 154-159, June 1993, University of Miami, Miami, Florida, 1993.

Benner, R., and R. E. Hodson, Microbial degradation of the leachable and lignocellulosic components of leaves and wood from *Rhizophora mangle* in a tropical mangrove swamp, *Mar. Ecol.-Prog. Ser., 23*, 221-230, 1985.

Bohnsack, J. A., Reef resource habitat protection: The forgotten factor, *Mar. Recreat. Fish., 14*, 117-129, 1992.

Buddemeier, R. W., and S. V. Smith, Coral reef growth in an era of rapidly rising sea level: Predictions and suggestions for long term research, *Coral Reefs, 7*, 51-56, 1988.

Buesa, R. J., Population biomass and metabolic rates of marine angiosperms on the northwestern Cuban shelf, *Aquatic Botany, 1*, 11-231, 1975.

Burroughs, R. H., Nonpoint sources as external threats to coastal water quality: Lessons from park service experience, *Coastal Manage., 21*, 131-142, 1993.

Calderon, E., Occurence of the mangrove *Pelliciera rhizophorae* on the Caribbean coast of Colombia with biogeographical notes, *Bull. Mar. Sci., 35*(1), 105-110, 1984.

Carlton, J. T., G. J. Vermeij, D. R. Lindberg, D. A. Carlton, and E. C. Dudley, The first historical extinction of a marine invertebrate in an ocean basin: The demise of the eel grass limpet *Lottia alveus, Biol. Bull., 180*, 72-80, 1991.

Connell, J. H., The influence of interspecific competition and other factors on the distribution of the barnacle *Cthamalus stellatus, Ecology, 42*(4), 710-723, 1961.

Connell, J. H., Diversity in tropical rain forests and coral reefs, *Science, 199*, 1302-1310, 1978.

Cowardin, L. M., V. Carter, F. C. Golet, and E. T. LaRoe, The classification of wetlands and deepwater habitats of the United States, U.S. Fish and Wildlife Service, FWS/OBS-79/31, 131 pp., 1992.

de Hartog, C., *The Seagrasses of the World*, 275 pp., North Holand Publ. Co., Amsterdam, 1970.

FAO, Marine fishery resources of the Antilles: Lesser Antilles, Puerto Rico, Hispaniola, Jamaica, and Cuba, Fish. Tech. Paper No. 326, 235 pp., Rome, 1993.

Fonseca, M. S., W. J. Kenworthy, and G. W. Thayer, Stemming the tide of coastal fish habitat loss, *Mar. Recreat. Fish., 14*, 141-147, 1992.

Goenaga, C., The distribution and growth of *Montastrea annularis* (Ellis and Solander) in Puerto Rican inshore platform reefs, Ph.D. thesis, 214 pp., University of Puerto Rico, Mayaguez, 1988.

Goenaga, C., and R. H. Boulon, The state of Puerto Rican and U.S. Virgin Island corals: An aid to managers, Special Report, Caribbean Fishery Management Council, Puerto Rico, 70 pp., 1991.

Gonzalez, G., Ecotourism in the Caribbean, *Business Puerto Rico*, 35-36, 1993.

Griffin, N., Maine weighs green urchin regulations, *National Fisherman, 74*(6), 22-23, 1993.

Hedgepeth, J. W., Treatise on marine ecology and paleoecology, I: Ecology, *Mem. Geol. Soc. Amer.*, 1296 pp., 1957.

Hendry, M., Sea-level movements and shoreline changes, in *Climatic Change in the Intra-Americas Sea*, edited by G. A. Maul, pp. 115-152, Edward Arnold Publ., New York, 1993.

Highsmith, R. C., Geographic patterns of coral bioerosion: A productivity hypothesis, *J. Exp. Mar. Biol., 46*, 77-96, 1980.

Hughes, T. P., Coral reef degradation: A long-term study of human and natural impacts. Proceedings, Global Aspects of Coral Reefs: Health, Hazards, and History, 208-213. June 1993. University of Miami, Miami, Florida, 1993.

Hutchinson, G. E., The concept of pattern in ecology, *Proc. Acad. Natur. Sci. Phila., 105*, 1-12, 1953.

Kaye, C. A., Shoreline features and Quaternary shoreline changes, Puerto Rico, *U.S. Geol. Survey Prof. Paper 3317-B*, 49-139, 1959.

Levine, E. A., Nitrogen cycling by the red mangrove *Rhizophora mangle* (L.) in Joyuda lagoon on the west coast of Puerto Rico, MS thesis, 103 pp., University of Puerto Rico, Mayaguez, 1981.

Lopez, J. M., A. W. Stoner, J. R. Garcia, and I. Garcia-Muniz, Marine food webs associated with Caribbean island mangrove wetlands, *Acta Cientifica, 2*(2-3), 94-123, 1988.

Lugo, A. E., and S. Brown, The wetlands of Caribbean islands, *Acta Cientifica, 2*(2-3), 48-61, 1988.

Maul, G. A., *Climatic Change in the Intra-Americas Sea*, 389 pp., Edward Arnold Publishers, London, 1993.

Mann, K. H., Macrophyte production and detritus food chains in coastal waters, *Memorie Ist. Ital. Idrobiol. 29* (Suppl.), 353-383, 1972.

Mann, K. H., Seaweeds: Their productivity and strategy for growth, *Science, 182*, 975-981, 1973.

Mann, K. H., and A. R. O. Chapman, Primary production of marine macrophytes, in *Photosynthesis and Productivity in Different Environments*, edited by J. P. Cooper, pp. 207-248, IBP Vol. 3, Cambridge University Press, New York, 1975.

Milne, L. J., and M. J. Milne, The eelgrass catastrophe, *Sci. Amer., 184*(1), 52-55, 1951.

Moffitt, J., and C. Cottam, Eelgrass depletion on the Pacific coast and its effect on the Black Brant, U.S. Department of Interior, U.S. Fish and Wildlife Service, Wildlife Leaflet, *204*, 1-26, 1941.

Müller-Karger, F. E., C. R. McClain, and P. L. Richardson, The dispersal of the Amazon's water, *Nature, 333*, 56-59, 1988.

Paine, R. T., Food web complexity and species diversity, *Amer. Natural., 100*, 65-75, 1966.

Paine, R. T., and R. L. Vadas, The effect of grazing in the sea urchin *Strongylocentrotus* on benthic algal population, *Limnol. Oceanogr., 14*, 710-719, 1969.

Peters, E. C., A survey of cellular reactions to environmental stress and disease in Caribbean scleractinian corals, *Helgolander Meeresunters, 37*, 113-137, 1984.

Porter, J. W., and O. W. Meier, Quantification of loss and change in Floridian coral reef populations, *Amer. Zool., 32*, 625-640, 1992.

Rogers, C. S., The response of a coral reef to sedimentation, Ph.D. thesis, 195 pp., University of Florida, Gainsville, 1977.

Rutzler, K., and C. Feller, Mangrove swamp communities, *Oceanus, 30*(4), 16-24, 1988.

Sanders, H. L., Marine benthic diversity: A comparative study, *Amer. Natural., 102*(925), 243-282, 1968.

Smith, S. V., and R. W. Buddemeier, Global change and coral reef ecosystems, *Ann. Rev. Ecol. Syst., 23*, 89-118, 1992.

Snedaker, S. C., Impact on mangroves, in *Coastal Plant Communities of Latin America*, edited by E. Seliger, pp. 123-133, Academic Press, New York, 1993.

Stevenson, J. C., and N. M. Confer, Summary of available information on Chesapeake Bay submerged vegetation, U.S. Fish and Wildlife Service, FWS/OBS-78/66, Washington, D.C., 1978.

Thayer, G. W., D. A. Wolfe, and R. B. Williams, The impact of man on seagrass systems, *Amer. Scien.*, *63*, 288-296, 1975.

Thayer, G. W, D. R. Colby, and W. F. Hettler, Utilization of the red mangrove prop root habitat by fishes in south Florida, *Mar. Ecol.-Prog. Ser.*, *35*, 25-38, 1987.

Torres-Rodriguez, M., 19th Symposium on the Natural Resources of Puerto Rico, October 14, 1993, San Juan, Puerto Rico, abstracts, pp. 16, 1993.

Vicente, V. P., Ecological effects of sea level rise and sea surface temperatures on mangroves, coral reefs, seagrass beds, and sandy beaches of Puerto Rico: A preliminary evaluation, *Science, 16*(2), 27-39, 1989b.

Vicente, V. P., Expected response of Caribbean coral reefs to disturbances associated with sea level rise, International Workshop, The Rising Challenge of the Sea, Margarita Island, Venezuela, (WMO/UNEP/IPCC), 1992a.

Vicente, V. P., Resource category I designation: The seagrass beds of Culebra Island, Puerto Rico, submitted to the U.S. Fish and Wildlife Service, 27 pp., 1992b.

Vicente, V. P., A summary of ecological information on the seagrass beds of Puerto Rico, in *Coastal Plant Communities of Latin America*, edited by E. Seliger, pp. 123-133. Academic Press, New York, 1992c.

Vicente, V. P., Characteristics of green turtle (*Chelonia mydas*) grazing grounds on some Caribbean islands, Proceedings, Sea Turtle Symposium, 1-5, February 1992, Jekyll Island, Georgia, 1992d.

Vicente, V. P., Locations of extinct corals in La Parguera: Ecological and economic impacts, 19th Symposium on the Natural Resources of Puerto Rico, October 14, 1993, San Juan, Puerto Rico, 1993.

Vicente, V. P., Structural changes and vulnerability of a coral reef in La Parguera, Puerto Rico, Proceedings, Global Aspects of Coral Reefs: Health, Hazards, and History, C39-C44, June 1993, University of Miami, Miami, Florida, 1993.

Vicente, V. P., and N. M. Carballeira, Studies on the feeding ecology of the hawksbill turtle *Eretmochelys imbricata* in Puerto Rico, NOAA TM, NMFS-SEFSC-302, 117-119, 1992.

Vicente, V. P., and J. A. Rivera, The depth limits of the seagrass *Thalassia testudinum* (Konig) in Jobos and Guayanilla Bays, *Carib. J. Sci.*, *17*(1-4), 79-89, 1982.

Wanders, J. B. W., The role of benthic algae in the shallow reef of Curacao (Netherlands Antilles), I. Primary productivity in the coral reef, *Aquatic Botany, 2*, 235-270, 1976.

WMO/UNEP (World Meteorological Organization/United Nations Environment Program), Climate change: The IPCC response strategies, 270 pp., Intergovernmental Panel of Climate Change, 1990.

17

Microbial Water Quality on a Caribbean Island (Martinique)

Monique Pommepuy, Annick Derrien, Françoise Le Guyader, Dominique Menard, Marie-Paule Caprais, Eric Dubois, Elizabeth Dupray, and Michele Gourmelon

Abstract

To assess microbial water quality, bacterial and viral studies were carried out in Fort-de-France Bay from 1992 to 1993. The objectives were to evaluate fecal contamination of rivers and sewage discharged into the bay and the behavior of fecal bacteria in seawater. In this study, molecular techniques such as the polymerase chain reaction (PCR) were used to detect the main pathogenic viruses encountered in hydric environments, i.e., Hepatitis A virus (HAV), rotaviruses (RV), and enteroviruses (EV), while bacterial contamination was evaluated by β-galactosidase activity using 4-methyl-umbelliferyl-β-D-galactoside (MUGal) and conventional techniques. Results showed that rapid techniques are successful in evaluating microbial contamination. In less than an hour bacterial analyses of fecal coliforms can be assessed; only two days are required for viral results by PCR. Martinique's rivers were often found to be highly contaminated by urban sewage. Viral RNAs were detected in 50% of the rivers, but no correlation between viral and bacterial contamination was noted in this study. An indication of the impact of sewage discharge in coastal areas was provided by fecal bacterial die-off rate studies. T90 was found to be less than 4 h in sunny weather and less than 30 h during the night or cloudy weather. The die-off rates (T90, which is the time necessary for bacteria to lose cultivability, in hours) found in Martinique are often shorter than in Europe, especially at night. This result can be attributed to local water conditions such as temperature, oligotrophic water, and quality of sunlight intensity.

Small Islands: Marine Science and Sustainable Development
Coastal and Estuarine Studies, Volume 51, Pages 284–297

Introduction

Martinique island in the lesser Antilles archipelago is reputed for tourism and its exceptional tropical climate. The development of air service has considerably increased the tourist population (Europeans and North Americans), and the native population tends to go to the beach more and more. Thus, new problems have appeared, such as disrespect of the environment and increasing pollution. Sewage treatment plants (secondary treatment) have been set up, but sewers are not always connected and raw waters are discharged directly. Hence, microbial contamination of surface waters is becoming an important problem, along with the sanitary impact of enteric bacteria behavior in the environment. The enteric bacteria dispersed in coastal marine areas depends completely on physical-chemical parameters such as temperature, salinity, nutrient concentrations, and turbidity. Depending on the water quality, these microorganisms can behave in completely different ways. Studies must be conducted in view of the economic impact of tourism on tropical coasts and purification difficulties. Bacterial water quality is routinely assessed using bacterial indicators, most frequently that of the fecal coliform test. However, current microbial standards may not always indicate viral pathogens.

In this study, molecular techniques such as the polymerase chain reaction were used to detect the main pathogenic viruses encountered in hydric environments, *i.e.*, hepatitis A virus (HAV), rotaviruses (RV), and enteroviruses (EV), while bacterial contamination was evaluated by β-galactosidase activity using 4-methyl-umbelliferyl-β-D-galactoside (MUGal) and conventional techniques. The die-off rates of fecal bacteria were studied in seawater to evaluate the impact of sewage outfall on coastal water quality.

Materials and Methods

Rivers and Sewage Contamination

Samples

Water samples (50 ml) were collected from 22 different rivers and wastewater sewage sites in Fort-de-France Bay (Figure 1). Samples were immediately shipped on ice to the laboratory. Each river was sampled nine times from June 1992 to March 1993. For the last sampling date (March 1993), points from 2 to 11 were collected on March 5, 1993 and points from 12 to 22 were collected on March 8, 1993.

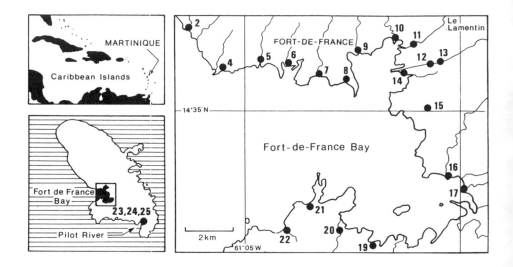

Fig. 1. Location of study area. Survival trials were conducted at (a) Fort-de-France Bay; (b) Arlet Bay; and (c) Le Robert Bay. River sampling points are indicated by ●.

Bacterial analyses

Fecal coliforms (FC) were enumerated at the river sites by the MPN MUGal microplaque technique [Hernandez *et al.*, 1991] using standard membrane filter techniques (Millipore HAWG 0.22 µm) with incubation on Mac Conkey agar (OXOID). Colony-forming units (CFU) were counted after 24 h of incubation at 44.5°C.

Fecal streptococci (FS) counts (D. group, Lancefield classification) were performed according to standard techniques (AFNOR no. 90411).

β-D-galactosidase (MU galactosidase) assay was used as a rapid technique to detect fecal coliforms as described by Berg and Fiksdal [1988] and slightly modified by Pommepuy *et al.* [submitted]. Water samples were filtered through 0.20 µm filter Nuclepore (PC). The filter was then aseptically placed in a 250 ml flask containing 13.5 ml sterile 0.05 M phosphate buffer (pH 7.9) and 0.05% sodium lauryl sulfate (SLS) (Sigma Chemical Co.); 9 ml β-D-galactosidase was added to each flask and to a sterile control flask, incubated in a shaking water bath at 44.5°C. Fluorescence intensity of sample aliquots was measured every 5 min. for 25 min. with a Sequoia Turner (excitation at 360 nm, emission at 450 nm) after addition of 100 µl 10 M NaOH to 2.5 ml samples in the cuvette. The enzymatic activity measured as production rate of MU (M min^{-1}) was determined by least squares linear regression.

Viral analysis

Extraction of viral RNAs: Water samples (400 µl) or DEPC water (distilled sterile water treated by diethyl pyrocarbonate) used as negative extraction control (400 µl) were treated as previously described [Le Guyader et al., in press]. After ethanol precipitation, the pellet was resuspended in 10 µl of DEPC water.

For virus detection after reverse transcriptase (RT), two rounds of PCR (one of them semi-nested) were performed.

For enterovirus, the primers used were derived from a conserved sequence in the 5 minute non-coding region of the poliovirus type 1 [Chapman et al., 1990; Rotbart, 1990]. Using the primer combination E1 (623-640) and E2 (446-465), PCR resulted in a 194 bp product; using primer combination E1 and E3 (584-603), a 157 bp product was obtained.

For HAV detection, the primers amplified the VP1 region [Robertson et al., 1989]. With the first round of PCR using primers H1 (2167-2192) and H2 (2389-2251), the product obtained was a 247 bp fragment; with the second round, primers H1 and H3 (2232-2251) yielded a 210 bp product.

For rotavirus detection, primers were specifically synthesized for gene segment 8 or 9 which encodes VP7 [Gouvea et al., 1990; Flores et al., 1990]: R1 (1-28) and R2 (376-392) were used to amplify a 392 bp fragment during the first round PCR and R2 used with R3 (51-71) amplified a 342 bp product for the second round.

RT PCR: for EV or HAV detection, 2 µl of each extract were used for cDNA synthesis in a 10 ml reaction with the downstream primer (E2 or H2) and a reverse transcriptase MMuLV (Stratagene) under conditions recommended by the manufacturer. For RV detection, the transcription was performed in the same conditions but with 10 µM of methylmercuric hydroxide (CH$_3$HgOH) (Strem-Chemicals).

Enzymatic amplification was performed with 3 µl of cDNA in 25 µl of PCR mix, according to Perkin Elmer conditions in a Perkin Elmer 9600 apparatus (Cetus). After denaturation for 2 min at 95°C, 30 cycles were performed as follows: 94°C for 30 s, 50°C for 30 s, 72°C for 30 s. The temperature was then maintained at 72°C for 3 min before being lowered to 10°C.

For the second round of PCR, 1 µl of the first amplification product (samples and controls) was further amplified under the same conditions as for the first PCR. The negative extraction control, a RT negative control, and a positive control were included in each amplification series.

PCR products were analyzed by 9% polyacrylamide gel electrophoresis and ethidium bromide staining. Controls were performed to ensure no contamination occurred during any of the experiments.

Survival Experiments

Survival trials were carried out in the field to determine the die-off rate (T90) of fecal coliforms. Culture preparation: the strain used was *Escherichia coli* H10407 from Evans isolated, provided by Pr Joly (Clermont-Ferrand, France). Cells were washed and the suspension was introduced in sterilized diffusion chambers. Survival experiments were carried out as described by Salomon and Pommepuy [1990]; chambers were immersed at depths of 0.50 m, 1 m or more depending on location, and supported by floating buoys; a dark chamber was also used as a control. Aliquots were periodically withdrawn in sterile tubes and reserved in cold boxes. Culturable bacteria were recovered on Drigalski agar (OXOID) incubated overnight at 37°C.

Light intensities were measured at different depths with a submersible quantum sensor (LICOR) in wavelength bands from 400 to 700 nm. Vertical attenuation coefficients were then calculated. Light intensities were recorded at 1 m under the surface during all experiments. Decay coefficient T90 (time needed for 90% of the bacteria to be unculturable) was calculated from bacterial results.

Results

Rivers and Sewage Contamination

Water samples were collected from 22 main rivers and wastewater sewage sites located around the bay. Fecal coliforms and streptococci analyses are presented in Table 1. Annual contamination variations of a polluted river (urban sewage--Trois Ilets, No. 19) and a slightly contaminated river (agricultural slope basin--Petit Bourg, No. 17) are shown in Figure 2. Fecal coliforms are often more numerous than fecal streptococci (10 to 100 times). Seasonal variations were observed (Table 1, mean and standard deviations). Bacterial counts allowed the 22 rivers to be classified in three groups:

TABLE 1. Bacterial River Contamination for Fort-de-France Bay

Samples	Fecal Coliform (/100 ml)		Fecal Streptococci (/100 ml)		
	Mean	Standard Deviation	Mean	Standard Deviation	# of replicates
1	$4.80 \ 10^4$	$1.01 \ 10^5$	$9.51 \ 10^3$	$1.55 \ 10^4$	7
2	$2.08 \ 10^4$	$3.32 \ 10^4$	$5.51 \ 10^3$	$1.00 \ 10^4$	7
3*	$7.28 \ 10^5$	$1.12 \ 10^6$	$3.29 \ 10^4$	$2.58 \ 10^4$	6
4	$2.95 \ 10^4$	$5.52 \ 10^4$	$1.27 \ 10^4$	$1.82 \ 10^4$	7
5*	$4.20 \ 10^5$	$7.19 \ 10^5$	$9.00 \ 10^4$	$1.54 \ 10^5$	13
6*	$2.60 \ 10^6$	$5.51 \ 10^6$	$3.00 \ 10^5$	$6.51 \ 10^5$	13
7	$3.32 \ 10^5$	$6.43 \ 10^5$	$2.81 \ 10^4$	$3.65 \ 10^4$	13
8*	$1.04 \ 10^5$	$1.76 \ 10^5$	$1.11 \ 10^4$	$1.66 \ 10^4$	13
9*	$8.41 \ 10^4$	$7.15 \ 10^4$	$3.47 \ 10^4$	$6.17 \ 10^4$	12
10*	$8.43 \ 10^4$	$8.31 \ 10^4$	$3.32 \ 10^4$	$5.64 \ 10^4$	7
11	$1.92 \ 10^4$	$1.80 \ 10^4$	$4.42 \ 10^3$	$3.20 \ 10^3$	7
12	$3.89 \ 10^4$	$7.72 \ 10^4$	$4.97 \ 10^3$	$1.10 \ 10^4$	7
13*	$8.52 \ 10^5$	$1.12 \ 10^6$	$3.77 \ 10^4$	$4.76 \ 10^4$	7
14	$1.03 \ 10^2$	$1.29 \ 10^2$	$2.89 \ 10^2$	$2.28 \ 10^2$	7
15*	$7.08 \ 10^6$	$1.88 \ 10^7$	$1.01 \ 10^5$	$1.73 \ 10^5$	7
16	$4.80 \ 10^6$	$1.24 \ 10^7$	$4.40 \ 10^5$	$1.12 \ 10^6$	7
17	$4.86 \ 10^3$	$6.31 \ 10^3$	$6.76 \ 10^2$	$3.15 \ 10^2$	7
18*	$1.70 \ 10^5$	$2.82 \ 10^5$	$1.42 \ 10^5$	$1.73 \ 10^5$	7
19*	$3.50 \ 10^7$	$9.42 \ 10^7$	$2.40 \ 10^5$	$3.77 \ 10^5$	7
20	$2.92 \ 10^3$	$3.17 \ 10^3$	$3.22 \ 10^2$	$3.03 \ 10^2$	7
21*	$1.99 \ 10^5$	$2.36 \ 10^5$	$2.16 \ 10^4$	$2.05 \ 10^4$	7
22*	$4.15 \ 10^5$	$8.43 \ 10^5$	$2.94 \ 10^5$	$4.57 \ 10^4$	7

*Sewage treatment plant.

1. 10^5-10^7 FC/100 ml: sewage treatment plant and rivers situated in the northern part of the bay (industrial area).
2. 10^3-10^5 FC/100 ml: most rivers.
3. $<10^3$ FC/100 ml: only two rivers located in the southern part of the bay.

In the first category cited, the ratio between the number of FC and FS was very high (>100), whereas in slightly polluted rivers this ratio was very low (<10).

To evaluate the interest of using rapid techniques, water samples were collected in March 1993. Fecal coliforms were investigated using enzyme activities of MU galactosidase. Logarithms of activities ranged from -3.70 μM MU min^{-1} in non-contaminated waters to 0.04 μM MU min^{-1} in polluted waters while fecal coliform counts by conventional techniques varied from 5 10^2 CFU/100 ml to 6.7 10^7 CFU/100 ml. Viral analyses of the same rivers were performed by RT nested PCR and showed eight positive water samples for HAV RNA, and seven for EV RNA and RV RNA.

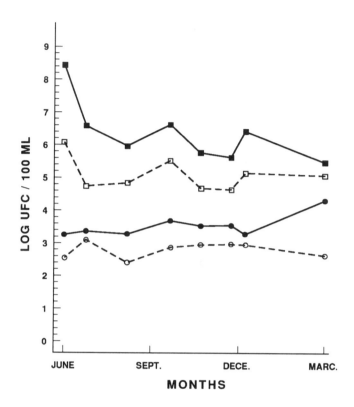

Fig. 2. Bacterial river contamination at Trois îlets River, point 19 (□ = fecal coliforms; ■ = fecal streptococci) and Petit Bourg River, point 17 (O = fecal coliforms; ● = fecal streptococci).

Among these samples, five were found to be positive for all viral RNA types tested, two contained RV and EV RNAs, two contained HAV and RV RNAs, one contained only EV RNA, and one contained only HAV RNA (Table 2).

Survival Experiments

Experiments were carried out in three areas: Fort-de-France Bay, close to the Lamentin river outfall (June 1992, turbidity: 4.9 mg/l, salinity: 35.2‰); Arlet Bay (June 1992, turbidity: 2.4 mg/l, salinity 34.75‰); and Le Robert Bay (December 1992, turbidity: 20.9 mg/l, salinity: 32‰ and March 1993, turbidity: 4.85 mg/l, salinity: 35.8‰).

TABLE 2. Microbiological results for river sampling points in Fort de
France Bay (March 1993)

Sample	Polymerase Chain Reaction			βgal (log)(a)	FC (/100 ml)(a)
	HAV	EV	RV		
2	+	+	+	-3.58	$3.5\ 10^3$
4	+	+	+	-3.23	$7\ 10^4$
5*	-	-	-	-3.07	$6\ 10^4$
6*	-	+	-	-0.02	$3\ 10^7$
7	+	+	+	-1.10	$3\ 10^5$
8*	+	+	+	-2.01	$2.6\ 10^6$
9*	+	-	+	-2.36	$4.8\ 10^5$
10*	+	-	+	-1.64	$8\ 10^5$
11	+	-	-	-2.17	$1\ 10^5$
12	-	-	-	-3.72	$1\ 10^4$
13*	-	+	+	-1.49	$1.8\ 10^6$
14	-	-	-	nd	$6\ 10^2$
15*	-	-	-	-1.12	$6.7\ 10^7$
16	-	-	-	-2.83	$5\ 10^4$
17	-	-	-	-3.11	$4.4\ 10^4$
19*	-	-	-	-0.04	$1.8\ 10^6$
20	-	-	-	nd	$5\ 10^2$
21*	-	-	-	-0.29	$3\ 10^6$
22*	+	+	+	-1.74	$2\ 10^5$

HAV = hepatitis A virus; EV = enterovirus; RV = rotavirus; βgal = galactosidase
assay (log. $\mu Mol.1^{-1}$); FC = fecal coliforms; * = sewage treatment plant; +(-) =
sample found positive (negative) after RT-nested-PCR; nd = not done; (a) = two
replicates.

Figure 3 shows an example of bacterial counts and light intensities. In the quartz
chamber (UV light receptive) bacterial counts decrease very quickly, while in Plexiglass
chambers (UV light blocked) counts decrease more slowly during the day, but continue
decreasing through the night. Black chambers (protected from sunlight) show a very low
count evolution due to local physical chemical conditions (salinity, temperature,
nutrients). Sunlight intensities received by bacteria in chambers were calculated from
light recording taking into account light attenuation due to chambers (quartz or
Plexiglass) and depth of immersion. A simple first order model was used to estimate the
bacterial die off (T90) which is the time (in hours) needed for 90% of the bacteria to lose
their culturability. As results in Table 3 show, we observed low T90 (1.85 or 2.5 h)
depending on light intensity ($10^6\ \mu E\ m^{-2}\ h^{-1}$). Variations of T90 were observed
depending on depth of immersion, location, *etc.* Dark chamber results indicated that
T90 was 7 h higher and more (29 h) were found when bacteria were protected from
sunlight.

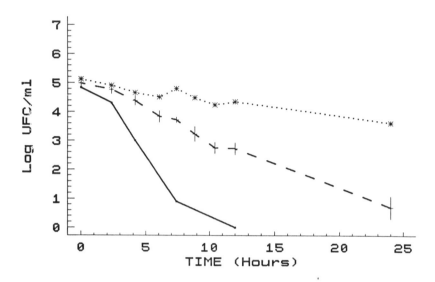

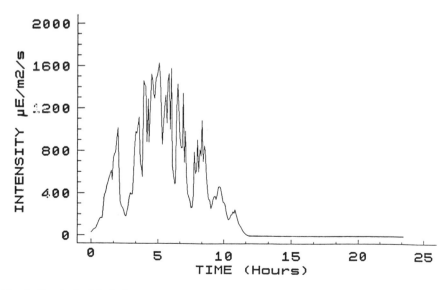

Fig. 3. Survival trials in Le Robert Bay, March 1993: (a) Bacterial decay: quartz chamber (---), plexiglass chamber (-+-), dark chamber(¨*¨); and (b) light intensities recorded at 1 m depth.

TABLE 3. Examples of T90 of *Escherichia Coli* in Three Locations

Date	Location	T° (°C)	Depth (m)	Chambers	Light Intensity $\mu Em^{-2}s^{-1}$	T90 (hours)	Salinity (‰)	Turbidity (mg/l)
June 1992	A	27.7	2	Light	nd	2.50	35.20	4.6
		28.0	5	Light	nd	9.50	35.60	4.6
		28.0	5	Light	nd	6.40	35.60	4.6
		28.2	6.50	Dark	$<10^4$	7.80	35.70	nd
June 1992	B	28.0	2	Light	$1.38\ 10^6$	1.85	34.70	3.9
		28.0	10	Light	$5.13\ 10^5$	7.10	34.75	nd
		28.0	10	Light	$3.65\ 10^5$	4.80	34.75	nd
		28.0	13	Dark	$<10^4$	6.50	34.75	nd
Dec. 1992	C	26.8	1	Light	$1.73\ 10^6$	1.85	32.00	0.7
		26.8	1	Light	$8.44\ 10^5$	4.50	32.00	0.7
		26.8	1	Light	$2.00\ 10^4$	8.30	32.00	0.7
		nd	3	Dark	$<10^4$	16.30	nd	nd
March 1993	C	27.0	0.5	Light	$3.82\ 10^6$	1.55	35.80	4.85
		27.0	1	Light	$1.01\ 10^6$	5.10	35.80	4.85
		nd	3	Dark	$<10^4$	29.00	nd	nd
		nd	3	Dark	$<10^4$	15.35	nd	nd

A = Fort-de-France Bay; B = Arlet Bay; C = Le Robert Bay; nd = not done.

Discussion

Martinique island, with a surface area of 1090 km², is the second archipelago of the Lesser Antilles. Numerous beaches and bays are crowded during weekends and the tourist season. Adrover [1993] indicated that some of the beaches were of poor bacterial quality. Significant and affirmative action was taken by the administration to improve water quality.

Bacterial water quality is usually assessed using conventional techniques to detect indicators (fecal coliforms and fecal streptococci). New techniques based on molecular detection (PCR) and on fluorescent substrate have been proposed recently [Kopecka *et al.*, 1993; Tsai *et al.*, 1993; Berg and Fiksdal, 1988; Augoustinos *et al.*, 1993].

In this study, both new and conventional assays were investigated. Standard methods are inexpensive and easy to use for the monitoring of bacterial quality. In addition, 24 hours or more are needed to obtain results with these methods. Using enzymatic assays, contamination can be evaluated in less than 1 hour. In our study, a fairly good relationship was found between fecal coliform counts and enzymatic activity: log MUG = 0.87 . log FC-7.73 (n=15 and r=0.89). Similar results were indicated by Fiksdal *et al.* [submitted] in European coastal waters. Relationship coefficients in Martinique differ from those in Europe because of different water quality (temperature, salinity, *etc.*). However, rapid techniques present some limits for routine use since their analyses have not been automated yet, and threshold values remain slightly higher (about 100 FC/100 ml).

The current microbial standards used as safety criteria for water are bacteria counts. It is well known that the presence of fecal bacteria can be linked to the presence of pathogenic microorganisms. In our study, viral RNAs were found in numerous analyses of rivers using semi-nested PCR. Our results showed that over 50% (10/20) of river samples contained viral RNA. They were positive for a sample volume as small as 400 µl which agrees with the findings of Kopecka *et al.* [1993] who reported EV RNA in 5 µl of polluted surface water. These authors demonstrated that concentration steps not only led to enrichment of viruses but also to the introduction of impurities and humic acids. Most urban sewage in Martinique is not depurated, which may explain why contamination levels for certain rivers were as high as those of the sewage treatment plant.

No correlation between viral and bacterial contamination was noted except for two rivers with low bacterial counts and without viral RNA. However, some river samples with fecal coliform counts as high as 10^7 CFU/100 ml were negative for all the three viral RNAs. This does not necessarily mean that the sample was free from human viruses but that inhibitors may have been present. In this respect, Abbaszadegan *et al.* [1993] have suggested that more purification steps are required. Lack of correlation between bacterial and viral contamination has been shown previously in wild shellfish [Le Guyader *et al.*, 1993] and in waters [Berg and Metcalf, 1978].

The impact of contaminated river water from sewage on the quality of coastal areas was investigated through survival trials and T90 studies. Various die-off rates were observed for different ambient conditions. For sunny weather, T90 values corresponded to those cited in the literature (2 h or less) [Bellair *et al.*, 1977; Martin and Bonnefont, 1986]. Slightly longer T90 was measured in cloudy weather by Salomon and Pommepuy [1990]. In light-proof chambers, the die-off rate of bacteria ranges from 7 to 30 h. Pommepuy *et al.* [1992] observed that water quality (osmoprotective compounds, nutrients) determines the decrease of culturable bacteria. On European coasts (Mediterranean Sea and English Channel) these authors found very high survival times in the dark (up to 7 or 9 days). The fact that T90 measured in the dark in coastal areas in Martinique is shorter than in Europe can be attributed to temperature and oligotrophic conditions: fecal bacteria would be more stressed than in European estuarine waters. Another difference was noted between these two areas: while bacterial counts stopped decreasing at sunset in Europe, on Caribbean islands they continue to decrease. We can hypothesize that high light intensity or certain wavelengths may be more efficient in rendering fecal bacteria non-culturable. On the other hand, the osmotic shock of oligotrophic waters added to sunlight stress can provoke drastic effects on culturable bacteria. These results were provided by Gourmelon et al. [1992] who observed that organic matter protects bacteria from the bactericidal effect of sunlight.

For engineers in charge of impact studies and coastal management, the T90 must be chosen carefully. The specific island climate features frequent cloudy weather with brief, intense periods of sunlight (12 h from sunrise to sunset but only about 8 h of effective and efficient illumination). Thus, T90 of several hours used by Lazure et al. [submitted] seems to be a realistic value for this part of the world. Our approach of assessing the impact of fecal bacterial river loads on survival rates presents a number of limitations. Some authors indicate that sunlight provokes the appearance of viable but non-culturable bacteria [Barcina et al., 1990; Dupray et al., 1993] but with no evidence that membrane fatty acids should be destroyed [Gourmelon et al., submitted]. These stressed bacteria remain in seawater with a very low metabolism [Rosack and Colwell, 1987]. Their behavior and remaining pathogenic power is currently unknown. Another limitation of our study is due that the behavior of viruses in the environment is very different from that of bacteria. Moreover, Le Guyader et al. [submitted], using polymerase chain reaction, demonstrated that viral RNA was detectable for a long period (140 days) in sterile seawater while Tsai et al. [1993] proved that, in the presence of natural seawater, viral RNA cannot be detected after 7 days. T90 viral die-off rates must be interpreted with caution.

Conclusions

Rapid techniques such as bacterial enzymatic activity or polymerase chain reaction have proved to be successful in evaluating culturable microbial contamination. In less than 1 h bacterial analyses of fecal coliforms can be assessed; only 2 days are required for viral results by PCR.

Martinique's rivers were found to be more often highly contaminated by urban sewage. Viral RNAs were detected in 50% of the rivers, with hepatitis A viruses, rotaviruses, and enteroviruses being representative of human fecal contamination. World-wide rotavirus is the major agent of severe diarrhea in children. While cases in Europe mainly occur during winter, on Caribbean islands pathological cases are reported throughout the year. The prevalence of seropositivity for HAV observed on Martinique, as for example in Venezuela [Pujol, 1990], can be correlated with social and economical standards, while in Europe a decline of seropositivity has been observed.

An indication of the impact of sewage discharge on coastal areas is provided by fecal bacterial die-off rate studies. T90 was found to be lower than in Europe, especially during the night or cloudy weather. However, very little is known about the behavior of pathogenic bacteria, and the fate of viruses is currently unknown. Only an epidemiological study can provide information on the real impact of pathogenic microorganisms discharged in coastal waters on native and foreign populations.

Acknowledgments. We thank local administrations for their help. Financial support was received from the French Research Ministry.

References

Abbaszadegan, M. S., A. S. Huber, C. P. Gerba, and I. L. Pepper, Detection of enteroviruses in groundwater with the polymerase chain reaction, *Appl. Environ. Microbiol.*, *59*, 1318-1324, 1993.

Adrover, M., Les problèmes de propreté des plages martiniquaises, *TSM, L'eau*, 5, 273-280, 1993.

Augoustinos, M. T., N. A. Grabow, B. Genthe, and R. Kfir, An improved membrane filtration method for enumeration of fecal coliforms and *E. coli* by a fluorogenic-glucuronidase assay, *Wat. Sci. Tech.*, *27*(3-4), 267-270, 1993.

Barcina, I., J. M. Gonzalez, J. Iriberri, and L. Egea, Survival strategy of *E. coli* and *E. faecalis* in illuminated fresh and marine systems, *J. Appl. Bacterial.*, *68*, 189-198, 1990.

Bellair, J. T., G. A. Pair-Smith, and J. G. Wallis, Significance of diurnal variations in fecal coliform die-off rates in the design of ocean outfalls, *J. WPCF, 77*(9), 2022-2030, 1977.

Berg, G., and T. G. Metcalf, Indicators of viruses in waters, in *Indicators of Viruses in Water and Food*, edited by G. Berg, pp. 267-296, Ann Arbor Science, Ann Arbor, Michigan, 1978.

Berg, J. D., and L. Fiksdal, Rapid detection of total and fecal coliforms in water by enzymatic hydrolysis of 4-methylumbelliferone-β-D-galactoside, *Appl. Environ. Microbiol. 4*, 2118-2121, 1988.

Chapman, N. M., S. Tracy, C. J. Gauntt, and U. Fortmueller, Molecular detection and identification of enteroviruses using enzymatic amplification and nucleic acid hybridization, *J. Clin. Microbiol.*, *28*, 843-850, 1990.

Dupray E., M. Pommepuy, A. Derrien, M. P. Caprais, and M. Cormier, Use of the direct viable count (DVC) for the assessment of the survival of *E. coli* in marine environments, *Wat. Sci. Tech.*, *27*(3-4), 395-399, 1993.

Fiksdal, L., M. Pommepuy, M. P. Caprais, and I. Midttun, Monitoring of fecal pollution in coastal waters by use of rapid enzymatic technique, *Applied Env. Mic.*, *20(5)*, 1581-1584, 1994.

Flores, J., J. Sears, I. P. Schael, L. White, D. Garcia, C. Lanata, and A. Z. Kapikian, Identification of human rotavirus serotype by hybridization to polymerase chain reaction-generated probes derived from a hyperdivergent region of the gene encoding outer capsid protein VP7, *J. Virol.*, *64*, 4021-4024, 1990.

Gourmelon, M., J. Cillard, and M. Pommepuy, Visible light damage on *Escherichia coli* in seawater: Oxidative stress hypothesis, *J. Appl. Bacteriol.,77*, 105-112, 1994.

Gourmelon, M., J. Cillard, M. Pommepuy, M. P. Caprais, G. Cahet, and M. Cormier, Toxicity of visible light on *Escherichia coli*: Involvement of reactive oxygen species, 6th Internat. Symp. Microbial Ecology, Barcelona, Spain, September 6-11, 1992.

Gouvea, V., R. I. Glass, P. Woods, K. Taniguchi, H. F. Clark, B. Forrester, and Z. Y. Fang, Polymerase chain reaction amplification and typing of rotavirus nucleic acid from stool specimens, *J. Clin. Microbiol.*, *28*, 276-282, 1990.

Hernandez, J. F., J. M. Guibert, J. M. Delattre, C. Oger, C. Charrière, B. Hugues, R. Serceau, and F. Singre, Evaluation of a miniaturized procedure for enumeration of *Escherichia coli* in seawater, based upon hydrolysis of 4-methylumbelliferyl-β-D glucuronide, *Wat. Res.*, *9*, 1073-1078, 1991.

Kopecka, H., S. Dubrou, J. Prevot, J. Marechal, and J. M. Lopez-Pila, Detection of naturally occurring enterovirus in waters by reverse transcription, polymerase chain reaction, and hybridization, *Appl. Environ. Microbiol.*, *59*, 1213-1219, 1993.

Lazure, P., J. C. Salomon, and M. Breton, Subtidal circulation in Fort-de-France Bay, *Coastal and Estuarine Studies*, 1994, (in press).

Le Guyader, F., V. Apaire-Marchais, J. Brillet, and S. Billaudel, Use of genomic probes to detect hepatitis A virus and enterovirus RNA in wild shellfish and relationship of viral contamination to bacterial contamination, *Appl. Environ. Microbiol.*, *59*, 3963-3968, 1993.

Le Guyader, F., M. L. Dincher, D. Menard, L. Schwartzbrod, and M. Pommepuy, Comparative study of the behavior of poliovirus in sterile seawater using RT-PCR and cell culture, *Mar. Poll. Bull.*, 1994, (in press).

Martin, Y., and J. L. Bonnefont, Conditions de décroissance en milieu marin des bactéries fécales des eaux usées urbaines, *Oceanis*, *12*(6), 403-418, 1986.

Pommepuy, M., J. F. Guillaud, E. Dupray, A. Derrien, F. Le Guyader, and M. Cormier, Enteric bacteria survival factors, *Wat. Sci. Tech.*, *25*(12), 93-103, 1992.

Pommepuy, M., L. Fiksdal, M. P. Caprais, and M. Cormier, Evaluation of bacterial contamination in an estuary using rapid enzymatic techniques, Water Quality International '94--poster presentation--abstract--Budapest, 44, 1994 IAWPC.

Pujol, F. H., I. Rodriguez, C. Borberg, M. O. Favorov, and H. A. Fields, Viral hepatitis serological markers among pregnant women in Caracas, Venezuela, 9th Internat. Congress of Virology, Glasgow, Scotland, August 8-13, 1993.

Robertson, B. H., V. K. Brown, and B. Khanna, Altered hepatitis A: VP1 protein resulting from cell culture propagation of virus, *Virus Res.*, *13*, 207-212, 1989.

Rosack, D. B., and R. R. Colwell, Survival strategies of bacteria in the natural environment, *Microbiol. Rev.*, *51*(3), 365-379, 1987.

Rotbart, H. A., Enzymatic RNA amplification of the enteroviruses, *J. Clin. Microbiol.*, *28*, 438-442, 1990.

Salomon, J. C., and M. Pommepuy, Mathematical model of bacterial contamination of the Morlaix estuary (France), *Wat. Res.*, *24*, 983-994, 1990.

Tsai, Y. L., M. D. Sobsey, L. R. Sangermano, and C. J. Palmer, Simple method of concentrating enteroviruses and hepatitis A virus from sewage and ocean water for rapid detection by reverse-transcriptase-polymerase chain reaction, *Appl. Environ. Microbiol.*, *59*, 3488-3491, 1993.

18

Fisheries of Small Island States and their Oceanographic Research and Information Needs

Robin Mahon

Abstract

Small island states are primarily tropical. Their fisheries are mainly artisanal or small-scale commercial, and can be viewed as shelf-based, mostly in coral reef habitats, or offshore oceanic. The former tend to be traditional, while the latter tend to have developed or expanded recently with extended jurisdiction. With regard to the fishery and fishery- related characteristics examined in this paper, small island states differ more from mainland states than they do among themselves in different regions of the world. Island states can be characterized as small and densely populated with relatively large areas of shelf and exclusive economic zone (EEZ). However, the yields of offshore pelagics (tunas/billfishes/mackerels) and shelf fishes (other fishes) are relatively low per unit area of EEZ and shelf, respectively, in comparison with mainland states. Lower yields of demersal fishes from island shelves than from continental shelves may be partly due to lower productivity; however, island EEZs should be able to produce similar yields of pelagic fishes to those of mainland countries.

The institutional and infrastructural basis for fisheries development and management is weak in most small island states, particularly in relation to the newly claimed EEZs. Research requirements in support of management are generally identified as being fishery socioeconomics and fish population dynamics (stock assessment). Oceanography

Small Islands: Marine Science and Sustainable Development
Coastal and Estuarine Studies, Volume 51, Pages 298–322
Copyright 1996 by the American Geophysical Union

is seldom cited as a high priority need. This may be partly due to the high cost of physical oceanographic research and the almost non-existent capability for this type of research in small island states. However, physical oceanographic information can provide valuable insights for fishery managers and developers, and can sometimes be accessed incidentally, as in the case history presented for the southeastern Caribbean. Physical oceanographic information of value for fisheries is usually acquired for other purposes. Better coordination among users and collectors could result in a significant increase in availability of physical oceanographic information to fisheries managers, with a minimal increase in the cost of acquisition.

Introduction

Small island developing states (SIDS) share many problems with developing states on large islands and mainland areas, but are distinctive in several ways [Beller *et al.*, 1990]. In general, their small size presents difficulties in providing the facilities and services considered normal in developed countries and, in some cases, even basic health and education services. Many are independent, but even those belonging to or associated with large developed or rapidly developing nations experience the difficulties described above. Most SIDS are tropical. Consequently, their marine environments are broadly similar and, as islands, life is closely associated with the sea.

Fishing and fisheries are usually important components of the culture in SIDS. Their contribution to the national product varies widely, but is usually significant. In this study, I will describe the main characteristics of fisheries in SIDS, evaluate the major constraints, and review the ways in which improved knowledge of oceanography could contribute to the development of sustainable fisheries.

Sources of Data

Most SIDS are found in one of three tropical geographic regions: the west Central Atlantic Ocean, including the Caribbean Sea; the Indian Ocean; and the South Pacific Ocean. There are a few SIDS in the east Central Atlantic (Cape Verde Islands, São Tomé and Príncipe) and the Mediterranean Sea (Cyprus and Malta).

Aspects of fisheries in SIDS and other countries were compared using data which I compiled for 171 islands and coastal mainland states. The states were grouped into four categories, as follows.

1. Small island developing states (SIDS), primarily those listed by UNDP [1993] for the Global Conference on Sustainable Development of Small Island Developing States. Several small island states belong to, or are associated with, developed

nations. Inclusion in category 1 is based on the extent of self government; for example, French territories are included in category 1, whereas French Departments are included in category 2. This category includes: American Samoa (U.S.), Anguilla (U.K.), Antigua and Barbuda, Aruba, British Virgin Islands (U.K.), Bahamas, Bahrain, Barbados, Bermuda (U.K.), Cape Verde Islands, Cayman Islands (U.K.), Comoros, Cook Islands (NZ), Cyprus, Dominica, Federated States of Micronesia, Fiji, French Polynesia (FR), Guam (U.S.), Grenada, Kiribati, Maldives, Malta, Marshall Islands, Mauritius, Montserrat, Northern Marianas (U.S.), Nauru, Netherlands Antilles (NE), New Caledonia (FR), Niue, Palau, São Tomé and Príncipe, Seychelles, Solomon Islands, St. Helena (U.K.), St. Kitts and Nevis, St. Lucia, St. Vincent and the Grenadines, Tokelau (NZ), Tonga, Trinidad and Tobago, Turks and Caicos (U.K.), Tuvalu, Vanuatu, Wallis and Futuna (FR), and Western Samoa.

2. Other island and island-based states (OIS). This category includes: Azores (PG), Brunei, Cuba, Dominican Republic, Falkland Islands (U.K.), Faroe Islands (DE), Guadeloupe (FR), Haiti, Hong Kong (U.K.), Jamaica, Madagascar, Martinique (FR), Mayotte (FR), Papua New Guinea, Philippines, Puerto Rico (U.S.), Reunion (FR), Singapore, Sri Lanka, St. Pierre and Miquelon (FR), Taiwan, and the U.S. Virgin Islands (U.S.).

3. Mainland coastal states, primarily those listed as poor or less developed states (PLDS) by the World Resources Institute [1992]. This category includes: Algeria, Angola, Bangladesh, Belize, Benin, Cambodia, Cameroon, China, Colombia, Congo, Costa Rica, Cote d'Ivoire, Djibouti, Ecuador, El Salvador, Equatorial Guinea, Ethiopia, French Guiana, Gabon, Gambia, Ghana, Guatemala, Guinea, Guinea-Bissau, Guyana, Honduras, India, Indonesia, Iran, Iraq, Jordan, Kenya, North Korea, Lebanon, Liberia, Libya, Malaysia, Mauritania, Mexico, Morocco, Mozambique, Myanmar, Namibia, Nicaragua, Nigeria, Oman, Pakistan, Panama, Qatar, Senegal, Sierra Leone, Somalia, Suriname, Syria, Tanzania, Thailand, Togo, Tunisia, Uruguay, Vietnam, Yemen, and Zaire.

4. Developed and rapidly developing states (DRDS) [World Resources Institute, 1992]. This category includes: Albania, Argentina, Australia, Belgium, Brazil, Bulgaria, Canada, Chile, Denmark, Egypt, Finland, France, Germany, Greece, Iceland, Ireland, Israel, Italy, Japan, South Korea, Kuwait, Monaco, Netherlands, New Zealand, Norway, Peru, Poland, Portugal, Romania, Saudi Arabia, South Africa, Spain, Sweden, Turkey, USSR, United Arab Emirates, United Kingdom, United States, Venezuela, and Yugoslavia.

The distinction between less developed and rapidly developing nations was guided by information provided by the World Resources Institute [1992], but was subjective as they did not provide lists of countries in these categories.

I include the following variables: land area; 1990 population [World Resources Institute, 1992; Hoffman, 1991; Shepard, 1990]; shelf area [World Resources Institute, 1992; Hoffman, 1991; Shepard, 1990; Mahon, 1990a]; EEZ area (km^2) [Fenwick, 1992; World

Resources Institute, 1992; Shepard, 1990; Mahon, 1990a]; average annual landings of
tunas, billfishes and mackerels (mt) for 1986-1990 [FAO, 1992]; average annual
landings of other marine fishes (mt) for 1986-1990 [FAO, 1992]; and average annual
per capita fish consumption for 1990 [World Resources Institute, 1992].

Comparisons among groups are displayed graphically by means of box and whisker
plots. The horizontal line in the box is the median value. The box shows the
interquartile range (contains 50% of the values) and the whiskers show the range of the
data. Differences among groups are tested by the non-parametric Kruskal-Wallis test,
the results of which are reported in Tables 1 and 2.

TABLE 1. Comparison of characteristics among categories of states (SID =
small island developing, OI = other island, PLD = poor and less developed,
DRD = developed and rapidly developing, K-W = Kruskal-Wallis statistic, P =
probability of K-W value).

| Characteristic | Sample Size | | | | K-W | |
	SID	OI	PLD	DRD	Statistic	P
EEZ area	38	12	62	40	5.2	0.161
EEZ/unit land area	38	12	62	40	90.8	<0.001
Fish consumption	11	8	62	36	23.1	<0.001
Land area	46	21	63	40	105.3	<0.001
Other fish catch	46	19	62	40	80.7	<0.001
Other fish/unit shelf	21	12	49	39	18.1	<0.001
Population	44	20	63	40	23.8	<0.001
Shelf area	21	12	50	39	36.3	<0.001
Shelf/unit land area	21	12	50	39	46.7	<0.001
Tuna catch	32	10	46	41	6.6	0.087
Tuna/unit EEZ	27	7	46	31	12.4	0.006

Physical Characteristics of Small Island States

Tropical small island states are the smallest states in the world in terms of both land
area and population (Figures 1a and 1b). Those in the west Central Atlantic are the
smallest (Figures 1c and 1d); they are generally of either volcanic or coralline origin.
The volcanic islands are generally steep and high with abundant rainfall and vegetation
(e.g., Mayotte and Reunion in the Indian Ocean, several islands of the Solomon Islands
and Vanuatu in the southwest Pacific Ocean, and Dominica and St. Vincent in the
western Atlantic Ocean). Consequently, agriculture is frequently important. Volcanic
islands frequently descend steeply into deep water and have relatively small areas of

TABLE 2. Comparison of characteristics of small island states in different regions (ECA = east Central Atlantic, IO = Indian Ocean, MED = Mediterranean, SP = South Pacific, WCA = west Central Atlantic, K-W = Kruskal-Wallis statistic, P = probability of K-W value).

Characteristic	Sample Size					K-W	
	ECA	IO	MED	SP	WCA	Statistic	Probability
	2	5	2	16	13	12.4	0.014
	2	5	2	16	13	3.9	0.422
	1	3	2	3	2	3.2	0.533
	3	5	2	20	16	1.2	0.873
	3	5	2	19	17	9.8	0.044
	0	4	2	2	13	1.2	0.752
	3	5	2	18	16	13.1	0.011
	0	4	2	2	13	4.8	0.190
	0	4	2	2	13	0.8	0.837
	3	5	2	15	7	4.0	0.406
	2	5	2	13	5	4.8	0.308

shelf. The coralline islands are usually low with minimal rainfall, vegetation, and soil (*e.g.*, Maldives in the Indian Ocean, Marshall Islands and Kiribati in the South Pacific, and Antigua and Bahamas in the western Central Atlantic). Agriculture is usually less important on low coralline islands than on volcanic islands and is often restricted to small portions of the island. In the South Pacific and, to a lesser extent, the Indian Ocean, they have often formed on submerged volcanoes as atolls with extensive central lagoons (*e.g.*, Cook Islands).

These small islands exhibit the full range of tropical coastal and marine habitats: estuaries, mangrove lagoons, seagrass beds, fringing reefs, patch reefs, and barrier reefs [Lewis, 1981], the slopes of the island platforms, deep bank reefs, and open ocean [see Longhurst and Pauly, 1987, for a review of tropical marine systems]. The extent to which these habitats can be found in association with the various islands depends to a large extent on whether the island is volcanic or limestone. The freshwater and sediment inputs of the rivers which occur on the high volcanic islands may inhibit nearshore coral reef development. In the coralline islands, coral reefs tend to be the most well developed.

Small Island Fisheries

The literature on tropical island fisheries is considerable, but is often in the "grey literature" of the various commissions and organizations that are active in the respective

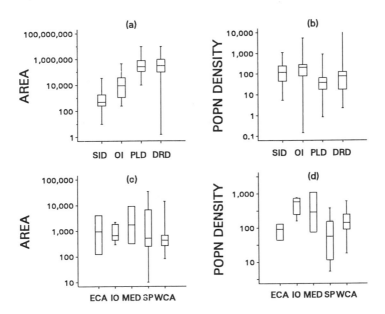

Fig. 1. The land area (km^2) and population density (individuals/km^2) of small island states compared with those of other countries (a and b) and among themselves for different regions (c and d) (SID = small island developing, OI = other island, PLD = poor and less developed, DRD = developed and rapidly developing) (ECA = east Central Atlantic, IO = Indian Ocean, MED = Mediterranean, SP = South Pacific, WCA = west Central Atlantic).

geographic regions. Such organizations include: the FAO West Central Atlantic Fishery Commission (WECAFC), the Organization of Eastern Caribbean States (OECS), the Gulf and Caribbean Fisheries Institute (GCFI), the South Pacific Commission (SPC), the Forum Fisheries Agency (FFA), the International Center for Living Aquatic Resource Management (ICLARM), the Indo-Pacific Tuna Development and Management Programme (IPTP), and the Regional Fisheries Management and Development Project for the South West Indian Ocean (SWIOP). A detailed review of island fisheries is beyond the scope of this paper. However, an overview is provided as a basis for discussing oceanographic needs of small island fisheries.

Accounts of tropical island fisheries suggest that they can be viewed in two broad categories: shelf-based and offshore oceanic [Sanders *et al.*, 1988; Forum Fisheries Agency, 1990; Shepard, 1990; Mahon, 1990a; Mahon and Mahon, 1990].

Island shelves, and thus the areas available for shelf-based fisheries, range widely in area (Figure 2a), but are generally smaller than the shelves of mainland countries. However, island shelf area per unit of land area is considerably larger than for mainland countries (Figure 2b). There is no significant difference in shelf areas of small islands among regions (Figures 2c and 2d). However, there are few data on island shelf areas from the South Pacific [J.-P. Gaudechoux, South Pacific Commission, New Caledonia, pers. comm.].

The resources available for offshore ocean fisheries are determined by the productivity of the ocean, and the extent of ocean area which belongs to the state. In the past decade, most states have claimed EEZs of up to 200 miles, or the greatest extent possible considering the proximity of other coastal states. Thus, many small oceanic states have claimed extensive areas of open ocean as EEZs [Fenwick, 1992]. Island EEZs are on

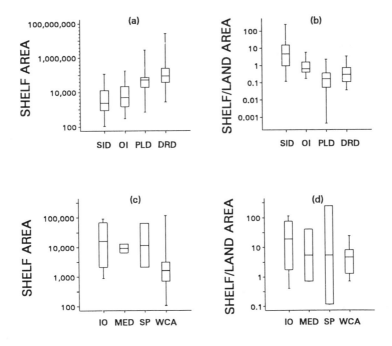

Fig. 2. The shelf area (km^2/1000) and shelf area per unit land area of small island states compared with those of other countries (a and b) and among themselves for different regions (c and d) (see Fig. 1 for abbreviation).

average the same size as those of the mainland countries, which have much larger land areas (Figure 3a). Consequently, the islands have significantly larger amounts of EEZ per unit land area (Figure 3b). EEZ areas and EEZ areas per unit land area do not differ significantly among island states in different regions. However, those of states in the west Central Atlantic and Mediterranean Sea tend to be smaller than those of island states in the other regions (Figures 3c and 3d).

The scale of shelf-based fisheries on small islands may range from subsistence (for personal consumption or barter), through artisanal (using traditional gear and vessels to catch fish for sale), to small-scale commercial (usually small owner-operated vessels using traditional or new technology). The boundaries between these categories, the terms for which are widely and variously used, are not distinct.

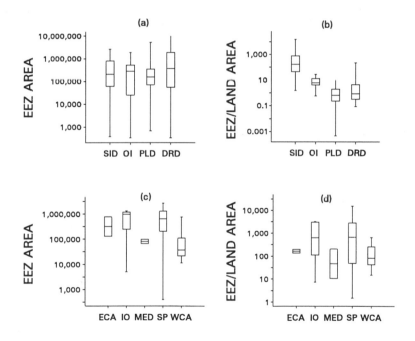

Fig. 3. The EEZ area (km^2) and EEZ area per unit land area of small island states compared with those of other countries (a and b) and among themselves for different regions (c and d) (see Fig. 1 for abbreviation).

There are several types of shelf-based fisheries. The most important of these is for reef associated fishes and is carried out in coral reef and reef-related habitats such as seagrass beds, mangroves, and lagoons. These fisheries, which are typically multispecies and multigear, using traps, handlines and spears, have been the subject of considerable study [Russ, 1991]. In islands with relatively small shelf areas, these resources are easily overfished.

Less important are the shelf-based fisheries for small pelagics such as herring-like fishes, scads and jacks (clupeids, *Decapterus* spp., *Caranx* spp.) [Dalzell and Lewis, 1989; Mahon, 1990a]. These species may be caught in the vicinity of coral reef habitats using various types of lift nets, often with lights to attract the fish. On some islands where permanent rivers drain into the sea, the substrate is primarily sand and rock deposited by the rivers. The sandy areas may support beach seine and lampara fisheries for schooling coastal pelagic species.

A specialized shelf-based fishery is that for deep-water snappers and groupers which typically occur along the slopes of island shelves and banks [Rodman, 1989; Mahon, 1990; Dalzell and Preston, 1992]. These species are more commercially valuable than the typical reef species and are often exported. Owing to the depths fished and the rough sea conditions frequently encountered at the shelf edge, this fishery is usually pursued from larger vessels than those used on the shelf itself. Typical gear includes vertical multi-hook lines deployed by hand or by electric reels, traps, longlines and, to a lesser extent, trammel nets. These fisheries have been the focus of recent expansion in most tropical island states. However, as they are restricted to shelf slopes and banks, the extent of fishable habitat is relatively small, and the resource can be easily over exploited.

Various specialized fisheries for crustaceans (primarily lobsters), gastropods (conch and *Trochus*), sea cucumbers (beche de mer), sea urchins, and decorative organisms (shells, black and stony corals) take place on the shelves of many islands and are often more commercially oriented than fin fish fisheries, as those species frequently have high export value.

Tropical oceans have been viewed as less productive than temperate oceans. However, they support significant stocks of commercially-valuable tunas and other large pelagic species. This is particularly so for the South Pacific, Indian, and east Central Atlantic Ocean regions. However, even in the west Central Atlantic, previously considered to be poor in this regard, there are considerable large-scale commercial fishing operations by developed and rapidly developing mainland coastal states (*e.g.*, Japan, United States, Venezuela).

In small island states, offshore oceanic fisheries for pelagic fishes may be pursued in the immediate vicinity of the island (<50 km) by artisanal and small commercial fishing operations (trolling, handlines, small-scale longlines) [IPTP, 1988; ICOD, 1991; Mahon, 1990a]. These are generally for large pelagic species such as tunas and billfishes, but may also be for smaller species, particularly flyingfishes [Gillett and Ianelli, 1991; Oxenford *et al.*, 1993]. However, safe and efficient exploitation of the valuable large pelagic (primarily tuna) resources offshore usually requires large commercial vessels and gear (*e.g.*, purse seines, longlines) typically used by the large industrial fishing nations [Shepard, 1990].

To examine the landings of fishes by small islands and other categories of states, the landings have been divided into two groups. The first group consists of tunas, billfishes, and mackerels, hereafter referred to as "tuna." These landings are considered to be indicative of the extent of development of commercial fishing for offshore pelagics. The second group consists of all other finfishes, and is referred to as "other fish." Landings of tunas have been assumed to relate to the EEZ area, whereas those of other fish have been assumed to relate to the shelf area. Invertebrate fisheries, which may be of considerable significance, have not been included in this analysis.

Total tuna catch is lower, but not significantly so, in island states than in mainland states (Figure 4a), and does not differ significantly among regions for island states (Figure 4c). However, tuna catch per unit of EEZ does differ significantly among category of state and is lowest in small island states (Figure 4b). Tuna catch per unit EEZ does not differ among regions for islands. The main reason for this is that the majority of tuna caught in the EEZs of small island states is taken by large-scale commercial fleets from other countries, illegally or under license. For small island states, the landings of "tuna" may not fully reflect the extent of offshore pelagic fisheries. If offshore pelagic landings are small, *e.g.*, <500 tons, they may not be reported separately to FAO, and small but significant landings of sharks, dolphinfish, jacks, *etc.* are not included in this category.

The total catch of "other fish" differs significantly among categories of states, being considerably lower in island states than in the other three categories (Figure 5a). This is clearly due to the smaller size of island shelf areas, but it is also due to lower catches per unit area because the difference is still evident when the catch per unit area of shelf is considered (Figure 5b). The total catch of "other fish" also differs significantly among regions for island states (Figure 5c), again primarily due to differences in shelf area, as there is no significant difference among regions in catch per unit shelf area (Figure 5d).

The above analysis indicates that yields per unit area from both shelf-based and offshore oceanic fisheries are significantly lower in small island states than in other categories of states. For shelf-based fisheries, this could largely reflect differences in productivity of coral shelves and continental shelves. Mainland shelf areas support the large trawl fisheries of the world, and are known to be highly productive. However, reviews of fisheries in the South Pacific and Indian Ocean suggest that there is considerable scope for development of shelf-based fisheries, with market development being the primary constraint. In the west Central Atlantic, overexploitation of shelf resources is more often the case, and yields may be lower than optimal for this reason [Neilson *et al.*, 1994].

For offshore oceanic fisheries, as indicated by tuna catches, the lower yields for the small island states are primarily due to low capacity of the fisheries. The proportion of reported "tuna" catch taken by small island states for each of the three major regions

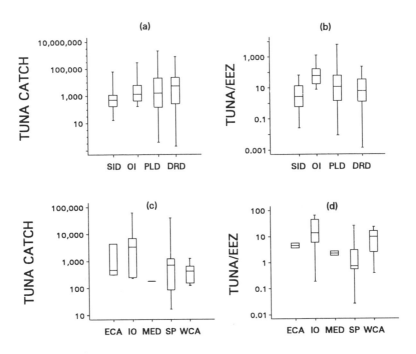

Fig. 4. The total landings of "tuna" (tunas, billfishes, mackerels) (mt) and "tuna" landings per unit shelf area (mt/km^2) (a and b) of small island states compared with those of other countries, and among themselves for different regions (c and d) (see Fig. 1 for abbreviations).

where small island states occur is relatively small (Table 3). Although the reported catch is for the entire region, not just the EEZs of the small island states, the small proportion taken by small island states reflects the extent to which large-scale commercial fleets from other countries dominate the tuna fisheries. Only Indian Ocean small island states, which have placed an emphasis on development of domestic tuna fishing capacity, take an appreciable proportion. The scale of fishing operation required to effectively access the offshore pelagic resources believed to be available in the newly acquired EEZs of small island states is frequently beyond the capabilities of local investors. These operations require substantial capital investment in vessels and shore-based facilities and secure access to markets. Most small island countries have been actively developing their capacity to access offshore pelagic resources on several fronts: improvement of small scale technology, joint ventures, and licensing of vessels from countries which have traditionally fished on the high seas [FAO, 1988; Lewis and Hampton, 1992].

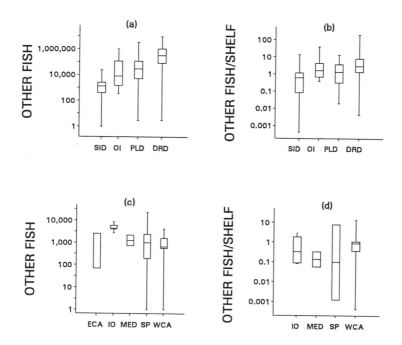

Fig. 5. The total landings of "other fish" (mt) and "other fish" per unit shelf area (mt/km²) (a and b) of small island states compared with those of other countries, and among themselves for different regions (c and d) (see Fig. 1 for abbreviations).

TABLE 3. Average total landings of "tuna" (tunas, billfishes, mackerels) (1986-1990) (mt/1,000), and landings by small island developing states (SID) in the Indian Ocean (IO), South Pacific (SP), and west Central Atlantic (WCA).

Region	Total	SID	Percentage
IO	315	74	23.5
SP	915	57	6.2
WCA	113	3	2.6

Improvement of small-scale technology involves the adaptation of longline gear to small scale vessels, introduction of commercial longline vessels in the 10-20 m length range, and exploration of the use of fish attracting devices (FADS). This approach has been most actively pursued by SIDS in the Indian Ocean and is reflected in the relatively high domestic landings of tuna in islands of that region. Similar activities have only recently begun to have an impact in the west Central Atlantic in the Lesser Antilles.

Licensing large commercial vessels has been the primary approach to obtaining benefits from large pelagic fish resources in the South Pacific, where the available resource has been sufficiently abundant to partially support the monitoring, control, and surveillance activities required for successful licensing.   Most small islands do not have the capability for surveillance of the large EEZs over which they have claimed jurisdiction. In the South Pacific, collaboration under the Forum Fisheries Agency has significantly improved the surveillance situation there [Doulman and Terawasi, 1990].   In the west Central Atlantic, similar efforts are underway with the Organization of Eastern Caribbean States (OECS).

Fishery Management and Development Issues

One of the main constraints to fisheries management and development in small islands is the institutional capability [Chakalall, 1986; Shepard, 1990].   According to Forum Fisheries Agency [1990], ICOD [1991], and Mahon and Boyce [1992], government fisheries departments in islands of the South Pacific and west Central Atlantic are typically small, and only senior staff have received post-secondary training in fisheries. Ardill [1983], Sathiendrakumar and Tisdell [1986], and papers in Sanders *et al.* [1988] suggest that a similar situation prevails in the western Indian Ocean. On a day-to-day basis, staff are primarily engaged in supervision and maintenance of facilities and extension activities with little time for planning and implementation of management and development.

In small island states, as in other developing countries, one of the main tasks of policy makers and planners is to define the objectives of management and development for the various fisheries [Mahon, 1990a]. The appropriate scale of development is of concern to small island states. Many individuals in rural areas who are dependent on subsistence or artisanal fishing can be negatively impacted by attempts to develop commercial fisheries. This can even be true of small-scale operations, which may be perceived as more efficient, better earners of foreign exchange, and "better" for the small island economy.

Resolution of potential conflicts between fishers operating at various scales and choice of appropriate scales of development for various fishery types requires knowledge of the socioeconomic and biological components of the system. Few fisheries departments in small island states have the research capability to address the issues described above. Most of the relevant research is carried out by the few research institutions that exist in those regions and, more recently, by regional organizations.

In small island states, limited land and marine shelf space brings fisheries into conflict with others sectors. Tourism, a significant contributor to many small island economies, is often closely linked to the marine environment, and can be negatively impacted by overexploitation of reef resources, as in the case of Bermuda [Towle *et al.*, 1991]. On the other hand, coastal construction for tourism, industry, and housing often results in destruction of coastal habitats which are important for fisheries production.

Fishery Research and Information Requirements

In theory, the needs of small islands for oceanographic information in support of fisheries management are similar to those of other countries. However, oceanographic research requirements of small island fisheries must be considered in the overall context of research and information requirements. Review of fishery research requirements of small island states reveals that oceanographic research is seldom listed as having high priority [Shepard *et al.*, 1989; Shepard, 1990; Forum Fisheries Agency, 1990; CFRAMP, 1993]. Information on the socioeconomic aspects of fisheries, population dynamics of the exploited stocks (stock assessment), and development of new fishing technology is given the highest priority. To a large extent, this prioritization reflects the real needs of fisheries management at the current stage of development in most island states, but is probably also due to the high cost of oceanographic research, its long-term nature, and to the virtual absence of oceanographic research capability in small island states [Freon *et al.*, 1991; Shepard, 1990]. Some examples for shelf-based and offshore oceanic fisheries are as follows.

Shelf Habitat Mapping

The productivity of tropical island shelves varies considerably with habitat type. Therefore, quantification of habitats can be an important component of estimating potential yield. Using the example of the Bahamas, Caddy and Piaggesi [1983] suggest that satellite imagery may be the most effective means of mapping habitats at the scale required for fisheries management. Owing to the high cost of satellite imagery and the expertise and equipment required to use it, there are only a few published instances where it has been used to map marine habitats of small island shelves [Armstrong, 1983; Lantieri, 1988]. Studies in progress in the Pacific SIDS by ORSTOM and other agencies indicate the recent recognition of the importance of this type of information for management of shelf resources.

Dispersal/Retention of Early Life History Stages

Most of the organisms on island shelves have planktonic early life history stages. Therefore, the stock structure and genetic discreteness of shelf-based resources has been a concern in regions where island states are in close proximity [e.g., Mahon, 1987]. The extent to which recruitment of young organisms to an island shelf depends on the reproductive output from that shelf, or on the output from another island shelf, depends on the extent to which nearshore circulation tends to favor retention rather than dispersal of the planktonic early life history stages [Brothers et al., 1983; Shapiro et al., 1988]. Knowledge of small-scale nearshore circulation is required to resolve these questions but is limited in most small islands.

Ciguatera Poisoning

Fish poisoning by ciguatera is a serious problem in many tropical areas. It occurs primarily in fishes associated with shelf habitats. Its occurrence is patchy, and appears to be persistent in localized areas of reef or bank. Fishers report that it may increase in prevalence after storms and other events which result in mixing of the shelf waters or damage to coral substrates. It is, therefore, possible that areas of local occurrence are those where mixing occurs continuously at a small-scale. As above, there is insufficient information on small-scale circulation on island shelves to address this question.

Marine Production Near Islands

Currents flowing past small islands produce eddies and mixing downstream of the island. This has been termed the "island mass effect" [Doty and Oguri, 1956]. This effect may have considerable importance for coastal and nearshore pelagic fisheries of islands through enhanced production and retention of early life history stages. Studies of the "island mass effect" are reported only for Hawaii, Bermuda, and Barbados [Sander, 1981], and have not extensively examined seasonal and interannual variability in the patterns.

Seasonality and Interannual Variability of Pelagic Resources

Pelagic resources are highly seasonal in most small islands. This is primarily because only those fish which pass close to the islands are available to the small-scale fleets. These resources also typically vary in abundance or availability from one year to the next, and also in regard to the time at which they are available for exploitation. An understanding of the seasonal and interannual variability in physical oceanography in the region can enhance understanding of the observed variability in the fishery.

Time-series of oceanographic variables which can be used in conjunction with time-series of fisheries data are seldom available in small island states, although several national and international oceanographic data centers (U.S. NOAA, National Oceanographic Data Center, French IFREMER Sismer) may compile such time-series and make them available to users. Accessing them requires expertise and equipment which is not usually available on small island states.

Oceanographic Correlates of Pelagic Fish Abundance

Certain mesoscale oceanographic features are known to be related to pelagic fish abundance, and may determine migratory paths which may be relevant to the variability discussed above [Olsen and Podesta, 1987]. These oceanographic features include areas of upwelling, fronts, and eddies. Information on mesoscale oceanographic features and their variability collected by ships of opportunity, research vessels, and satellites are seldom available in small island states. As with the oceanographic time-series and satellite imagery for habitat mapping discussed above, accessing these data from developed country sources is often beyond the financial, equipment, or expertise capability of small island states.

Real-Time Information Needs

Fishing operations often require real-time information on weather and sea conditions, particularly artisanal and small-scale fishers for whom safety at sea is a serious consideration. Offshore fishing vessels can also use real-time information on the location of mesoscale ocean features, such as eddies and fronts, which are known to be areas of high local abundance of fish, to determine appropriate locations for fishing. Satellite data on ocean temperature can provide some of the latter type of information. However, accessing this information requires specialized on-board equipment which is frequently unavailable for small-scale vessels in developing states.

As would be expected, the literature on fisheries in small islands includes very few physical oceanography studies or instances in which physical oceanographic research conducted for other purposes has been used in fisheries management. Given the scanty application of physical oceanography in fisheries management and development in small islands, it is difficult to evaluate the extent to which it could contribute in this area. However, the recent extension of economic jurisdiction to extensive ocean areas is likely to require that managers of these areas be better informed about the physical oceanographic conditions therein.

Fisheries Oceanography in Southeastern Caribbean Islands: A Case History

For managers to pay increased attention to acquisition of oceanographic information, technicians must provide specific examples of the ways in which this information can be of value. Therefore, in the following section, I provide a case history of the ways in which oceanographic information has contributed to an understanding of the offshore pelagic fisheries in the islands of the southeastern Caribbean.

Studies of fisheries for offshore pelagics in the southeastern Caribbean have raised many questions on which oceanographic information could be expected to shed some light. Despite the scarcity of such information for this area and the virtual absence of physical oceanographic research capability, it has been possible to piece together an informative picture of some aspects of physical oceanography which relate to fisheries.

In the eastern Caribbean islands, fisheries for offshore pelagics are most prominent from St. Lucia south (Figure 6). This has been explained as due to the more extensive shelves of the northern islands, their consequent preoccupation with demersal resources, and the greater travel distance required to get off their shelves to fish pelagics. It has been assumed that pelagics are available in the northern islands in similar abundance to the southern islands and that it would simply be a matter of developing fisheries for them.

It is common knowledge among island fishers that pelagic fishes exploited by the southern islands are most common in the (greenish) waters affected by the outflow of the Amazon River, which reaches the eastern Caribbean at certain times of year and, in particular, around floating objects which originate from South America. Recent studies based on Coastal Zone Color Scanner (CZCS) imagery (Nimbus 7) indicate that the northern extent of the influence of the Amazon River appears to be in the vicinity of St. Lucia [Müller-Karger et al., 1989] (Figure 6). Thus, differences in the importance of pelagic fisheries in northern and southern islands may have a basis in the respective physical oceanographic conditions, and fishery developers may wish to take an exploratory approach to developing small-scale domestic fisheries for large pelagics in the northern Lesser Antilles.

Within the group of southern islands there are considerable differences in the relative importance of various species in the catches of large pelagics. In Grenada, yellowfin tuna and sailfish are prominent. In the other islands, dolphinfish and wahoo are more prominent. Another recent study using CZCS imagery indicates the extent to which the outflow of the Orinoco River affects the islands of the southeastern Caribbean. The outflow regularly extends northward to Grenada [Müller-Karger, 1990], but only occasionally in the second half of the year does it reach St. Vincent and rarely Barbados or St. Lucia (Figure 6). Thus, the differences in fisheries within this small area appear to correspond to differences in physical oceanographic conditions, although the mechanisms are unknown. Although the CZCS imagery required for this study had been available for several years and fishery researchers in the southeastern Caribbean were aware of its existence, it was not until the countries perceived a fisheries crisis and appealed to FAO that the CZCS imagery could be accessed, using expertise from the United States [FAO, 1992].

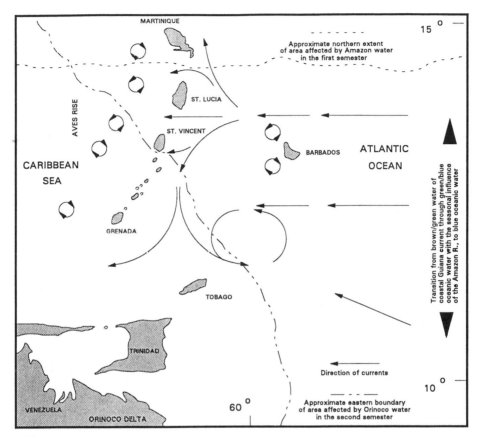

Fig. 6. A synthesis of physical oceanographic characteristics of possible significance to fisheries in the southeastern Caribbean.

The patterns of mesoscale circulation in the southeastern Caribbean, which can be pieced together from the various oceanographic surveys in the region over the past two to three decades, also provide insight into the distribution patterns of pelagic fishes known from the fisheries. Ingham and Mahnken [1966] described areas west of St. Vincent where flocks of seabirds were associated with high densities of schooling fishes. They noted that plankton production in these areas was higher than in the Atlantic Ocean east of the island chain. They assumed these areas to be eddies which formed downstream of the islands by the currents passing through the channels between them. However, these eddies are now known to form primarily on the eastern edge of the Aves Rise [Molinari *et al.*, 1981] (Figure 6). They appear to range in diameter from 20-100 km and move to the west. Although they are known as areas of increased productivity, they are not commonly fished by small-scale fishers for safety reasons, as the Aves Rise is 150-200 km downcurrent of the island chain. The recent acquisition of well-equipped, small-scale commercial vessels makes it possible for fishers to access these areas.

The preferred fishing area for flyingfish, a species of considerable importance to the southeastern islands [Oxenford et al., 1993], is east of the island chain, particularly in the area between Tobago, Grenada, and Barbados. The current flow through the eastern Caribbean has generally been depicted as to the west. However, closer examination of the literature reveals considerable mesoscale complexity east of the island chain. Water moving to the west appears to "pile up" against the island arc in the vicinity of St. Vincent and the Grenadine Bank. Some of this water flows directly through the passages between the islands, and some forms currents which flow northward and southward parallel to the island chain before flowing through the passages [Mazeika, 1973].

The southward flow appears to result in a cyclonic eddy in the area between Tobago, Grenada, and Barbados (Figure 6). Surveys in two winter periods, 1977 and 1979, show similar eddy patterns, with a large eddy or gyre in the area between Barbados and Tobago [Mazeika et al., 1980]. Febres-Ortega and Herrera [1976] found a cyclonic gyre east of Tobago in August 1972, and Johannesson [1971] found an area of current moving southeastward in a hydrographic section from Barbados to Tobago in August. The high densities of flyingfish known to occur in this area may be there because of the increased productivity and/or retention of early life history stages by the eddy just described. Nothing is known of the oceanographic conditions in this eddy, relative to those in the waters of the North Equatorial and Guiana Currents from which it is derived.

Pelagic fishes in the southeastern Caribbean are highly seasonal and exhibit considerable interannual variability [Mahon et al., 1990] (Figures 7 and 8). The interannual variability can result in catch rates which fluctuate by a factor of two or more, and is, therefore, of considerable concern to fishery managers. Very low catch rates in 1986 prompted an examination of the causes of this variability through analysis of correlations of fisheries data with oceanographic environmental variables [FAO, 1991]. These efforts were only marginally successful because direct physical oceanographic variables were not available; therefore, correlation analysis had to be carried out using atmospheric climatic variables [Mahon, 1990b]. The results were inconclusive but suggestive of mechanisms which might be more evident if oceanographic data could be used.

Knowledge of physical oceanography of relevance to fisheries in the southeastern Caribbean has been acquired from studies carried out for non-fishery purposes. Nevertheless, it has been possible to gain an enhanced understanding of the fishery patterns through the use of these data.

Conclusions

Small island states are primarily tropical. Their fisheries are mainly artisanal or small-scale commercial. They can be viewed as shelf-based, mostly in coral reef habitats, or

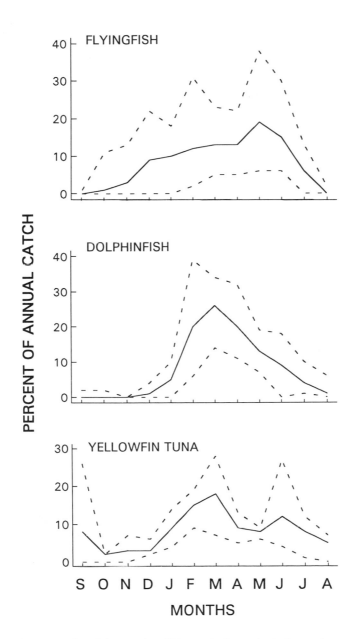

Fig. 7. The seasonality of three major offshore pelagic species in the southeastern Caribbean.

offshore oceanic. The former tend to be traditional, while the latter tend to have developed or expanded recently with extended jurisdiction.

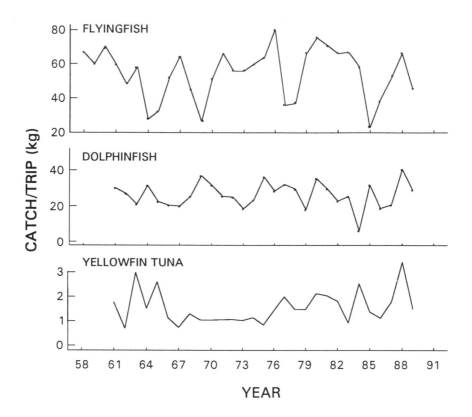

Fig. 8. The interannual variability of three major offshore pelagic species in the southeastern Caribbean.

With regard to the fishery and fishery-related characteristics examined in this paper, small island states differ more from mainland states than they do among themselves in different regions of the world. Island states can be characterized as small and densely populated, with relatively large shelves and EEZs. However, the yields of offshore pelagics (tunas, billfishes, mackerels) and shelf fishes (other fishes) are relatively low per unit area of EEZ and shelf, respectively, in comparison with mainland states. Lower yields per unit area of shelf-based fisheries on island shelves than from continental shelves may be partly due to productivity, but indicates the need for management and, in certain cases, development.

Island EEZs should be able to produce similar yields of large pelagic fishes as those of mainland countries. Despite the proximity of major upwelling areas to continents, and the resulting high fish yields, these yields are usually highest for small clupeid fishes such as sardines, pilchards, and anchovies. Large pelagics, characterized by tunas, are more widely distributed in tropical oceans. In the three major regions where small island states predominate, their proportion of the total tuna catch is small, with most of it being taken by large commercial distant water fleets.

The institutional and infrastructural basis for fisheries development and management is weak in most small island states, particularly in relation to the newly claimed EEZs. Research requirements in support of management are generally identified as being of fishery socioeconomics and fish population dynamics (stock assessment). Oceanography is seldom cited as a high priority. This may be partly due to the high cost of physical oceanographic research and the minimal research capability in small island states.

Physical oceanographic information can, however, provide valuable insights for fisheries managers and developers and can sometimes be accessed incidentally, as in the case history presented for the southeastern Caribbean. Physical oceanographic information relevant to fisheries is usually acquired for other purposes. However, most of the information required by other economic sectors is nearshore, e.g., for the siting of effluent outfalls, shoreline erosion, beach sediment distribution, siting of moorings, and location of ship channels. Better coordination among users and collectors could result in a significant increase in availability of nearshore physical oceanographic information to fisheries managers, with a minimal increase in the cost of acquisition. Acquisition of offshore oceanographic information will probably continue to be primarily from developed country sources. However, collaboration among island states in establishing the necessary facilities and expertise could make this information much more readily available for fisheries and other users.

Acknowledgments. I would like to thank Molly Mahon for assistance with compiling the data, and Sharon Almerigi for comments on the manuscript.

References

Ardill, J. D., Fisheries in the southwest Indian Ocean, *Ambio, 12,* 341-344, 1983.

Armstrong, R. A., Report to the Caribbean Fishery Management Council on the marine environments of Puerto Rico and the Virgin Islands: Automated mapping and inventory using LANDSAT data, Caribbean Fishery Management Council, Hato Rey, Puerto Rico, 37 pp., 1983.

Beller, W., P. d'Ayola, and P. Hein, *Sustainable Development and Environmental Management of Small Islands,* 419 pp., Parthenon Publishing, New Jersey, 1990.

Brothers, E. B., D. M. Williams, and P. F. Sale, Length of larval life in 12 families of fish at "One Tree Lagoon," Great Barrier Reef, Australia, *Mar. Biol., 76,* 319-324, 1983.

Caddy, J. F., and D. Piaggesi, Mapping of shelf areas for fisheries yield estimation, *FAO Fish. Rep., 278* (suppl.), 296-308, 1983.

CFRAMP, Large pelagic, reef, and deep-slope fish assessments, CARICOM Fishery Research Document No. 10, 76 pp., 1993.

Chakalall, B., Perspectives and alternatives for fisheries development in the Lesser Antilles, *Proc. Gulf Carib. Fish. Instit., 37*, 154-165, 1986.

Dalzell, P., and A. D. Lewis, A review of the South Pacific tuna bait fisheries: Small pelagic fisheries associated with coral reefs, *Mar. Fish. Rev., 51*(4), 1-10, 1989.

Dalzell, P., and G. L. Preston, Deep reef slope fishery resources of the South Pacific, South Pacific Commission, IFRP, Technical Document No. 2, 299 pp., 1992.

Doty, M. S., and M. Oguri, The island mass effect, *J. Cons. Int. Explor. Mer., 22*, 33-37, 1956.

Doulman, D. J., and P. Terawasi, The South Pacific regional register of foreign fishing vessels, *Marine Policy, July*, 324-332, 1990.

FAO, Studies of the tuna resource in the EEZs of Maldives and Sri Lanka, FAO/UNDP Bay of Bengal Programme, BOBP/REP/41, 144 pp., 1988.

FAO, Factors responsible for low catches of large pelagic fishes in the eastern Caribbean, FAO FI:TCP/RLA/8963, Terminal Statement, 11 pp., 1991.

FAO, Microfiche, fisheries statistics time-series ver. 2.0. FAO, Fisheries Department, Rome, Italy (FAO Fisheries Yearbooks on diskette, with software system for searching), 1992.

Febres-Ortega, G., and L. E. Herrera, Caribbean Sea circulation and water mass transports near the Lesser Antilles, *Bol. Inst. Oceanogr. Univ. Oriente, 15*, 83-96, 1976

Fenwick, J., *International Profiles on Marine Scientific Research: National Marine Claims, MSR Jurisdiction, and U.S. Research Clearance Histories for the World's Coastal States*, 202 pp., Woods Hole Oceanographic Institution, Woods Hole, Massachussetts, 1992.

Forum Fisheries Agency, Current fisheries development research activities in South Pacific Forum countries, in *The Forum Fisheries Agency: Achievements, Challenges, and Prospects*, edited by R. Herr, pp. 88-156, Institute of Pacific Studies, Suva, Fiji, 1990.

Freon, P., B. Gobert, and R. Mahon, La recherche halieutique et les pecheries artisanales dans la Caraibe insulaire, in *La Recherche Face a la Peche Artisanale*, edited by J.-R. Durand, J. Lemoalle, and J. Weber, pp. 195-222, International Symposium, Montpellier, France, July 3-7, 1989, ORSTOM-IFREMER, 1991.

Gillett, R., and J. Ianelli, Aspects of the biology and fisheries of flyingfish in the Pacific Islands, FAO/UNDP Regional Fisheries Support Program, RAS/89/039, 35 pp., 1991.

Hoffman, M. S., *The World Almanac and Book of Facts, 1992*, 960 pp., Pharos Books, New York, 1991.

ICOD, A general overview of inshore fisheries in the South Pacific, Consultancy Report, International Center for Ocean Development, 145 pp., Halifax, Canada, 1991.

Ingham, M. C., and C. V. W. Mahnken, Turbulence and productivity near St. Vincent Island, British West Indies: A preliminary report, *Carib. J. Sci., 6*(3-4), 83-87, 1966.

IPTP, Report of the expert consultation on stock assessment of tunas in the Indian Ocean, FAO/UNDP, Indo-Pacific Tuna Development and Management Programme, IPTP/88/GEN/14, 87 pp., 1988.

Johannessen, O. M., Preliminary results of some oceanographical observations carried out between Barbados and Tobago, March/April 1968, *Symposium on Investigations and Resources of the Caribbean Sea and Adjacent Regions*, 95-112, 1971.

Lantieri, D., Use of high resolution satellite data for agricultural and marine applications in the Maldives, FAO Technical Report, TCP/MDV/4505, 47 pp., 1988.

Lewis, A. D., and J. Hampton, Tuna research and monitoring in the South Pacific, *NAGA, 15*(1), 4-7, 1992.

Lewis, J. B., Coral reef ecosystems, in *Analysis of Marine Ecosystems*, edited by A. R. Longhurst, pp. 127-158, Academic Press, New York, 1981.

Longhurst, A. R., and D. Pauly, *Ecology of Tropical Oceans*, 407 pp., Academic Press, New York, 1987.

Mahon, R. (Ed.), Report and proceedings of the expert consultation on shared fishery resources of the Lesser Antilles region, FAO Fisheries Report No. 383, 278 pp., 1987.

Mahon, R., Fishery management options for Lesser Antilles countries, FAO Fishery Technical Paper 313, 126 pp., 1990a.

Mahon, R., Seasonal and interseasonal variability of the oceanic environment in the eastern Caribbean: With reference to possible effects on fisheries, FAO FI: TCP/RLA/8963 Field Document 5, 45 pp., 1990b.

Mahon, R., and S. L. Boyce, CARICOM Fisheries Resource Assessment and Management Program baseline survey of fisheries divisions in participating countries, CARICOM Fishery Report No. 7, 72 pp., 1992.

Mahon, S., and R. Mahon, OECS Island Fisheries: An Overview for Students and Fishermen, Food and Agriculture Organization, Rome, Italy, 35 pp., 1990.

Mahon, R., F. Murphy, P. Murray, J. Rennie, and S. Willoughby, Temporal variability of catch and effort in pelagic fisheries in Barbados, Grenada, St. Lucia, and St Vincent: With particular reference to the problem of low catches in 1989, FAO FI: TCP/RLA/8963 Field Document 2, 74 pp., 1990.

Mazeika, P. A., Circulation and water masses east of the Lesser Antilles, Deutsche Hydrogr. Zeit. 26, 49-73, 1973.

Mazeika, P. A., D. A. Burns, and T. H. Kinder, Mesoscale circulation east of the southern Lesser Antilles, J. Geophys. Res., 85(5), 2743-2758, 1980.

Molinari, R. L., M. Spillane, I. Brooks, D. Atwood, and C. Duckett, Surface currents in the Caribbean Sea as deduced from Lagrangian observations, J. Geophys. Res., 86(C7), 6537-6542, 1981.

Müller-Karger, F. E., A Coastal Zone Color Scanner (CZCS) analysis of the variability in pigment distribution in the southeastern Caribbean region, FAO: TCP/RLA/8963 Field Document No. 4, 38 pp., 1990.

Müller-Karger, F. E., C. R. McClain, T. R. Fisher, W. E. Esaias, and R. Varela, Pigment distribution in the Caribbean Sea: Observations from space, Prog. Oceanog., 23, 23-64, 1989.

Neilson, J. N., K. A. Aiken, and R. Mahon, Potential yield of reef and slope fishes: A review of methods with special reference to the Caribbean, Proc. Gulf Carib. Fish. Instit. 46, 1994.

Olsen, D. B., and G. P. Podesta, Oceanic fronts as pathways in the sea, in Signposts in the Sea, edited W. F. Herrnkind and A. B. Thistle, pp. 1-14, Proceedings, Multidisciplinary Workshop on Marine Animal Orientation and Migration, Dept. of Biological Sciences, Florida State University, Tallahassee, Florida, 1987.

Oxenford, H. A., R. Mahon, and W. Hunte (Eds.), The Eastern Caribbean Flyingfish Project, 187 pp., OECS Fishery Report No. 9, 1993.

Rodman, M., Deep Water, Development, and Change in Pacific Village Fisheries, 173 pp., Westview Press, London, 1989.

Russ, G. R., Coral reef fisheries: Effects and yields, in The Ecology of Fishes on Coral Reefs, edited by P. F. Sale, pp. 601-635, Academic Press, New York, 1991.

Sander, F., A preliminary assessment of the main causative mechanisms of the "island mass" effect of Barbados, Mar. Biol., 64, 199-205, 1981.

Sanders, M. J., P. Sparre, and S. C. Venema, Proceedings, Workshop on the Assessment of the Fishery Resources of the Southwest Indian Ocean, FAO/UNDP: RAF/79/065/wp/41/88/E, 277 pp., 1988.

Sathiendrakumar, R., and C. Tisdell, Fishery resources and policies in the Maldives: Trends and issues for an island developing country, Marine Policy, October, 279-293, 1986.

Shapiro, D. Y., D. A. Hensley, and R. S. Appeldoorn, Pelagic spawning and egg transport in coral reef fishes: A skeptical overview, Env. Biol. Fish., 22, 3-14, 1988.

Shepard, M. P., Fisheries research needs of small island countries, International Center for Ocean Development, 71 pp., Halifax, Canada, 1990.

Shepard, M. P., A. Wright, and S. T. Fakahau, Reassessment of South Pacific fisheries research needs, Forum Fisheries Agency Report No. 89/91, 22 pp., 1989.

Towle, E., R. Carney, and R. Mahon, Report of the Commission of Inquiry into the future of fisheries management and the future of the marine environment in Bermuda, Government of Bermuda Printing Office, 46 pp., Hamilton, Bermuda, 1991.

UNDP, Preparations for the Global Conference on Sustainable Development of Small Island Developing States on the basis of General Assembly resolution 47/189 and taking into account other relevant General Assembly resolutions: Overview of the activities of the United Nations System, UNDP GENERAL A/CONF.167/PC/6, 36 pp., 1993.

World Resources Institute, *World Resources 1992-1993*, 385 pp., Oxford University Press, New York, 1992.

19

Towards Integrated Coastal Zone Management in Small Island States

Gillian Cambers

Abstract

Integrated Coastal Zone Management (ICZM) is still a relatively new concept within the smaller eastern Caribbean islands. The problems facing the implementation of ICZM are discussed; these include the absence of quantitative coastal inventories of natural resources, the weakness of physical planning mechanisms, shortages of equipment, and the difficulty of reconciling the short term political time frame and the longer term environmental time scale. The paper recommends increased cooperation between professionals in regional research institutions and government CZM agencies to develop applied research projects to solve specific problems.

Introduction

This paper discusses Integrated Coastal Zone Management (ICZM) within small island states from the coastal manager's perspective and thus in many ways presents a very practical approach. The paper is based on the smaller English-speaking islands of the eastern Caribbean, also known as the Lesser Developed Countries (LDCs); however, many of the observations and conclusions may be applied to larger Caribbean islands and islands in other parts of the world.

Small Islands: Marine Science and Sustainable Development
Coastal and Estuarine Studies, Volume 51, Pages 323–328
Copyright 1996 by the American Geophysical Union

Background to Integrated Coastal Zone Management in the Caribbean

ICZM has been defined as follows:

> "Integrated Coastal Zone Management is a dynamic process in which a coordinated strategy is developed and implemented for the allocation of environmental, socio-cultural, and institutional resources to achieve the conservation and sustainable multiple use of the coastal zone" [CAMPNET, 1989].

ICZM is still a relatively new concept in these islands where traditionally there is a sectoral approach to government such that areas such as water resources, fisheries, sea defenses, and planning are dealt with by separate agencies with very little communication amongst themselves. The ICZM framework follows a geographical approach such that a coastal zone is defined which may extend from the seaward limit of the territorial sea to a certain distance inland. Within this coastal zone, government agencies and the public work together within an integrated framework to utilize coastal resources in a sustainable manner. One agency has to take the lead coordinating role. Coordination is an essential part of ICZM, since the different agencies remain in place; however, they work together through mechanisms such as inter-agency consultative committees. The change-over from a traditional sectoral approach to an ICZM approach is a very difficult one to make within existing government frameworks. As a result, no eastern Caribbean island has yet achieved full ICZM, although several islands are working towards this goal, *e.g.*, Barbados and British Virgin Islands [Cambers, 1993].

ICZM is not an exact science; essentially, it provides the background for a rational decision-making process for the allocation and use of coastal resources. Within small island states ICZM has to balance four equally important viewpoints: scientific, social, economic, and political. However, in reality these four viewpoints do not have equal status. Often the scientific database is insufficient to fully justify this viewpoint. In addition, in a small developing country, new development is almost always viewed as beneficial from a socio-economic viewpoint.

The political viewpoint is, perhaps, the most important in these small island states. In these countries, all decisions, whether large or small, are made by the political directorate. Thus, whether the proposed development is for placing one mooring buoy, removing four mangrove trees, or for a major hotel and marina, the decision whether or not to proceed will be made by a politician. This is different to the CZM model developed in the United States and its island territories, where there is usually a statutory body which will review the development applications and make the decisions. In these small island countries the coastal manager or committee essentially only makes a recommendation to a minister who then makes the decision. Since this situation is unlikely to change in the foreseeable future, the question must be asked whether small island states such as the Caribbean LDCs will ever achieve ICZM.

Within the Caribbean islands, the scientific database is often inadequate. Coastal managers are forced to make decisions or recommendations in the absence of the necessary data. In most cases, they do not have the option of deferring the decision until the data is collected. This is especially so when a development decision is being reviewed.

One of the major data deficiencies is the lack of quantitative natural resource inventories, which will also be discussed in Chapter 21. To illustrate this point with an example, a particular marina proposal may involve the dredging of a large area of seagrass. An Environmental Impact Assessment (EIA) may have concluded that the loss of this area of seagrass was of no major impact to the country because there were many other similar areas of seagrass. Unless the coastal manager has at hand a complete inventory of the geography, ecology, and health of seagrasses around his island, inevitably the statement in the EIA will be accepted at face value.

The existence of good quantitative inventories of natural resources is of immense importance to the coastal manager as he tries to balance various options and alternatives in the management process. An example from Guyana illustrates this point. The entire coastline of 450 km used to consist of mudflats and mangroves. As land was reclaimed for agriculture, earthen banks and, eventually, seawalls were constructed. The situation is now such that much valuable infrastructure is protected with seawalls that have been poorly maintained and now require considerable financial outlay. A detailed mapping and inventory of coastal mangroves in 1970 and 1990 was prepared [Pastakia, 1991]. This has shown that very few mangroves remain in front of the 125 km of coastline protected with seawalls. However, there are considerable mangrove resources left on the 325 km of undefended and lesser defended coasts. The inventory information clearly suggests a two phased management approach for Guyana's coastal zone, one approach for the "seawall" coasts and a different approach for the "mangrove/earthen bank" coasts.

Another example from the British Virgin Islands clearly illustrates the usefulness of inventories to the coastal manager. Here mangroves were mapped and then prioritized based on biological and socio-economic criteria. Mangroves were then categorized as critical, very important, and less important [Blok-Meeuwig, 1990]. Such an inventory, which covers an entire country, provides a coastal manager with a very useful tool for decision making. However, in many of the eastern Caribbean islands the basic inventory information covering each country's natural resources does not exist.

The EIA process is still in its very early stages in eastern Caribbean islands. While in recent years there have been several training workshops on EIA in the islands, few EIAs have actually been carried out. This is partly because it is not mandatory to conduct an EIA for a new development, although this situation is slowly changing.

One of the major constraints to ICZM in the small islands is the weakness of the planning agencies. While economic planning is on a sound footing, physical planning has always been weak. Physical planning agencies are critical to the ICZM process, especially in the early stages when there is no coastal legislation. Existing planning legislation often provides a mechanism for implementation of some initial CZM policies. Physical planning agencies in the islands have been in existence for several decades; however, they usually concentrate on development control rather than physical planning for development. Thus, there is very little zoning and in most of the islands any type of development can take place in any location provided certain conditions such as road allowance, building density, *etc.*, are fulfilled.

The islands also lack the basic equipment which would assist them in the implementation of ICZM. The policy of most lending institutions has been to concentrate on technical assistance and training. While these are very important and much needed forms of assistance, the near impossibility of obtaining equipment such as computers, laboratory equipment, boats, and vehicles is a very serious constraint. For instance, it is theoretical to talk about sea level rise in the smaller eastern Caribbean islands when there are no permanent tide gauges installed on these islands.

The time to develop and achieve ICZM may take as long as 30 years. Figure 1 shows some of the important milestones to be achieved in developing ICZM. The government's perception that there is a need for a special form of management for the coastal zone is an essential starting point. Rarely has an island country in the eastern Caribbean set out to establish a CZM program. Usually a specific problem such as severe coastal erosion or rapidly declining fish stocks generates a certain response that provides the nucleus from which a CZM program can grow.

Discussion

There is considerable scope for increased cooperation between academic institutions and government agencies within the framework of ICZM, particularly in the field of data collection, analysis, and interpretation. However, it is important to have close dialogue so that the research provides the answers to the problems being faced. It is also necessary for the researcher to understand the government agency's needs and the process whereby those needs can be fulfilled and implemented. Within ICZM there is no purely scientific answer; social, economic, and political viewpoints have to be considered.

While obviously there will always be a need for pure research, since this is the foundation on which science progresses, it is advocated that more applied research should be directed towards specific problem solving. This will require close cooperation between the scientific and government professionals. There are so many problems facing coastal managers in the islands for which they have no specific answers, *e.g.* how

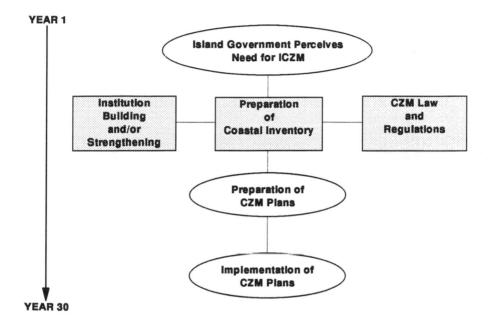

Fig. 1.   Milestones in ICZM.

much sand can safely be mined from an accreting beach; does scuba diving for conchs and lobsters damage the stocks; what is the impact of a particular sewage outfall on a nearby coral reef. Solving problems such as these will involve short- and long-term data collection, analysis, and interpretation projects. The answers may be provided by direct academic research or through students' dissertations and postgraduate work.   The preparation of natural resource inventories can also be included in these projects.

Island countries without a university or research institution require special attention since these are often omitted from the research field because of extra costs and other logistical problems. However, these are often the countries that are most in need of the assistance.

Technical cooperation between developing countries (TCDC) is a useful mechanism which is underutilized in the Caribbean region. There is experienced local expertise available in specific fields in certain countries which could provide regional benefit within TCDC projects. Similarly, a system of attachments to government agencies with progressive CZM programs would benefit small island states.

Perhaps the major problem facing these small island states in the achievement of ICZM is how to rationalize the political time scale (four to five years) with the natural resources time scale (10+ years). It is necessary to develop tools which will convince senior administrators and politicians of the need to consider the longer term viewpoint when making decisions related to the coastal zone.

Recommendations

The recommendations are listed as follows:

1. To develop closer links between professionals in the Caribbean academic institutions and government CZM agencies and to develop applied research projects to provide solutions for specific problems facing coastal managers.
2. To develop a cooperative approach between the CZM agencies, academic institutions, and lending institutions to prepare quantitative coastal resource inventories.
3. To encourage technical cooperation between developing countries.
4. To develop tools which will convince politicians and senior administrators of the need to consider the longer term viewpoint.
5. To strengthen physical planning in the small island states.
6. To encourage lending institutions to make equipment more readily available.

References

Blok-Meeuwig, J., Mangrove systems of the British Virgin Islands resource mapping and assignment to protection categories, Tech. Rept. No. 5, Conservation and Fisheries Department, Ministry of Natural Resources and Labour, Government of the British Virgin Islands, p. 45, 1990.

Cambers, G., Coastal zone management in the smaller islands of the eastern Caribbean: An assessment and future perspectives, Proceedings, 8th Symposium on Coastal and Ocean Management, New Orleans, Louisiana, pp. 2345-2353, American Society of Civil Engineers, 1993.

CAMPNET, The status of integrated coastal zone management: A global assessment, Workshop Summary Report, Coastal Area Management and Planning Network, Charleston, South Carolina, July, 4-9, 1989, p. 13, Rosenstiel School of Marine Science, University of Miami, 1989.

Pastakia, C. M. R., A preliminary study of the mangroves of Guyana, Final Report, Article B 946/89, contract no 8912, p. 32, 1989.

20

Water Supply and Sewerage in a Small Island Environment: The Bahamian Experience

Richard V. Cant

Abstract

The Bahama islands comprise several hundred low-lying limestone islands which are well suited to and heavily dependent upon the tourism industry. Unfortunately, water supplies and liquid waste disposal present serious problems in such an environment and these have impacted the economic development of the islands. Specific problems include the availability and distribution of freshwater resources. Water systems are proving to be difficult and costly to develop, and few residential communities can afford the full cost of water supplied by alternate methods such as reverse osmosis. Groundwater resources in this environment are also very prone to man's abuse and are exceedingly vulnerable to pollution. Mistakes have been made which have resulted in serious, long-term damage. Liquid and solid wastes are proving difficult to manage and dispose of, and appropriate technology needs to be applied where conventional methods are found to be unsatisfactory.

Experience has shown that adequate legislation and regulation are required to control and protect water resources, and there needs to be an institutional structure that can administer and enforce the fair use of these resources. In small island states all those involved in the environment and water supply sector have to work diligently to make the public aware of the issues involved, and the potential consequences of inaction, so that the public will accept whatever measures are needed to safeguard the future.

Small Islands: Marine Science and Sustainable Development
Coastal and Estuarine Studies, Volume 51, Pages 329–340
Copyright 1996 by the American Geophysical Union

Introduction

The Bahamas typifies a carbonate environment in which there are extensive shallow marine banks, deep intervening troughs, and a scattered distribution of low-lying limestone islands. Because there are no true rivers draining and eroding the islands and because it is a totally carbonate environment, the Commonwealth is blessed with crystal clear seas and beautiful white beaches and, due in part to this, in combination with a mild climate and close proximity to North America, the Bahama islands have benefited from a healthy tourist industry for many years. Although this environment is attractive to visitors from colder climates, it is also one in which water resources are limited; it is one in which it is extremely difficult to dispose of various wastes that are created by tourism and urbanization on a long-term basis.

Water supply and sewerage are two critical concerns that present major constraints in achieving sustained economic growth in islands of this type. The Bahamas provides a good example for studying just what problems will emerge in this situation, as well as the types of solutions needed to achieve goals that are acceptable for the future.

The Environment

The Commonwealth of the Bahamas (see Figure 1) consists of 14 shallow marine banks that emerge locally above sea level, and are separated from one another by deep water. The banks, which are usually less than 9 m deep, stretch over a distance of 900 km, and appear like submerged plateaus with steep sides, displaying considerable size variation. By far the largest is the Great Bahama Bank. The islands occur unevenly but are usually on the margins of the larger banks and in the center of the smaller ones. Many islands are long (100 miles, 160 km), narrow (1 mile, 2 km), and orientated from northwest to southeast, with central ridges extending to a maximum height of 62 m. The larger islands are generally flat and tend to be partially inundated by the sea on their protected leeward shores. In all, there is a land mass of about 13,878 km, but only 16 islands have areas greater than 50 km^2.

The population of 260,000 is concentrated largely in Nassau, the capital, where 67% reside on a small island known as New Providence. The remainder are dispersed as indicated in Table 1, which also provides information on the area of the islands listed and population densities. In total there are more than 700 small islets and cays in the region. The majority of them are uninhabited.

With regards to climate, the Bahamas is broadly characterized by subtropical temperatures, and can be classified as marine tropical being affected by regular easterly trade winds. There is a warm, wet summer from May to October, with a dry cool winter from November to April. Rainfall varies considerably from north to south and from year

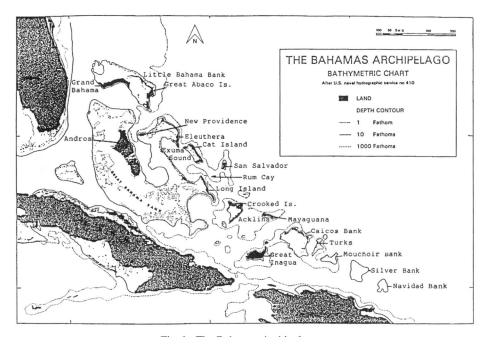

Fig. 1. The Bahamas Archipelago.

to year [Clarkson, 1971]. To the north the average annual rainfall is 1580 mm, declining to 630 mm in the south. In winter periodic depressions from North America tend to dominate weather conditions with northerly winds; however in summer the western Atlantic is the most important control on weather, and this includes the effects of hurricanes.

Geology and Aquifers

The plateaus of the Bahama Banks are made up largely of carbonate-evaporite deposits that accumulated as layered, horizontally-aligned sedimentary units. These deposits extend from sea level down to a depth of about 8 km. There is no proven indication of tectonic warping or folding; however, the region is subsiding differentially, the highest rates being located in the Central Bahamas.

The surface geology of the Bahamas comprises Pleistocene limestones, largely shallow marine deposits, together with littoral and wind-blown material; less common are marshland sediments and lake or pond deposits. Soils and caliche commonly occur with any of the above sediment types. Typically, the limestone surface displays a young karst topography but in places there are large sink holes known as "blue holes," some of

TABLE 1. Percentage Distribution of Population by Island, 1980 and 1990 Censuses. Island Area and Population (1990 Census).

Island	1980 Pop.	1980 Percent. Distribution	1990 Pop.	1990 Percent. Distribution	Area (sq. miles)	Density (population per sq. mile, 1990)
All Bahamas	209,505	100.00	254,689	100.00	5,382	47.32
Abaco	7,271	3.47	10,061	3.95	649	15.50
Acklins	618	0.29	428	0.17	192	2.22
Andros	8,307	3.97	8,155	3.20	2,300	3.54
Berry Islands	509	0.24	634	0.25	12	52.83
Bimini Islands	1,411	0.67	1,638	0.64	9	182.00
Cat Island	2,215	1.06	1,678	0.66	150	11.18
Crooked Island	553	0.26	423	0.17	93	4.54
Eleuthera	8,331	3.98	8,017	3.15	187	42.87
Exuma	3,670	1.75	3,539	1.39	112	31.59
Grand Bahama	33,102	15.80	41,035	16.11	530	77.42
Harbor Island	1,133	0.54	1,216	0.48	3	405.33
Inagua	924	0.44	985	0.39	599	1.64
Long Island	3,404	1.62	3,107	1.22	230	13.50
Mayaguana	464	0.22	308	0.12	110	2.80
New Providence	135,437	64.65	171,542	67.35	80	2144.27
Ragged Island	164	0.08	89	0.03	14	6.35
Rum Cay	78	0.04	53	0.02	30	1.76
San Salvador	747	0.36	486	0.19	63	7.71
Spanish Wells	1,167	0.56	1,291	0.51	10	129.10

*including Long Cay.

which are as much as 100 m wide and 150 m deep. In coastal areas, the solid geology may be overlain by uncemented sands of Holocene age.

Cemented marine deposits make up the bulk of the larger islands and represent widespread flat banks that were once shallowly submerged. Eolianites occur most commonly as long linear ridges, and multiple ridge systems indicate several phases of dune accretion. Littoral deposits are surprisingly common and occur in low-relief concentric ridges, or as less obvious ridge-and-swale structures overlying marine flat lands.

Close inspection reveals that the marine deposits are most commonly bioturbate or burrowed lagoonal sediments, or well-sorted sands with cross-current sedimentary structures. The eolianites are cross bedded, frequently include soils and root casts, and may have high depositional dips. Beach deposits are usually well sorted and finely laminated, and display constant low-angled depositional dips.

The geology relevant to the occurrence of fresh water ranges from the surface to a depth of about 40 m; and is termed the Lucayan Limestone [Beach and Ginsburg, 1980].

Coring has revealed that this formation commonly consists of marine deposits occurring as several distinct beds, separated by soil-covered erosion surfaces; the sequence, no doubt, results from sea level fluctuations during the Pleistocene. Each bed has cementation differences, and cavern or fissure development [Little *et al.,* 1977], which together with the varying effects of soil horizons, impart a strong influence on aquifer hydraulics. Vertical variations in permeability and porosity are often very pronounced, and can help sharply define the limits of fresh water lense development [Cant and Weech, 1986, Vacher and Bengtsson, 1989].

Another important aquifer consists of uncemented superficial Holocene sand which occurs to a maximum depth of about 10 m. Such coastal sands have low permeabilities but can yield fresh water in places where rock would not; their occurrence is, therefore, important. Sometimes they are partially cemented and their water-bearing characteristics become somewhat changed.

Ground Water Resources

The low-lying, young Karstic terrain of the Bahamas, coupled with the limited size of the islands, precludes the development of rivers. Rainfall infiltrates rapidly to the water table. Surface water bodies usually occur in topographic lows and are no more than exposures of the water table lying at or just above sea level. Often surface water bodies are brackish or hypersaline, particularly in the southeast, where evaporation far exceeds precipitation. In Inagua, modified ponds are used on a large scale to produce salt by solar evaporation.

Ponding above the water table often occurs following heavy rain but usually lasts only for a matter of days. Where ridges or hills are able to channel or concentrate runoff, flooding occurs in response to a raised water table; which can stay higher than usual for two or three weeks, but will eventually returns to normal as a result of increased coastal discharge.

Infiltrated rain water that percolates to the water table is capable of accumulating in significant quantities and forms freshwater lenses, the name being based on their theoretical shape. They accumulate in rock or sand, and are maintained partially as a result of the lower density of the fresh water which causes it float to on the seawater beneath, and also as a result of the seaward movement of groundwater which tends to push the salt water out. The freshwater lenses are, therefore, entrapped rainfall and they are known to develop to a maximum thickness of 38 m. If the rock is very permeable, the lenses will not form; instead, brackish water results from tidal mixing.

Freshwater lenses are the only exploitable groundwater resources in the Bahamas and their natural occurrence is limited by the amount of effective rainfall recharge and by the retention characteristics of the rock. Thick widespread lenses are often found in the larger islands. Lenses are subject to the movements of underlying seawater, from which they are separated by a 1-5 m thick transition zone. Salt water intrusion easily takes place as a result of overpumping or excess evapotranspiration. Tidal and barometric pressure effects can be observed at the water table, and at the freshwater/salt water interface. The lens water table usually is slightly higher than sea level, the highest recorded mean elevation being +1.2 m.

Where lenses occur, a state of approximate equilibrium exists because of fixed geological controls, stabilized sea level, limited area for growth, and fairly uniform rates of recharge. Man-induced changes affect the state of these water resources more than any other cause, and the ease and speed of infiltration from the surface also makes them very easily polluted.

The groundwater resources of the major islands in the Bahamas were quantified and studied in detail by the Bahamas Land Resources Survey between 1969 and 1975. The results of these investigations are summarized in Little *et al.* [1977]. Other private and government-sponsored investigations have taken place since the Bahamas Land Resources Survey and these have helped to complete the picture for many of the smaller islands where there were gaps in the previous knowledge. Cant and Weech [1986] summarized the occurrence of fresh water in the Bahamas and, using this information, showed how lens development is controlled by island size, shape, climate, and geology. Table 2 shows how the freshwater resources are distributed over the 13 larger Bahamian islands.

TABLE 2. Comparison of Freshwater Resources on the 13 Largest Bahamian Islands.

Island	Freshwater Lense (acerage)	Maximum Volume Available Daily (million gallons)	Water Available IIG/D Person (1990 census)	Total Population (1990 census)
Abaco	116,280	79.10	7,859	10,061
Acklins	15,783	4.36	10,178	428
Andros	338,585	209.92	25,742	8,155
Bimini	395	0.17	75	2,272
Cat Island	14,774	6.80	4,050	1,678
Crooked Island	5,923	1.74	4,104	423
Eleuthera	16,599	8.13	773	10,524
Exumas	6,586	2.90	819	3,539
Grand Bahama	147,884	93.17	2,270	41,035
Great Inagua	3,571	0.86	870	985
Long Island	9,301	2.88	928	3,107
Mayaguana	2,340	0.65	2,096	308
New Providence	17,503	9.63	56	171,542
Total	695,524	420.31	59,820	254,057

Development and Management of Water Supplies

Groundwater resources in the Bahamas have always been easy to exploit, and regular usage dates back to the earliest settlers. Today, water is still privately obtained by bucket from shallow hand-dug wells in the less developed areas. Such wells usually contain less than 1 m of water and are practicable where the freshwater lens is very thin. Other methods such as hand, petrol, or electric pumping systems may lift water to overhead storage, thereby providing running water for domestic usage. People who require larger quantities use several dug wells or a network of drilled boreholes.

Besides dug and drilled wells, public supply of groundwater is obtained from mechanically cut trenches, pits, and seasonal freshwater marshes. Under certain circumstances, long shallow trenches are a better source than wells, particularly where salt water upconing is a serious problem. Abstraction methods include the use of wind, diesel, and electrically powered pumps, and compressed-air-operated systems.

Where fresh groundwater is not available, rain water catchments are normally used. Alternatively, where costs allow, water is obtained by flash desalination, reverse osmosis, and also by marine transport of groundwater from other islands. The latter method provides up to 40% of the water presently supplied in New Providence. Piped supplies have been available in New Providence since the late 1920's and in some other islands since the 1940's.

Presently about 88% of the population has access to a piped water supply. This high figure reflects the fact that supplies are available in most of the heavily populated islands; however, there are many smaller communities that do not have supplies. Some islands have no piped supplies available because there is no fresh water available, and in many locations the piped supplies are brackish because this is all the resource is capable of producing.

Most water systems operating in the Bahamas are owned and managed by the Water and Sewerage Corporation. This is a government-owned corporation that was formed in 1976. This corporation operates about 60 separate water systems distributed over 26 inhabited islands. About 10 million Imperial gallons are supplied each day. A number of privately-owned water supply companies operate in New Providence and on other islands. The largest of these is the Grand Bahama Utility Company which provides water in the city of Freeport, Grand Bahama.

Water rates vary from place to place. In some of the more remote locations the water is provided free of cost to the local residents. Water rates are generally lower where local groundwater is all that is utilized ($2.30 to $5.42/1000 Imperial gallons), but it increases where groundwater is blended with barged water, or water produced by reverse osmosis ($10.80 to $19.20/1000 Imperial gallons), and is highest where salt water reverse osmosis is the only source of supply ($20 to $30/1000 Imperial gallons).

There are many existing laws and regulations that impinge upon the use of water; however, there are no regulations that control the use of groundwater or enable this resource to be appropriately administered. Efforts are now in hand to achieve this.

Sewerage and Effluent Disposal

Sewer collection systems serve approximately one fifth of the capital Nassau. On the other islands sewer systems are limited to a few small subdivisions and some private developments and hotels. Septic tanks are used most commonly on the major islands although these do not always conform to the building code and, therefore, may not function in the manner that they should. In less developed areas, pit latrines may be used and there are some places where direct discharge into the sea is still used as a means of disposing of wastes. The use of septic tanks is usually combined with a drain field or disposal well.

Where main sewerage exists, the wastes are normally treated to primary or secondary levels and the effluent is then discharged in a deep disposal well. Many different types of deep disposal wells are utilized, and dispose of a wide variety of liquid wastes [Cant, 1992]. The wells that are used for large volumes of effluent are normally cased down to about 200 m and are open below that depth. Golf courses in tourist areas usually require considerable volumes of irrigation water. In such situations the wastewater from hotels is usually treated and reused on a nearby golf course. The waste disposal methods used in the Bahamas are presently far from satisfactory, and studies have shown that the groundwater underlying urbanized areas has relatively high levels of pollution [Weech, 1993]. There is also evidence of seawater pollution, particularly in some enclosed harbors, and this is now giving rise to cause for concern, particularly where these harbors are important tourist destinations or may be involved in the seafood industry.

Discussion

Development Difficulties

While it may be simple to provide water to an individual household in a Bahamian-type island, either by use of a private well or a rain catchment system, it has become evident that such sources of supply cannot meet the needs of a developing and viable community. The water demands of a small hotel, marina, or even a small business are usually excessive for a private system for a number of reasons. These include the fact that the well or wells will become saline when developed on a small scale, and also the fact that rain catchment and storage are not practical where demands are high and where there is a long dry season such as occurs in the Bahamas. Alternate sources such as reverse osmosis or even barging need to be considered; however, such options are excessively costly when implemented on a small scale. Large plants produce cheaper water than small plants and, therefore, all such systems are best developed on a centralized or community basis. This centralization has been difficult to implement because only a small proportion of those making up the water demand on islands in this category are able to pay the true cost of water. These persons often prefer to pay lower water rates, even if poor water quality results, and they will obtain essential drinking water from other sources; often bottled water or rain water which will only be needed in small amounts. These persons are usually the local residents.

Businesses catering to the needs of tourism or industry require proper water supplies and if these are not available the businesses are then forced into the position of providing the utility themselves. This problem acts as a definite constraint to growth; however, new developments built to cater exclusively to the tourism industry have no problem in developing the quality water supplies required as these costs can be incorporated into the total package. The result of this type of imbalance is that new developments often do not wish to incorporate the needs of the existing community. Since this disparity will result in differences in standards of living in a small island environment, it is left to government agencies to ensure that the new development actually subsidizes the water needs of the established community. Normally this is achieved by a tiered rate structure that favors the small consumer. However, in the present day context of heavy competition between tourist destinations and the subsequent need to lower room rates, there is now pressure to reduce this form of subsidy at the cost of the community itself. In addition to the above, the Bahamas has problems in developing water supplies because many of the demand centers are poorly located with respect to the resources. Settlements are often located on the coast or on small offshore islands, as are the tourist developments. Costly pipelines are, therefore, needed to bring the water to these locations. Also, there are many logistical problems in supplying water on a large number of islands. Technical and managerial capabilities have to be duplicated, besides the water installations themselves, and all costs are increased by this difficult situation.

Resource Vulnerability

Oceanic island groundwater resources are extremely vulnerable, particularly so where the islands are low-lying limestone islands. This vulnerability stems from several factors. The main ones are:

1. Such a resource can be overexploited easily, and when this occurs freshwater will be replaced by seawater. Nearly all small islands provide evidence of the above in the form of abandoned or saline wellfields. Salt water upcones beneath an overpumped well making it unfit for use.
2. The ease by which man can change the conditions in which the fresh lens stabilized, *e.g.*, it is a common practice to cut canals in islands to provide valuable waterside lots. This practice, of course, results in facilitating the mixing of fresh and saline water and can cause the total loss of the resource. In Grand Bahama one canal system was cut right across the center of the island where there was a 13 m

thick lens, and this action has resulted in the direct loss of 7.5 billion gallons of freshwater held in storage [Cant *et al.,* 1990].

3. Flat, low-lying islands are extremely prone to inundation, either as a result of storms or changes in sea level. Any form of saline inundation will damage groundwater resources and, in situations where water levels are rising, there is danger that the water table will rise close to or above the land surface resulting in full potential evapotranspiration that will, in due course, destroy the resource.

4. Many forms of pollution threaten the resource. Limestone islands are particularly vulnerable for several reasons:

    a. The water table is near to the surface.

    b. The rock is usually very porous.

    c. Good thick soils are normally lacking; therefore, many beneficial biochemical reactions that could nullify the effects of certain elements and compounds are lost.

    d. Internal drainage is the normal situation; therefore, everything disposed of or decomposing on the ground will be introduced into the groundwater environment.

    e. Suitable disposal sites are very scarce in a small island environment. Sink holes, swamps, and marshes are often used as dump sites but these are probably the worst of all possible options. Also, islands of this type are very often undeveloped and do not have mains sewerage. This situation results in widespread problems because of the use of septic tanks and other such means of sewage disposal. Domestic wastes are introduced directly to the water table.

Once groundwater is polluted it is extremely difficult and costly to clean up. Efforts have been in hand to clean up one fuel spill in Nassau for more than 12 years and there is still a significant volume of hydrocarbon in the subsurface.

Legislation and Regulation

Small island states very often do not have the necessary institutional structure nor the laws or regulations that are required for the protection, management, and fair use of their groundwater resources. In the Bahamas, existing laws usually relate to water supply or the environment in general, and efforts are required to address this shortfall. In remote islands where the inhabitants are fairly self-reliant, there may be some objection to the government administering what is often regarded as a private possession.

It is very important that freshwater resources be established as a national resource, and that an institutional structure be created that can administer the proper use of this resource. The institutional structure should be empowered to issue permits, licenses, and penalties as necessary. A water law consultant is usually the best person to draft what is needed and this should be structured to suit existing laws and the constitution, if there is one. Changes in legislation will normally respect existing uses, and it may be beneficial to develop an organizational distinction between policy making, water management, and the provision of water supplies and sanitation services. The policy-making body should include representatives from all the relevant government agencies, as well as the private sector. Delays in implementing legislative control can be extremely damaging.

Wastewater and Effluent Disposal

All wastes create problems in small islands, and developing safe and reliable means of disposal is an ongoing challenge. Mains sewerage coupled with full treatment is an ideal that is difficult to realize. Usually it takes a serious epidemic to attune the public to the danger of improper waste disposal, and a great deal of lobbying and effort is required to get a third world government to invest the large sums of money that are needed to develop this particular sector.

Surrounded by sea, the island mentality dictates that liquid wastes go directly into the sea, or, indirectly, via the subsurface. As an area develops, this simple approach to the problem becomes unsatisfactory. If an island's economy is tourism based, it usually involves marine activities and these can only be sustained if the sea is kept in pristine condition. Achieving this will require some form of waste treatment and an effluent disposal system that does not involve coastal waters. Deep sea outfalls can be used but they are not regarded as being satisfactory in the Bahamas. Instead, deep well disposal is used, as this method is well suited to the geological characteristics of a carbonate bank. Zones of high transmissivity are used as receiving zones, and in these it is assumed that wastes are rapidly mixed, diluted, and dispersed, thereby nullifying any threat that they pose to the environment. To employ this method on a sustained and successful basis it is imperative that the geology of the subsurface be fully understood and also the hydrodynamics of this hidden environment. Disposal wells should be designed to take advantage of specific geological features, and there should be a full range of specifications established to control well depth, well head installations, casing requirements, pretreatment, and any subsequent monitoring. Deep well disposal is a viable option and a relatively low cost one. However, it can only be applied if it is done so with all the necessary knowledge and regulatory controls.

Visiting yachts and cruise ships have created pollution problems in coastal waters in many parts of the world. Finding a solution to the cruise ship problem should not be difficult. However, large numbers of small vessels are much harder to regulate and the tendency to gather in small safe harbors, which may be fairly enclosed, does present a health risk. There is presently no simple solution to this problem, but boating visitors are an important part of island tourism and a solution has to be found.

Education

The value and vulnerability of groundwater resources in a small island dictate that the public should be fully aware of the situation and know exactly what threatens this essential requisite to life. Conserving and protecting this resource is the responsibility of everyone. In order to carry the message to the general public, the agencies involved in dealing with health, the environment, and water need an ongoing campaign to publicize all aspects of the issue. School children should be the main target of this campaign but it should also include talks to various clubs and associations, articles in the press, and television exposure. If the country does not have the capability of such a program, there are many international agencies and professional affiliations that can assist and some can provide funding for this. In many islands ignorance has caused irreparable damage to its water resources, and this damage continues today.

Research

Because small islands have a tendency to introduce problems that are specifically case related, there is a need for detailed research that may not be appropriate in other environments and is, therefore, often lacking. A hydrologist working in the Bahamas finds it impossible to use work done elsewhere to solve his problems and, therefore, has to try to develop his own solutions. This will involve research which may be difficult to finance locally and, therefore, international funding needs to be sought. Small island states find it difficult to have the necessary knowledge in the wide range of disciplines that can impact on their sustained economic development.

References

Beach, D. K., and R. N. Ginsburg, Facies succession of Plio-Pleistocene carbonates, northwestern Great Bahama Bank, *Amer. Assoc. Petrol. Geol. Bull., 64*, 1634-1642, 1980.

Cant, R. V., Geological implications of deep well disposal in the Bahamas, in *Natural Hazards in the Caribbean*, edited by R. Ahmad, Special Issue No. 12, *J. Geol. Soc. Jamaica*, 1992.

Cant, R. V., and P. S. Weech, A review of the factors affecting the development of the Ghyben-Hertzberg lenses in the Bahamas, *J. Hydrology, 84*, 1986.

Cant, R. V., P. S. Weech, and E. Hall, Salt water intrusion in the Bahamas: A case study of the Grand Lucayan Waterway, Grand Bahama, Bahamas, Proceedings International Symposium on Tropical Hydrology and 4th Caribbean Islands Water Resources Congress, San Juan, Puerto Rico, July 23-27, 1990.

Clarkson, J. R., Rainfall distribution over the Bahamas, unpublished report produced for the Meteorological Office, Nassau, Bahamas, 1971.

Little, B. G., D. K. Buckley, R. V. Cant, W. T. Gillis, P. W. T. Henry, A. Jefferies, J. D. Mather, J. Stark, and R. N. Young, Land resources of the Bahamas: A summary, Land Resources Study 27, Land Resources Division, M.O.D., Surbiton, Surrey, 1977.

Weech, P. S., Bahamas, Paper presented at Regional Workshop on Water Quality Monitoring in the Caribbean, July 5-9, 1993, Port of Spain, Trinidad, 1993.

21

Coastal and Marine Environments of Pacific Islands: Ecosystem Classification, Ecological Assessment, and Traditional Knowledge for Coastal Management

Paul F. Holthus

Abstract

Coastal and nearshore marine environments in the tropical Pacific islands are subjected to severe and extensive impacts from a variety of sources of degradation and destruction. In spite of the importance of coastal and marine environments to the small islands of the Pacific, there is only limited scientific information available on these areas. Integrated coastal zone management (ICZM) is urgently required on the small islands of the Pacific to ensure the sustainable use, development, and conservation of coastal and nearshore marine environments. Information on the biological, ecological, geological, and oceanographic characteristics of these environments is an essential basis to ICZM. Although much research is needed to provide information for the scientific management of coastal and marine environments in the Pacific, these areas will not be able to be managed with scientific rigor for some time, due to the limited amount of research underway. However, a substantial amount of information on coastal and marine environments can be readily obtained through several approaches which are being employed in the Pacific and may have relevance to other small islands.

In the tropical island Pacific, three complementary approaches are being used to rapidly and cost effectively obtain information on coastal and marine environments as a basis for ICZM and sustainable development:

Small Islands: Marine Science and Sustainable Development
Coastal and Estuarine Studies, Volume 51, Pages 341–365
Copyright 1996 by the American Geophysical Union

1. Inventorying marine ecosystems using a comprehensive, hierarchical classification of the ecosystems of the tropical island Pacific which allows ecosystems to be inventoried on a systematic, regionally valid, and comparable basis in order to determine conservation priorities on a local, national, or regional scale.
2. Undertaking rapid ecological assessments of nearshore marine areas to obtain information on the biological and physical characteristics as a basis for coastal management planning and other aspects of sustainable development.
3. Collecting traditional knowledge of marine ecosystems from Pacific island marine resource users which hold a wealth of traditional knowledge on their marine environments, much of which is invaluable to the management of marine ecosystems.

All of these methods can be applied to evaluating the coastal and marine biodiversity and conservation needs of individual areas.

Introduction

The small islands of the tropical Pacific are characterized by: (1) small land masses dispersed across an immense ocean area; (2) rapidly growing populations; (3) economic and cultural dependence on the natural environment, especially coastal and marine ecosystems; (4) vulnerability to natural disasters; (5) a high level of ecosystem and species diversity and endemicity; and (6) a diverse mixture of cultures, languages, and customs. The island countries and territories of the South Pacific region consist of only 550,000 km^2 of land with 5.2 million inhabitants scattered across some 29 million km^2 of the Pacific Ocean (Figure 1). However, the figures drop to merely 87,587 km^2 of land area and 2.2 million people if Papua New Guinea (PNG) is excluded. Among the 22 island states in the region there are four with land areas of less than 100 km^2, 11 with 100 to 1,000 km^2, and another seven with land areas ranging from 2,935 to 27,556 km^2.

There are four principle types of islands in the South Pacific region: continental, volcanic, raised limestone, and low islands [Thomas, 1963; Dahl, 1980]. The first three types are all "high" islands, while low islands are built of coral reef sediment and rubble which accumulates to elevations of only a few meters above sea level on reef platforms, such as the ring of reef which forms an atoll. Virtually all of the islands in the region are entirely coastal in character. That is, all parts of the island influence, or are influenced by, processes and activities occurring on coastal lands and in nearshore waters.

Coastal and marine ecosystems are of critical importance to Pacific island peoples, cultures, and economies. The coastal and nearshore marine areas of all islands in the tropical Pacific are the location of the vast majority of human habitation, the focus of subsistence and commercial, agricultural, and fisheries activity, and the target of most economic development [Connell, 1984; Dahl, 1984]. This combination of factors is

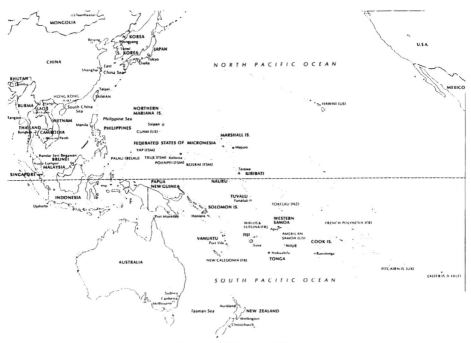

Fig. 1.The tropical island pacific region.

increasingly resulting in coastal habitats being destroyed or degraded, natural resources being overexploited, and growing conflicts in coastal resource usage.   Common problems in the nearshore environment include terrestrial sedimentation, sewage pollution, waste dumping, overharvesting of living resources, destructive fishing techniques, and dredging or filling of nearshore areas [Brodie *et al.*, 1990]. The coastal areas of the Pacific, especially low elevation islands, are also subject to the destructive effects of natural hazards from extreme events such as cyclones. Global warming now threatens to exacerbate these hazards through accelerated sea level rise and other possible changes to oceanographic conditions.

The island Pacific region contains a range of oceanic and nearshore marine environments.   Coastal and marine environments (hereinafter "marine environments") are used here to generally refer to the components, functions, and processes of the shoreline and intertidal area and the nearshore waters and benthos around small islands. Marine biological diversity generally decreases from west to east across the Pacific Ocean, with the regional center of high diversity in the southwest Pacific [Holthus and Maragos, 1992]. Within this broad pattern are ecosystems important to rare, threatened, or endangered species and a few known centers of endemism. Relatively little is known about the large ocean water masses of the Pacific and their deep sea features and benthic communities.   Shallow and nearshore marine areas support coral reef, lagoon, mangrove, and seagrass systems and contain a diverse flora and fauna which are critical to the cultural, subsistence, and economic life of Pacific islands. Marine protected areas (MPAs) are an important tool for the conservation of marine biological diversity; however, there are few MPAs in the island Pacific region and the marine biological diversity of no part of the region can be considered effectively protected [Thomas *et al.*, 1992].

Coastal management problems are widespread in the region and in some areas require urgent attention. The potential for sustainable development of coastal and marine resources is being permanently lost or compromised at the same time as their value to the rest of the world (*e.g.*, for biomedical purposes) is only beginning to be explored. Much of the degradation of coastal habitats and depletion of resources in the island Pacific could be avoided, reduced, or mitigated through integrated coastal management and planning (ICZM), *i.e.*, comprehensive, multi-sectoral, integrated planning and management for the sustainable development, multiple use, and conservation of coastal and marine environments [Holthus, 1991a]. ICZM is an essential approach for addressing the special environmental and development situation of small islands. The need for ICZM has been recognized globally, and at the regional and national level in the Pacific.

Need for Scientific Information on Marine Environments

Information on the biological, ecological, geological, and oceanographic characteristics and processes of marine environments is an essential component of ICZM. Unfortunately, in spite of the importance of coastal and marine environments to the region, this kind of information is usually lacking for most of the small islands of the island Pacific. Much research is needed to provide a sound scientific basis for the management of marine environments in the Pacific. Scientific research is often a slow and expensive process, and is more so in the island Pacific region due to the complexity of tropical marine environments and the scattered and isolated location of the islands. In addition, the economic realities facing the developing countries in the region mean very little of the needed research is being undertaken.

The marine environments of small islands in the Pacific will not be able to be managed with scientific rigor for some time and will have to rely on less than adequate information for many years to come at the same time as development pressures and impacts are increasing. However, a substantial amount of information important to ICZM, the management of marine resources, and sustainable island development can be readily obtained through several complementary approaches. These approaches, which are being applied in the Pacific and may have relevance to other small islands, are to:

1. Inventory ecosystems using a regional ecosystem classification, which means marine ecosystems can be evaluated, and conservation priorities set, on a systematic, comparable basis for any country, or for the region as a whole.

2. Undertake semi-quantitative field assessments of marine environments with multi-disciplinary teams covering a broad geographic range of coastal and nearshore areas, which rapidly provides extensive information for coastal management planning.
3. Systematically collect relevant traditional knowledge on marine environments, which immediately provides a wealth of detailed information important to ICZM, marine resources management, and sustainable development.

Each of these approaches is examined in more detail in the following sections.

Classification and Inventory of Marine Ecosystems

Ecosystem Classification for the Tropical Island Pacific

A regionally accepted framework for systematically identifying the presence of ecosystems, and inventorying their occurrence, is necessary as a basis for identifying areas which are a priority for conservation on a country, regional, or global scale. It has not been possible to determine the full range of ecosystems in each country or territory of the island Pacific, or in the region as a whole, without a complete ecosystem checklist based on a systematic classification. In spite of this, there have been efforts to inventory terrestrial biodiversity and recommend areas for conservation, but there has been no comparable effort for the vast majority of the tropical island Pacific--the marine ecosystems [Dahl, 1988].

In response to the need to inventory ecosystems on a systematic, regionally valid and comparable basis, the South Pacific Ecosystem Classification System (SPECS) was developed as a comprehensive, hierarchical, and scientific classification of the ecosystems of the tropical island Pacific region. Scientific and island country experts familiar with classifying natural systems and/or familiar with the ecosystems of the tropical island Pacific participated in workshops to develop the initial classification system, building on existing efforts. This process has resulted in: (1) a revised biogeography of the region [Stoddart, 1992]; (2) a freshwater ecosystem classification system [Polhemus et al., 1992]; and (3) a description of the development of both the terrestrial classification system and the marine classification system [Fosberg and Pearsall, 1993; Maragos, 1992].

The marine component of the SPECS combines biogeographical, morphological, and geophysical characteristics. The classification has a simplified biogeographic component, with only continental and/or oceanic subdivisions, following from Stoddart [1992]. The basic structure of the marine ecosystem classification is presented in Figure 2. At its higher levels, the classification system divides the marine environment into ocean and benthic components. The ocean area is subdivided into the open ocean (pelagic) and the nearshore ocean (neuritic), each of which has several water mass types. The bottom (benthic) area is subdivided into continental and non-continental (oceanic)

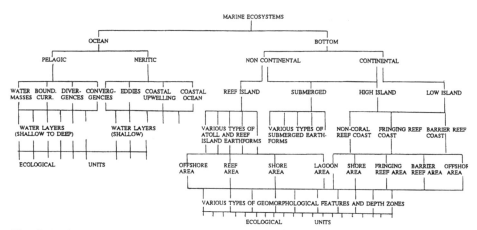

Fig. 2. Basic structure of the marine component of the South Pacific ecosystem classification system [Maragos, 1992].

classes. The next level divides islands into either high island and atoll/reef island earthform types. The oceanic area also has submerged earthform types.

Further levels of classification of marine ecosystems are outlined in Annex 1, which contains the full classification and continues to be refined. The lowest levels of classification are termed ecological units, which are not listed in the classification. A listing of examples of ecological units is provided in Table 1. Further work is required to fully develop the lowest levels of SPECS, as field surveys and ground truthing will generally be needed to identify dominant organisms in the ecological units. The marine ecosystem classification system developed for the tropical island Pacific may represent the most concerted effort to analyze and classify the marine environment of most of the world's small tropical islands. The classification may thus be applicable or adaptable, with some revision, to other small tropical island areas (*e.g.*, Caribbean Sea, Indian Ocean).

Preliminary Inventory of Pacific Island Ecosystems

It was anticipated that SPECS would be tested and refined through the preliminary inventory of ecosystems in a subset of Pacific island areas. Such a preliminary inventory was undertaken for the U.S. affiliated islands of the tropical Pacific: Republic of the Marshall Islands, Federated States of Micronesia (FSM), Territory of Guam, Commonwealth of the Northern Mariana Islands, Republic of Palau and Territory of American Samoa [Holthus *et al.*, 1993]. A total of 127 main islands or island clusters were evaluated by applying SPECS to an analysis of the available literature, maps, and aerial photographs for each of the islands. The number of principle islands in each country or territory, and their island type, is summarized in Table 2. The actual number of islands covered by the inventory was higher due to the lumping of island clusters (*e.g.*, the Rock

TABLE 1. Examples of Ecological Units to be Added to the Lowest Level of the Marine Component of the South Pacific Ecosystem Classification System [from Maragos, 1992].

Classification Element	Ecological	Dominant or Conspicuous
Open ocean, divergence, epipelagic	aggregations	*Thunnus albacares*
Open ocean, boundary current, epipelagic	Albacore schools	*Thunnus alalunga*
Nearshore ocean, coastal ocean, epipelagic	schools	*Katsuwonus pelamis*
Open ocean, water mass, epipelagic	aggregations	*Thunnus obesus*
Open ocean, coastal ocean, epipelagic	Eastern little tuna	*Euthynnuis affinis*
Bottom continental, high island, sand beach	community	*Ocypode sp.*
Bottom continental, high island, boulder beach	community	*Littorina sp., Nerita sp.,*
Bottom, non-continental, marine bench	community	*Echinometra mathaei*
Bottom, continental, lagoon fringing reef	Mangrove forest	*Brugiera gymnorhiza*
Bottom, non-continental, reef area	Microatoll corals	*Porites lutea*

Islands of Palau; the high and low islets surrounding Pohnpei Island) under a single island database record.

The type and level of information available for each island varied greatly, and for most of the islands information was very limited. The best information base available for an island was a combination of: (1) large scale maps (usually 1:25,000 scale); (2) recent low altitude, color air photos; and (3) a recent field survey report. This combination of sources was rarely available. For many islands, the only information source was a small scale hydrographic chart.

The following aspects of each island were systematically evaluated using SPECS during the ecosystem inventory:

1. Island shape and dimension.
2. Coastal ecosystems, including coastal type, coastline type, reef perimeter, shore area, reef islet area, and shoreline.
3. Marine ecosystems, including fringing, barrier and perimeter reef areas (*e.g.*, outer reef slope, reef top, and lagoon reef slope), and lagoon areas.
4. Terrestrial ecosystems, including native vegetation, non-native vegetation, and non-marine wetlands.

TABLE 2. Island Type and Numbers of Islands Inventoried [from Holthus et al., 1993].

	FSM					Marshall Islands	Palau	Guam	Northern Mariana Islands	American Samoa	Total
	FSM Total	Kosrae State	Pohnpei State	Chuuk State	Yap State						
High Islands											
Continental	0	0	0	0	0	0	2	0	0	0	2
Volcanic (basalt)	16	1	1[1]	14[2]	0	0	0	0	9	5	30
Raised Limestone	1	0	0	1	0	0	3[4]	0	3	0	7
Mixed	1	0	0	0	1[3]	0	2	1	2	0	6
Atoll/Reef Islands											
Atoll (many deep passes/open)	9	0	1	4	4	17	0	0	0	0	26
Atoll (few/one deep passes)	11	0	3	4	4[5]	6	2	0	1	1	20
Atoll (no deep passes)	13	0	5	3	5	6	1	0	0	0	20
Atoll (completely land ringed)	0	0	0	0	0	0	0	0	0	1	1
Reef island (with water body)	1	0	0	1	0	3	0	0	0	0	4
Reef island (without water body)	5	0	0	2	3	2	4	0	0	0	11
Total	57	1	10	29	17	34	14	1	14	7	

[1] Excludes high islets and barrier reef islets in Pohnpei lagoon.
[2] Only includes principle high islets in Chuuk Island lagoon.
[3] Includes all Yap Proper islands.
[4] The 504 individual Rock Islands not included.
[5] Includes Faurulep Atoll (type unknown).

The existing Island Directory database [Dahl, 1991] provided the framework for compiling and presenting the detailed ecosystem inventory information, and a Pacific ecosystem database record for each island was developed to hold the results of the preliminary inventory [Holthus *et al.*, 1993].

The inventory of marine ecosystems focused on intertidal and nearshore marine areas and was primarily based on the interpretation of maps and aerial photographs, as there were few field survey reports available on the marine ecosystems of most of the islands investigated. However, the few marine field survey reports which were available did provide "ground truth" information for the inventory of marine ecosystems on other islands. For a few islands, detailed, comprehensive information was available in coastal, shoreline, or coral reef atlases and survey reports or in island-specific summaries in the Coral Reef Directory [UNEP/IUCN, 1988]. Overall, the level of information available for each island was highly variable.

Interpretation and Application of Preliminary Ecosystem Inventory Results

Recognizing that conservation action is predominately undertaken at the country level, the preliminary inventory results were analyzed to indicate which islands within each country are more important for conservation (Table 3). Islands on which many kinds of ecosystem types were identified were considered to have a "rich" diversity of ecosystems for that country. Islands that supported ecosystems which had been identified on only one or a few islands in a country were considered to contain "rare" ecosystems.

Relative 'richness' or 'rareness' of ecosystems may provide an initial basis for determining the relative importance and interest of an island for conservation action or further investigations, although the determinations are likely to be biased towards those islands for which a more thorough and detailed data set was available. The results of the preliminary inventory reveal that there are generally more islands rich in shore, marine, and marine wetland ecosystems than there are islands "rich" in terrestrial and non-marine wetland ecosystems (Table 3). However, the number of islands in each country with "rare" shore, marine, and marine wetland ecosystems was variable in comparison to terrestrial ecosystems.

The adequacy of the information base used in this preliminary inventory was analyzed to indicate the validity of the results and to determine needs for further ecosystem inventory efforts. Those islands ranked "low" in information availability or adequacy were considered as a high priority for additional inventory efforts (Table 3).

Application of Ecosystem Classification and Inventory Information

The inventory of marine ecosystems using a regionally accepted classification system can identify representative, rare, and unique ecosystems at a variety of scales. The inter

TABLE 3. Number of Islands with Important Ecosystems and Number of Islands with "Low" Levels of Information

	FSM	Marshall Island	Palau	Guam	Northern Mariana	American Samoa
Islands rich in terrestrial and non-marine wetland eco-systems.	14	14	4	3	3	4
Islands with rare terrestrial and non-marine wetland eco-systems.	20	15	3	4	4	6
Islands rich in shore, marine, and marine wetland eco-systems.	27	20	13	7	7	3
Islands with rare shore, marine, and marine wetland ecosystems.	20	8	6	5	5	6
Islands "low" in information availability/adequacy.	42	10	*	11	11	2

*Information base is adequate due to a recent "rapid ecological assessment" field inventory [Maragos, et

pretation and application of inventory results may be undertaken on the scale of an island, a country, a region, or, if the classification system interfaces with a global ecosystem classification, on a global scale. The marine ecosystem classification system developed for the tropical island Pacific may be applicable to the small tropical islands of the Indian Ocean and Caribbean Sea.

The results of an inventory of marine ecosystems employing a hierarchical classification system will allow coastal and marine sites to be selected for a variety of purposes on a number of these scales. For example, for a given island, the inventory would provide the basis to select sites for long term marine ecosystem monitoring for local anthropogenic impacts. On a broader scale, the inventory could be used to select sites for long-term monitoring of impacts associated with global change as part of an international program. An inventory of marine ecosystems, especially using a system based on morphological features, can also provide valuable information to aid in determining the potential vulnerability of different shoreline areas to sea level rise and thereby aid in setting priorities for response planning [Holthus et al., 1992].

At the level of an individual island, an inventory of marine ecosystems would provide a valuable information base for many aspects of development planning, such as environment impact assessment (EIA) and oil spill contingency planning. The distribution of marine ecosystems is also an essential component of the information base needed for ICZM, particularly for the development of zoning and the identification of

protected areas.   The classification system potentially allows protected areas to be
selected as representative, rare, or unique on an island, national, regional, or global
scale. As part of the preliminary inventory of ecosystems in U.S. affiliated islands, a
range of criteria were identified which may be used to evaluate the selection of
ecosystem types and sites for conservation (Table 4) [Holthus *et al.*, 1993].

TABLE 4. Proposed Criteria for Selecting Ecosystem Types and Sites for Conservation [from Holthus *et al.*, 1993].

1. Biological Diversity (Richness)	4.Ecological Importance
a. Richness of Habitats	5.Naturalness
b. Richness of Taxa	6.Threats to the Ecosystem
c. Richness of Native Taxa	7.Ecosystem Fragility and Resilience
d. Richness/Fidelity of Endemic Taxa	8.Size
e. Richness of Rare or Threatened Taxa	9.Economic Importance
	10.Social and Cultural Importance
2. Ecosystems as Critical Habitat	11.Scientific Importance
a. Critical Habitat: Life Stages or Processes	12.Conservation Feasibility (Manageability)
b. Refugia	
3. Biogeographic Importance	
a. Rare Ecosystems	
b. Unique Ecosystem Aspects	
c. Representative Ecosystems	
d. Biogeographic Scale	

Rapid Ecological Assessment of Marine Environments

Rapid Ecological Assessment Needs and Methodology

The lack of information on the structure, components, characteristics, and status of
marine ecosystems around most of the tropical islands in the Pacific and the accelerating
pace of habitat degradation and destruction has created an urgent need for methods to
rapidly obtain and assemble this kind of information for management planning
[Holthus, 1991a]. As a result, methods for the rapid ecological assessment of marine
environments were developed in Hawaii and have been refined through application in
U.S. affiliated Pacific islands (American Samoa, parts of the FSM and the Marshall
Islands, Palau) [Maragos and Elliot, 1985].   This form of coastal resource survey
continues to be employed in these areas and has been used elsewhere in the region (*e.g.*,
parts of French Polynesia, Western Samoa and PNG).

The rapid ecological assessment of Pacific island marine environments usually begins
with a reconnaissance field evaluation of the nearshore marine area to be covered and a
review of maps, aerial photographs, and previous studies. Low-altitude color aerial
photographs, in conjunction with large-scale topographic and bathymetric maps, are an
excellent tool for pre-field work analysis of the area and for selecting field stations.

However, often there is little existing literature, only decades-old maps, and a few small-scale black and white aerial photographs.

A survey of the biophysical characteristics and resources of the area is undertaken by a multidisciplinary team of experts, which is usually composed of individuals with expertise in one or several of the following fields: corals, reef geomorphology, fish, marine turtles, algae, seagrass, and marine macroinvertebrates (*e.g.*, urchins, sea cucumbers, giant clams, starfish). Local, in-country experts form part of the team as much as possible and also participate in a training capacity. The survey is semi-quantitative in nature, to allow a large number of sites to rapidly be assessed. Field sites are selected to provide an inclusive sampling of the range of the ecosystems present and provide a wide geographic coverage of the area.

The underwater surveys are undertaken by either snorkeling or using scuba gear, depending on the depth of the area investigated. Each team member records information on underwater paper regarding the relative abundance of the group of organisms he/she is responsible for, employing a 1-5 point scale (*i.e.*, rare, occasional, common, abundant, dominant). Data on other characteristics or features (*e.g.*, percent bottom cover by benthic organisms, substrate type, geomorphology) is recorded in notes and sketches. The use of data sheets, pre-printed with species lists or other information, helps to speed up data recording underwater. Data collection is augmented by using underwater camera and video, which assist in the assessment of the ecological conditions and environment of each site visited.

Interpretation and Application of Information From Rapid Ecological Assessments

The information gathered in rapid ecological assessments is usually compiled into: (1) a summary report which describes the structure, components, characteristics, and status of the marine environment and resources for the area; and (2) an atlas which presents the features and characteristics which are readily mapped (*e.g.*, substrate type, seagrass beds, survey site location, *etc.*). These documents are often the only comprehensive report of the biological and physical characteristics of the marine environment around the islands which have been surveyed and have ready applications to conservation and management needs.

At the broadest scale, the ecological assessment information has been applied to economic development planning in some of the islands concerned. More often, the surveys have provided the information basis, and sometimes the impetus, for the development of coastal management plans, including zoning plans [Holthus, 1985; Dahl, 1989]. In particular, ecological assessment results can be used to identify candidate protected areas [Thomas *et al.*, 1989; Maragos *et al.*, 1994a, 1994b]. The results of these kinds of field surveys have been almost immediately put to use in supporting EIAs on many of the islands investigated. The documents resulting from the survey often contain the only environmental information on the nearshore marine environment of areas proposed for development projects. As EIA is not a mandated procedure in most of the island countries of the Pacific, the time and resources to undertake environmental investigations as part of development project planning are almost never available. However, rapid ecological assessments of marine environments usually provide enough information for resource management agencies to direct development projects away from sensitive, unique, critical, or otherwise important marine areas or propose mitigation measures to reduce impacts [Maragos *et al.*, 1983; Carpenter and Maragos, 1989].

Traditional Knowledge of Marine Environments

Gathering Traditional Knowledge Information

The development of an adequate information base on the type, location, and status of marine resource use is a necessary prerequisite to effective coastal management planning. Much of the harvesting of marine resources in the island Pacific is community-based subsistence and artisanal resource use, based on practices which have been employed for centuries. Pacific island marine resource users hold a store of traditional knowledge on a wide range of aspects of their marine environments, which is not restricted to resource harvesting. Unfortunately, Pacific islanders' rich knowledge of marine environments is rapidly being lost at the same time as this extensive information is of increasing value to resource managers in small island countries [Johannes, 1978].

The wealth of traditional knowledge held by Pacific islanders often includes detailed information on: (1) the distribution and abundance of marine resources, particularly food resources; (2) the types, levels, location, and timing of resource harvest efforts; (3) the life history and behavior of important coastal organisms, particularly reef fish and sea turtles; (4) the location and status of critical coastal habitats and threatened or endangered species; (5) physical oceanography (*e.g.*, current patterns and seasons); (6) important cultural features and aspects of marine resource use; and (7) former or existing traditional resource management practices [Johannes, 1981a, 1981b, 1981c]. Traditional marine ecosystem tenure and usage rights may be of particular interest to ICZM efforts [Hviding, 1991]. In addition, the age and experience of traditional knowledge holders often means they have personally witnessed the pattern and level of marine habitat destruction or degradation and overharvesting of marine organisms around their island. Further, they may have a unique perspective on potential ideas for marine resource management.

All of the above kinds of information are invaluable to the modern management of marine resources for sustainable island development. The research to obtain this kind of information would be expensive and time consuming. However, using the methods of ethno-biology, the traditional knowledge of indigenous experts can become available to the scientific community and resource managers. This methodology has been adapted to collect traditional knowledge relevant to coastal management planning in a timely manner [Holthus, 1991b]. In particular, when obtaining traditional knowledge for management applications, it is necessary to specifically seek out the geographic context of the information (*i.e.*, the location or distribution of features or activities) if the information is to be of maximum use. In this process, traditional information is obtained through village group interview meetings with knowledgeable fishermen and women, usually conducted in the local language with the assistance of a translator.

At the group interviews, participants are asked a series of questions on the location of various kinds of resources and resource uses, which they are asked to indicate on a map. The questions cover a range of topics. Knowledge on marine areas of special biological significance (*e.g.*, fish spawning aggregation sites, fish migration paths, turtle nesting, and feeding areas) is particularly important information which is often otherwise unavailable without expensive research. Because the harvest of marine resources is such a large and diverse activity, it is necessary to subdivide the discussion on fishing and other resource harvesting activities. Information on the location, timing (*e.g.*, lunar period, season), and status of each kind of fisheries activity is usually specified for each different group of fish or marine invertebrates harvested and/or by the harvesting methods used. Cultural significant areas (*e.g.*, archeological sites, areas of cultural value) are also an important topic of knowledge discussed during the interviews.

Interpretation and Application of Traditional Knowledge

The information gathered in the structured interview meetings is usually compiled into: (1) a summary report which describes the types and status of resources and the types and methods of resource uses; and (2) an atlas which presents the geographic distribution of the features, resources, and resource uses which are readily mapped (*e.g.*, fish spawning sites and migration paths, turtle nesting beaches, fishing areas, cultural sites). The report and atlas often provide the only systematic assemblage of traditional knowledge on the marine environments of that island which is easily available for conservation and resource management planning. These documents are usually combined with those of the rapid ecological assessment to present a comprehensive package of information on the marine environments of the island.

Relevant traditional knowledge which has been gathered and presented in this manner is applied to management planning for sustainable development in very much the same way as described for rapid ecological assessment. However, it is important to note that the traditional knowledge which has been gathered often contains information on: (1) biological processes (*e.g.*, fish spawning sites, turtle nesting beaches) critical to the local survival of important organisms; and (2) subsistence fishery activities critical to the health and welfare of rural or marginalized portions of the population which are dependent upon these activities and have few options. These kinds of information are particularly important to planning for the conservation and sustainable development of marine environments of small islands.

Conclusions and Recommendations

The inventorying of marine ecosystems using a regional classification system, the undertaking of rapid ecological assessments of nearshore marine areas, and the collecting of traditional knowledge of marine environments from indigenous experts are three approaches being successfully employed in the island Pacific to obtain marine science information in a timely and cost effective manner. These approaches, often used in concert, are providing a substantial amount of information for ICZM in support of the sustainable development of the small islands of the Pacific. However, the marine environments around most of the islands remain largely unknown, while the pressures and impacts of development and resource use continue to increase and expand. There is much that can be done to expand the application of the approaches described here to further support the development and implementation of ICZM in small islands of the tropical Pacific and elsewhere.

Specific recommendations are as follows:
1. Further the application of the South Pacific Ecosystem Classification System (SPECS). There are several areas in which the work of developing and applying SPECS to the islands of the tropical Pacific needs to be continued.
   a. Finalize the marine component of the classification system through additional efforts to develop the detailed ecological unit level of classification, complete a final working version of the marine classification for publication, and make it available to scientists and resource managers.
   b. Develop a handbook or field manual version of the SPECS, complete with diagrams and photographs of ecosystem types, which could be used for training and educational purposes and field use, particularly by in-country environment and natural resource agency officers.
   c. Conduct a preliminary inventory of ecosystems (using literature, maps, and aerial photography) for the rest of the island Pacific region as a basis for determining conservation priorities in these areas and at a regional level.
   d. Undertake more detailed field inventories using the classification system for those islands and areas which are determined to be a high priority for conservation (e.g., inventory existing MPAs to determine which ecosystems are already under some form of protection (ostensibly) at a national and regional scale).
   e. Further develop and refine the application of the ecosystem inventory and classification as a means to indicate the vulnerability of various shoreline types to sea level rise (e.g., determine the relative vulnerability of the various combinations of atoll reef and shoreline geomorphological characteristics).
2. Apply SPECS to other areas and issues. The potential application of the marine component of SPECS to other small tropical island areas should be explored and the interface of SPECS with global marine ecosystem classifications should be developed to allow information generated by its application in the Pacific to contribute to global programs.

a. Conduct a workshop on marine ecosystem classification in the Indian Ocean and the Caribbean Sea to evaluate and adapt the classification system developed for the Pacific to the marine environments of those areas.

b. Test the applicability of the marine classification system developed for the tropical island Pacific by conducting pilot preliminary inventories of islands in the Indian Ocean and Caribbean Sea.

c. Clarify the interface of the marine component of SPECS with global classifications of marine environments used by agencies with global mandate for the marine environment (*e.g.,* UNESCO/IOC, UNEP/OCA/PAC, GEF).

d. Use the marine component of SPECS to aid in systematically identifying sites in the island Pacific region for global projects (*e.g.,* the long-term global monitoring of marine environments for the effects of climate change).

3. Expand the use of rapid ecological assessments of marine environments. The rapid evaluation of nearshore marine areas should be continued to provide baseline information for coastal management planning.

a. Undertake rapid ecological assessments of the marine environments adjacent to all main urban centers and around all "primary" islands for each country (*i.e.,* the island which is the center of government, industry, commerce, population, and development for the country).

b. Undertake rapid ecological assessments of the marine environments adjacent to all proposed major developments (*e.g.,* resort hotels, airports, ports, coastal roads, coastal industry, or agriculture) as a support for EIA.

c. Develop in-country capabilities to undertake the rapid ecological assessment of marine environments through a program to train appropriate individuals in marine resource and environment agencies.

4. Develop a program for the collection of traditional knowledge of marine environments. The wealth of traditional knowledge of marine environments held by indigenous scientists is rapidly being lost and cannot be retrieved at a later time. A program to collect, compile, and make use of this store of information invaluable to the sustainable use of marine environments is a high priority for the island Pacific region.

a. Develop a program to systematically seek out traditional knowledge of marine environments on islands where development activities and social/cultural change are most pronounced, as this is where the information is most needed and is most rapidly disappearing.

b. Systematically seek out traditional knowledge of marine environments on islands where development activities and social/cultural change have had the least impact, as this is where the information is most extensive and intact.

c. Conduct a regional workshop to stimulate and coordinate efforts to collect and compile traditional knowledge of marine environments and develop further approaches for the application of this information to ICZM.

d. Develop capabilities to compile traditional marine knowledge in each country and make the information available for education and for application to ICZM.

References

Brodie, J., C. Arnould, L. Eldredge, L. Hammond, P. Holthus, D. Mowbray, and P. Tortell, Review of the state of the marine environment: South Pacific Action Plan Regional Report, UNEP Regional Seas Reports and Studies, No. 127, UNEP, 59 pp., 1990.

Carpenter, R. A., and J. E. Maragos (Eds.), *How to Assess Environmental Impacts on Tropical Islands and Coastal Areas*, Training Manual, South Pacific Regional Environment Program, 345 pp., 1989.

Connell, J., Islands under pressure: Population growth and urbanization in the South Pacific, *Ambio, 13*(5-6), 306-312, 1984.

Dahl, A. L., Regional ecosystems survey of the South Pacific area, SPC Tech. Paper 179, South Pacific Commission, 99 pp., 1980.

Dahl, A. L., Oceania's most pressing environmental concerns, *Ambio, 13*(5-6), 296-301, 1984.

Dahl, A. L., *Review of the Protected Areas System in Oceania*, IUCN, 239 pp., 1988.

Dahl, A. L., Island directory, UNEP Regional Seas Directories and Bibliographies, No. 35, UNEP, 573 pp., 1991.

Dahl, C., Developing a coastal resource management plan for Kosrae State, FSM, pp. 40-47, Proc., 4th Conf. on Parks and Protected Areas in the South Pacific Region, Vol. II, edited by P. E. J. Thomas, 146 pp., 1989.

Fosberg, F. R., and S. H. Pearsall, III, Classification of non-marine ecosystems, *Atoll Res. Bull., 389*, 38 pp., 1993.

Holthus, P. F., Reef resource conservation and management planning for Pohnpei Island (Caroline Archipelago, Micronesia), Proceedings, 5th International Coral Reef Congress, 4, 231-236, 1985.

Holthus, P. F., Strategies for improving coastal resource management in the South Pacific region, in *Regional Cooperation on Environmental Protection of the Marine and Coastal Areas of the Pacific Basin*, edited by J. Pernetta, pp. 91-96, UNEP Regional Seas, Reports and Studies, No. 134, UNEP, 196 pp., 1991a.

Holthus, P. F., Coastal resource use in Pacific Islands: Gathering information for planning from traditional resource users, *Trop. Coast. Resource Manage., 5*(1-2), 11-14, 1991b.

Holthus, P. F., and J. E. Maragos, Marine biological diversity and marine protected areas in the South Pacific region: Status and prospects, Proc., 4th World Parks Congress and GEF/World Bank Workshop on Marine Biodiversity Conservation (Caracas, Venezuela, February 1992), 1992.

Holthus, P., M. Crawford, C. Makroro, E. Nakasaki, and S. Sullivan, Case study on vulnerability to sea level rise: Majuro Atoll, Marshall Islands, SPREP Reports and Studies Series, No. 60, South Pacific Regional Environment Program, 107 pp., 1992.

Holthus, P., P. Brennan, S. Gon, L. Honigman, and J. Maragos, Preliminary classification and inventory of ecosystems of U.S. affiliated islands of the tropical Pacific, Report, U.S. Fish and Wildlife Service and The Nature Conservancy Pacific Program (draft), 1993.

Hviding, E., Traditional institutions and their role in contemporary coastal resource management in the Pacific Islands, *NAGA, ICLARM Quarterly, 14*(4), 3-6, 1991.

Johannes, R. E., Traditional marine conservation methods in Oceania and their demise, *Ann. Rev. Ecol. System., 9*, 349-364, 1978.

Johannes, R. E., *Words of the Lagoon: Fishing and Marine Lore in the Palau District of Micronesia*, 245 pp., Univ. of California Press, Berkeley, 1981a.

Johannes, R. E., Working with fishermen to improve coastal tropical fisheries and resource management, *Bull. Mar. Sci., 31*(3), 673-680, 1981b.

Johannes, R. E., Making better use of existing knowledge in managing Pacific island reef and lagoon ecosystems, SPREP Topic Review No. 4, South Pacific Regional Environment Program, 10 pp., 1981c.

Maragos, J. E., A marine ecosystem classification system for the South Pacific region, in *Coastal Resources and Systems of the Pacific Basin: Investigations and Steps Toward Protective Management*, pp. 253-299, UNEP Regional Seas, Reports and Studies, No. 147, UNEP, 308 pp., 1992.

Maragos, J. E., and M. E. Elliot, Coastal resource inventories in Hawaii, Samoa, and Micronesia, *Proc., 5th International Coral Reef Congress, 5*, 577-582, 1985.

Maragos, J. E., A. Soegiarto, E. D. Gomez, and M. A. Dow, Development planning for tropical coastal ecosystems, in *Natural Systems for Development: What Planners Need to Know*, edited by R. A. Carpenter, pp. 229-298, MacMillan, New York, 485 pp., 1983.

Maragos, J. E., A. K. Kepler, R. L. Hunter Anderson, T. J. Donaldson, S. H. Geermans, K. J. McDermid, N. Idechony, S. Patris, C. Cook, B. Smith, R. Smith, and K. Z. Meier, Synthesis Report: Rapid Ecological Assessment of Palau: Part I, June 1992, Natural and Cultural Resource Survey of the Southeast Islands of Palau, Prepared by CORIAL and the Nature Conservancy--Asia/Pacific Region for the Bureau of Resources and Development, Republic of Palau, 62 pp., 1994a.

Maragos, J. E., C. Birkeland, C. Cook, K. DesRochers, R. D. Rosa, T. J. Donaldson, S. H. Geermans, M. Guilbeaux, H. Hirsch, L. Honigman, N. Idechong, P. S. Lobel, E. Matthews, K. J. McDermid, K. Z. Meier, R. Meyers, D. Otobed, R. H. Richmond, B. Smith, and R. Smith, Marine and Coastal Areas Survey of the Main Palau Islands: Part I, Rapid Ecological Assessment Synthesis Report, Prepared by CORIAL and The Nature Conservancy - Asia/Pacific Region for The Department of Resources and Development, Republic of Palau, 124 pp., 1994b.

Polhemus, D. A., J. Maciolek, and J. Ford, An ecosystem classification system of inland waters for the tropical Pacific Islands, *Micronesia, 25*(2), 155-173, 1992.

Stoddart, D. R., Biogeography of the tropical Pacific, *Pacific Sci., 46*(2), 276-293, 1992.

Thomas, P. E. J., F. R. Fosberg, L. S. Hamilton, D. R. Herbst, J. O. Juvik, J. E. Maragos, J. J. Naughton, and C. F. Streck, Report of the northern Marshall Islands natural diversity and protected areas survey, SPREP Report, South Pacific Regional Environment Program, 133 pp., 1989.

Thomas, P., P. Holthus, and I. Reti, Pacific regional review of national parks and protected areas, in *Chapter 11: Regional Reviews*, Proc., 4th World Parks Congress (Caracas, Venezuela, February 1992), 1992.

Thomas, W. L., Jr., The variety of physical environments among Pacific Islands, in *Man's Place in the Island Ecosystem*, edited by F. R. Fosberg, pp. 7-37, Bishop Museum Press, Honolulu, 485 pp., 1963.

UNEP/IUCN, Coral reefs of the world, Vol. 3: Central and western Pacific, UNEP Regional Seas Directories and Bibliographies, IUCN/UNEP, 329 pp., 1988.

Annex 1: Marine Ecosystems of the Tropical Insular Pacific

A. MARINE ECOSYSTEMS
A.A. OCEAN
A.A.A. OCEAN-OPEN (PELAGIC)
1. Divergences (upwelling zones)

1.1 Surface Microlayer
1.2 Epipelagic (0-200 m)
   1.2.1 Shallow
      1.2.1.1 Surface Waters
      1.2.1.2 Mixed Waters
      1.2.1.3 Thermocline Waters
   1.2.2 Deep
      {Light Level: Photic, Aphotic}
1.3 Mesopelagic (200-1000 m)
1.4 Bathypelagic (1000-4000 m)
1.5 Abyssopelagic (4000-7000 m)
1.6 Hadalpelagic (>7000 m)
2. Convergence (downwelling zone)
   (as in 1.)
3. Boundary Current
   (as in 1.)
4. Water Mass
   (as in 1.)
A.A.B. OCEAN-NEARSHORE (NERITIC)
1. Eddy
   1.1 Surface Microlayer
   1.2 Epipelagic
      (as in 1.2)
2. Upwelling Zone - Coastal
   (as in 1.)
3. Ocean - Coastal
   (as in 1.)
4. Convergence
   (as in 1.)
A.B. BOTTOM (BENTHIC)
A.B.A. CONTINENTAL SHELF (NON-OCEANIC)
1. Earthform - High Island
   {Island Geology: Continental; Volcanic; Limestone; Mixed}
   1.1 Coast - Non-Coral Reef
      1.1.1 Area - Shore
         1.1.1.1 Coastline - Undifferentiated (scale: 10 km)
            1.1.1.1.1 Shoreline - Sediment
               1.1.1.1.1.1 Beach - Boulder/Cobble
               1.1.1.1.1.2 Beach - Sand/Gravel
               1.1.1.1.1.3 Beachrock
               1.1.1.1.1.4 Boulder/Cobble Field
               1.1.1.1.1.5 Sandflat
               1.1.1.1.1.6 Mudflat
               1.1.1.1.1.7 Bar and Spit
               1.1.1.1.1.8 Mangrove

        1.1.1.1.2   Shoreline - Solid Substrate
         1.1.1.1.2.1 Cliff - High (ht >10 m)
         1.1.1.1.2.2 Cliff - Medium (ht 2-10 m)
         1.1.1.1.2.3 Cliff - Low (ht <2 m)
         1.1.1.1.2.4 Sea Stacks
         1.1.1.1.2.5 Talus
         1.1.1.1.2.6 Bench/Ramp - Marine
         1.1.1.1.2.7 Notch/Cave - Marine
        1.1.1.1.3   Shoreline - Artificial
         1.1.1.1.3.1 Seawall/Revetment/Bulkhead
         1.1.1.1.3.2 Landfill/Causeway/Groin
         1.1.1.1.3.3 Fishpond/Fishtrap/Shipwreck
      1.1.1.2 Coastline - Cove (scale: 10 km)
         (as in 1.1.1.1)
      1.1.1.3 Coastline - Bay (scale: 10 km)
         (as in 1.1.1.1)
         {Salinity: Marine; Estuarine}
      1.1.1.4 Coastliine - Coastal Lagoon/Lake/Pond (scale: 10 km)
         (as in 1.1.1.1)
         {Connectedness:
          Subtidal connection/subtidal lagoon;
          Intertidal connection/subtidal lagoon;
          Intertidal lagoon (Barachois)}
         {Salinity: Marine; Estuarine}
      1.1.1.5 Coastline - Peninsula (scale: 10 km)
         (as in 1.1.1.1)
      1.1.1.6 Coastline - Irregular/Discontinuous/Islets
         (as in 1.1.1.1)
    1.1.2 Area - Nearshore Bottom
      {Steepness/Slope gradient}
      1.1.2.1 High Islet
         (as in 1.1.1.1)
1.2 Coast - Fringing Reef
  1.2.1 Area - Shore
      (as in 1.1.1)
  1.2.2 Area - Fringing Reef
      {Orientation: windward, leeward}
    1.2.2.1 Reef Top
      {Reef Top Width}
      1.2.2.1.1   Reef Flat Surface Features
      1.2.2.1.1.1 Reef Pavement
      1.2.2.1.1.2 Sand/Rubble/Reef Rock
      1.2.2.1.1.3 Mud/Silt Flats
      1.2.2.1.1.4 Sand/Gravel Sheet or Flats
      1.2.2.1.1.5 Cobble/Boulder Field

1.2.2.1.1.6 Rubble/Boulder Tract
1.2.2.1.1.7 Coral - Scattered/Microatolls
1.2.2.1.1.8 Algal Bed
1.2.2.1.1.9 Seagrass Bed
1.2.2.1.1.10 Algal Ridge
1.2.2.1.1.11 Surge Channel
1.2.2.1.2   Subtidal Features
1.2.2.1.2.1 Hoa (inter-islet Channel)
1.2.2.1.2.2 Moat and Depression
1.2.2.1.2.3 Reef Pool(depth <5 m)
1.2.2.1.2.4 Reef Hole(depth >5 m)
1.2.2.1.2.5 Submerged/Incomplete Reef Top
1.2.2.1.2.6 Dredge Pit/Quarry/Channel/Basin
1.2.2.1.3   Supratidal Features
1.2.2.1.3.1 Storm Block
1.2.2.1.3.2 Gravel/Boulder Ridge or Rampart
1.2.2.1.3.3 Beachrock
1.2.2.1.3.4 Conglomerate/Reef Limestone Platform
1.2.2.1.3.5 Aeolianite
1.2.2.1.3.6 Coral/Algal Dam and Spillway
1.2.2.1.3.7 Mangrove
1.2.2.1.3.8 Islet
1.2.2.1.3.9 Fishpond/Fishtrap/Shipwreck
1.2.2.1.4   Passes/Perimeter Openings
               {No. Passes} {Depth/Width}
               {Amount of Perimeter} {% of Perimeter}
1.2.2.1.4.1 Pass - Shallow (depth <10 m)
1.2.2.1.4.2 Pass - Deep (depth >10 m)
1.2.2.1.4.3 Submerged Perimeter (depth <10 m; width >2 km)
1.2.2.1.4.4 Open Perimeter (depth > 10 m; width >2 km)
1.2.2.1.4.5 Pass - False or Channel
1.2.2.2   Reef Slope - Outer
1.2.2.2.1   Reef Slope with Terrace
1.2.2.2.1.1 Groove and Spur
1.2.2.2.1.2 Tunnel (room and pillar)
1.2.2.2.1.3 Platform - Submarine
               {Furrows: with; without}
1.2.2.2.1.4 Cliff - Submarine
1.2.2.2.1.5 Notch/Cave - Submarine
1.2.2.2.1.6 Boulder Detritus
1.2.2.2.2   Reef Slope without Terrace
1.2.2.2.2.1 Buttress and Valley
1.2.2.2.1.2 Cliff - Submarine
1.2.2.2.1.3 Notch/Cave - Submarine
1.2.2.2.1.4 Boulder Detritus

1.2.2.3 Reef Islet Coastline
          (as in A.B.B.2.1.1.2 and 2.1.1.3)
    1.2.3 Area - Nearshore Bottom
          (as in 1.1.2)
1.3 Coast - Barrier Reef
    1.3.1 Area - Shore
          (as in 1.1.1)
    1.3.2 Area - Lagoon Fringing Reef
          (as in 1.2.2)
    1.3.3 Area - Lagoon
          {Area} {Depth}
          {Special Aspects: sub-lagoon(s); perched}
        1.3.3.1 Lagoon Reef
          1.3.3.1.1 Patch Reef/Pinnacle
                    {No. of Patch Reefs/Pinnacles}
          1.3.3.1.2 Patch Reef with Islet
          1.3.3.1.3 Reticulate Reef
        1.3.3.2 Reef Islet Coastline
                (as in 1.2.2.3)
        1.3.3.3 Lagoon Floor
          1.3.3.3.1 Shallow Floor (<10 m deep)
          1.3.3.3.2 Algal (Halimeda) Mound
          1.3.3.3.3 Sediment Sink - Deep Floor
        1.3.3.5 Lagoon High Islet
                (as in 1.1.1.1, 1.2.2.1-1.2.2.2)
    1.3.4 Area - Barrier Reef
          {Orientation: windward, leeward}
        1.3.4.1 Reef Top
                (as in 1.2.2.1)
        1.3.4.2 Reef Slope - Outer
                (as in 1.2.2.2)
        1.3.4.3 Reef Slope - Lagoon
          1.3.4.3.1 Reef Slope w/Terrace
            1.3.4.3.1.1 Terrace - Sediment
            1.3.4.3.1.2 Terrace - Sediment with Coral Mounds
            1.3.4.3.1.3 Terrace - Reef
            1.3.4.3.1.4 Terrace - Mixed Sediment/Reef
          1.3.4.3.2 Reef Slope without Terrace
            1.3.4.3.2.1 Reef Slope - Sediment
            1.3.4.3.2.2 Reef Slope - Mixed Sediment/Reef
        1.3.4.4 Reef Islet Coastline
                (as in 1.2.2.3)
    1.3.5 Area - Nearshore Bottom
          (as in 1.1.2)

2. Earthform - Atoll/Reef Island (height <10 m)
    (as in A.B.B.2.)
3. Earthform - Submerged
    3.1 Reef/Shoal - Nearshore (including fringing reef)
    3.2 Reef/Shoal/Bank - Mid-Shelf
    3.3 Reef/Shoal/Bank - Outer Reef (including outer barrier reef)
    3.4 Plain - Nearshore
    3.5 Plain - Offshore
    3.6 Canyon
    3.7 Slope - Continental

A.B.B. OCEANIC (NON-CONTINENTAL)
1. Earthform - High Island
    1.1 Coast - Non-Coral Reef
        1.1.1 Area - Shore
            (as in A.B.A.1.1.1)
        1.1.2 Area - Nearshore Bottom
            (as in A.B.A.1.1.2)
        1.1.4 Area - Deep Bottom
            1.1.4.1 Bathyal (200-4000 m)
            1.1.4.2. Abyssal (4000-7000 m)
            1.1.4.3 Hadal (>7000 m)
    1.2 Coast - Fringing Reef
        1.2.1 Area - Shore
            (as in A.B.A.1.1.1)
        1.2.2 Area - Fringing Reef
            (as in A.B.A.1.2.2)
        1.2.3 Area - Nearshore Bottom
            (as in A.B.A.1.1.2)
        1.2.4 Area - Deep Bottom
            (as in 1.1.4)
    1.3 Coast - Barrier Reef
        1.3.1 Area - Shore
            (as in A.B.A.1.1.1)
        1.3.2 Area - Lagoon Fringing Reef
            (as in A.B.A.1.2.2)
        1.3.3 Area - Lagoon
            (as in A.B.A.1.3.3)
        1.3.4 Area - Barrier Reef
            (as in A.B.A.1.3.4)
        1.3.5 Area - Nearshore Bottom
            (as in A.B.A.1.1.2)
        1.3.6 Area - Deep Bottom
            (as in 1.1.4)

2. Earthform - Atoll/Reef Island
  2.1 Atoll - Many Deep Passes/Open
      {Perimeter Length}
    2.1.1 Area - Reef Islet
      2.1.1.1 Coastline - Reef Islet
            {Linear Ocean Shore} {% of Reef Perimeter}
            {Approx. No. of Reef Islets}
            {Reef Islet Barachois/Anchialine Pond: with; without}
        2.1.1.1.1 Reef Islet - Large
        2.1.1.1.2 Reef Islet - Long/Narrow
        2.1.1.1.3 Reef Islet - Series
        2.1.1.1.4 Reef Islet - Isolated
      2.1.1.2 Coastline - Outer/Ocean
        2.1.1.2.1 Shoreline - Sediment
        2.1.1.2.1.1 Beach - Boulder/Cobble
        2.1.1.2.1.2 Beach - Sand/Gravel
        2.1.1.2.1.3 Beachrock
        2.1.1.2.1.4 Boulder/Cobble Field
        2.1.1.2.1.5 Sandflat
        2.1.1.2.1.6 Bar and Spit
        2.1.1.2.1.7 Mangrove
        2.1.1.2.2 Shoreline - Solid Substrate
        2.1.1.2.2.1 Cliff - Low (ht <2 m)
        2.1.1.2.2.2 Ramp
        2.1.1.2.2.3 Conglomerate/Reef Limestone Bench
        2.1.1.2.3 Shoreline - Artificial Features
                (as in A.B.A.1.1.1.1.3)
      2.1.1.3 Coastline - Inner/Lagoon
        2.1.1.3.1 Shoreline - Sediment
        2.1.1.3.1.1 Beach - Boulder/Cobble
        2.1.1.3.1.2 Beach - Sand/Gravel
        2.1.1.3.1.3 Beachrock
        2.1.1.3.1.4 Boulder/Cobble Field
        2.1.1.3.1.5 Sandflat
        2.1.1.3.1.6 Mudflat
        2.1.1.3.1.7 Bar and Spit
        2.1.1.3.1.8 Mangrove
        2.1.1.3.2 Shoreline - Solid Substrate
                (as in 2.1.1.2.2)
        2.1.1.3.3 Shoreline - Artificial
                (as in A.B.A.1.1.1.1.3)
      2.1.1.4 Coastline - Coastal Lagoon/Lake/Pond
              (as in A.B.A.2.1.1.4)
    2.1.2 Area - Lagoon
       (as in A.B.A.1.3.3; excluding 1.3.3.5)

2.1.3 Area - Perimeter Reef
(as in A.B.A.1.3.4)
2.1.4 Area - Nearshore Bottom
(as in A.B.A.1.1.2; excluding 1.1.2.1)
2.1.5 Area - Deep Bottom
(as in 1.1.4)
2.2 Atoll - Few/One Deep Pass(es) (depth >5 m)
(as in 2.1.)
2.3 Atoll - No Deep Pass
(as in 2.1.)
2.4 Atoll - Completely Land-ringed
(as in 2.1.)
2.5 Reef Island - with Water Body
(as in 2.1.)
2.6 Reef Island - without Water Body
2.6.1 Area - Reef Islet Shore
(as in 2.1.1; excluding 2.1.1.3)
2.6.2 Area - Fringing Reef
(as in A.B.A.1.2.2)
2.6.3 Area - Deep Bottom
(as in 1.1.4)

3. Earthform - Submerged
3.1 Atoll-Reef (upper surface depth <20 m)
3.1.1 Near Surface (<200 m)
3.1.2 Bathyal (200-4000 m)
3.1.3 Abyssal (4000-7000 m)
3.1.4 Hadal (>7000 m)
3.2 Table-Reef (depth <20 m)
(as in 3.1)
3.3 Shoal (depth <20 m)
(as in 3.1)
3.4 Bank (depth 20-200 m)
(as in 3.1)
3.5 Seamount (depth >200 m)
3.6 Guyot (depth >200 m)
3.7 Ridge
3.8 Plain
3.9 Trench
3.10 Fracture
3.11 Submarine Volcano
3.12 Geothermal Vent

22

Coastal Management, Oceanography, and Sustainability of Small Island Developing States

Paul H. Templet

Abstract

New emphasis by the United Nations on sustainable development as a result of UNCED has led to an increased need for developing nations to implement environmental planning and management programs. An important management approach for coastal areas, and by inference island states, is that of integrated coastal management (ICM) which seeks to resolve conflicts among coastal uses, encourage appropriate coastal development, and discourage development which is detrimental to coastal resources, *i.e.* to reduce man-made risk. As islands and countries develop, they undergo a development transition in which man-made risks initially increase. The primary role of environmental management is to minimize risk to people and ecosystems, our life support systems, and thus enhance sustainability. The use of science, and oceanography in particular, in developing management programs for island states reduces risk and promotes sustainability.

The American Samoa Coastal Program is presented as one model of a management program which may have application to developing island nations. A cultural-based model which relies heavily on science stands the best chance of being useful to developing nations, and recommendations are made on ways to develop such programs. Lessons learned from the development process indicate that spatial presentation of technical data, among others, is helpful in gaining acceptance by decision makers. The role of science and oceanography in ICM program development is discussed, and suggestions on making both more relevant to coastal planners are presented.

Small Islands: Marine Science and Sustainable Development
Coastal and Estuarine Studies, Volume 51, Pages 366–384
Copyright 1996 by the American Geophysical Union

Introduction

One outcome of the United Nations Conference on Environment and Development (UNCED) in Rio was the recommendation that integrated coastal management (ICM) be undertaken by nations as a means of promoting sustainable development. This recommendation is especially pertinent to small island states since their coastline/area ratios are the world's highest and they tend to be very dependent on their coastal areas for economic development potential and environmental amenities. Subsequent actions by the UN have strengthened the concept of coastal planning and management as a major tool in achieving sustainable development. Integrated coastal management refers to the combined or coordinated management of two or more coastal sectors (*e.g.*, fisheries, ports, tourism, *etc.*) within a defined region or zone [Sorensen *et al.*, 1984]. ICM is typically concerned with resolving conflicts among many coastal uses, determining the most appropriate use of coastal resources, and discouraging development which may be damaging to coastal resources and hence reduce sustainability. "The focus on resources management reflects the growing awareness among developing nations that renewable natural resources constitute foundation blocks needed to support the construction of economic and social development programs" [Sorensen *et al.*, 1984]. Thus, sustainable development, *i.e.*, development which contributes to the present generations' welfare without detracting from the welfare of future generations, can be enhanced with ICM.

The purpose of this chapter is to investigate the role of the ICM within the larger context of risk reduction necessitated by the stress of economic development on ecosystems. To accomplish this purpose a case study of the development of the American Samoa Coastal Management Program [ASCMP, 1980] is presented along with a discussion of the role of oceanography in coastal planning and linkages between planning, management, and sustainability which may be useful to small islands. The author developed an approved coastal management program and plan which was subsequently implemented in American Samoa [Templet, 1986]. Some of the lessons learned there may have application in other islands as they seek to plan and manage their coastal areas in an integrated manner.

The Need for Planning and Management

The Environment and Economic Sustainability

The economy, which is defined here as any activity designed to increase wealth or well being, is a system nested in the larger environment and is dependent on the environment as a source of resources and as a place to dispose of the inevitable waste (Figure 1). Thus, the environment provides these essential services which the economy must have to survive. For example, the environment provides "free" services to development in the

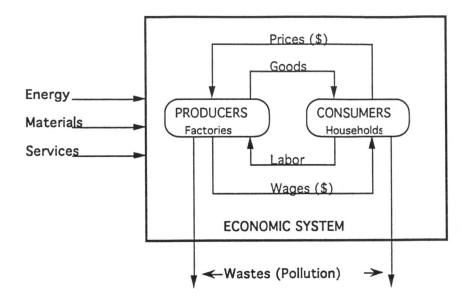

ENVIRONMENT

Fig. 1. The relationship of economy to environment.

form of clean water and air, timber, fish, waste treatment, and others. If the environment is degraded by the development, its ability to provide those services may be impaired and the development itself may suffer. Poorly planned tourist development has often led to polluted water, over crowded beaches, reef exploitation and decline, and a reduction in fisheries. But these were the very amenities which the tourists were seeking in the first place. Over time the quality of the development will decline, and with it the economic benefits. Planning and management can help prevent the economic decline by maintaining and enhancing the environment so that it continues to contribute to economic development over the long term.

One principle concern of developing countries is job creation and employment opportunities. Thus, it is helpful to examine the relationship between jobs and a clean environment. Job creation and economic development are frequently viewed as issues independent of, and even contradictory to, environmental and resource conservation. In reality, economic systems function within a larger ecosystem context. Economic development will be forced to be consistent with the nature of a region's ecosystem over the long-term unless ecosystem services can be imported. The direction, speed, and scale of development will ultimately be constrained by ecosystems. Rapid mining of ores in rainy, mountainous regions creates immediate jobs, but mine pollution will inhibit future replacement industries, such as silvaculture, recreation, or fishing. Industrial discharges in excess of the assimilative capacity of the ecosystem inhibit the economic potential of a region as the ecosystem base is destroyed. This effect is particularly important to small islands.

The traditional view is that enhancements to environmental conditions are costly to economic development. In other words, environmental and economic risks are inversely related. This perspective denies a more plausible relation between ecosystems and economies. Rather, the loss of an ecosystem base, or increased environmental risk and reduced carrying capacity of the ecosystem, reduces the long-term economic welfare secured from the ecosystem. Hence, environmental and economic risks are complementary.

The Bank of America [1993] has explained the mechanisms for the complementary relationship and proposed three ways in which the environment contributes to the economy.

1. The quality of life is improved. Providing an improved environment is similar to providing any consumer good or service: it fulfills the needs of the citizens and improves the quality of life.
2. Resources are better managed. Environmental regulations allow for the side effects of using resources. For example, if a factory pollutes a river, and water quality is not enforced, the water in the river may be unsuitable for use by other factories and cities downstream. Regulating pollution enhances overall economic efficiency and productivity.
3. Long run growth is maintained. Environmental protection will sustain long-term economic growth. Short-term exploitation of resources can result in unsustainable growth since current market prices do not take future resource limitations into account. Only specific environmental protection can insure resources will be available for sustainable economic growth.

Development Transitions

It is generally accepted that nations undergo transitions when they move from undeveloped through developing to developed status. For example, the demographic transition [Miller, 1993] explains that an undeveloped nation has high death and birth rates and thus the population is stable. As development occurs the death rate declines due to improved sanitation and health care and population increases due to the differences in the continued high birth rate and the lower death rate. This is the situation in the developing nations. As the country passes into the developed category, the birth rate also declines and the population stabilizes once more, although at a higher level. The economic improvements secured through development are retained and are able to adequately support a larger population. This is the current situation in the developed nations of the world.

There appears to be another transition through which nations must pass as they develop, one which involves levels of risk [Edgerton *et al.*, 1988] and affects sustainability. Figure 2 shows the risk transition model which is proposed here as a thinking tool to help us understand what sustainability means. In undeveloped nations there is a high level of "natural risk," *i.e.*, risk due to disease, animals, natural hazards, ignorance of the workings of the ecosystem, *etc.*, which declines when the nation begins to develop and uses information (science, technology, education, *etc.*) to reduce the risks. The man-made risks (*e.g.*, resource depletion, pollution, stress, crime, *etc.*) are low in undeveloped nations since little development has occurred. The culture is sustainable though in an undeveloped state and does not have the benefits of development. As development occurs, the man-made risks increase with increased industrialization and as massive amounts of energy are made available to man through exploitation of environmental resources. Economic well being also increases with development, even as the risk is increasing. Upon successfully reaching the developed status, man-made risks are expected to decline as we learn to use technology within environmental constraints while economic benefits are expected to remain high, although they may decline somewhat. The nation strives to be sustainable but now at a level of prosperity which is the benefit of development and which has low risk. Our economy and environment are only sustainable when risk has been reduced below a certain level; however, that level is still unknown. One of the benefits of sustainability is that both types of risk are minimized and indeed, this may be a condition for sustainability. A task of science is to determine an acceptable level of risk while retaining the fruits of development. This model admits to increased industrial risk in developing countries but the level must be kept within bounds if the country is to successfully make the transition to a developed country. Another task of science is to find ways that countries can

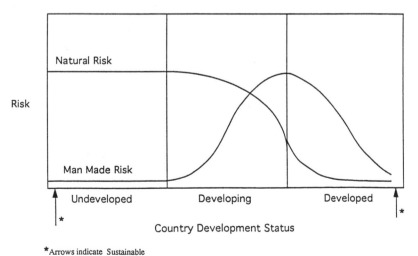

*Arrows indicate Sustainable
Status

Fig. 2. Risk transition upon development.

develop without incurring the excessive depletion of resources and subsequent pollution levels that occurred previously as countries developed. There is evidence that as man-made environmental risk increases beyond a threshold level, people become poorer and the probability that development is beneficial to all begins to decrease [Templet et al., 1994]. A country in this situation has little chance of successfully completing the transition to developed status.

American Samoa Coastal Management Program and Its Development—A Case Study[1]

In the United States, coastal management establishes broad national problem-solving goals and promotes diverse decentralized solutions at the state level. The key factor, flexibility, which makes CZM feasible in the United States, is also the primary reason such a program can serve as a model for other countries seeking to solve similar resource management problems [Mitchell, 1982; Sorensen et al., 1984; National Research Council, 1982]. For a comprehensive perspective on managing coasts and islands see [Towle, 1985]. One test of the applicability of the U.S. model in a different sociocultural setting involves American Samoa.

American Samoa (14°S, 170°W) is a complex mixture of traditional cultural practices and modern western institutions. This tropical South Pacific island territory has an enduring cultural heritage, a strong village life, extended family (aiga) ties, and a communal land system, all presided over by a hierarchy of familial chiefs (matai). Concurrently, American Samoa uses a cash economy, single-purpose government agencies, and competitive schools and business practices, modeled after their U.S. counterparts, to conduct daily territorial affairs. Although American Samoa is an American territory with all the advantages and disadvantages of having the U.S. as a bureaucratic model, it also has elements of being a separate nation. Its political leaders frequently interact with leaders of other Pacific island nations on regional issues. It contends with its own immigration problems, and it has substantial control over its adjacent 200-mile fisheries zone. In short, American Samoa is a microcosm of the sociocultural situation in which most developing countries find themselves; they intermingle their own traditional culture with external influences from other countries. The parallel authority embodied in traditional and borrowed culture must generally be in agreement before important decisions can be made or activities performed.

Sorensen   et al. [1984] reported on possible institutional arrangements for coastal resource management in 76 developing countries. Each country can be classified by a typology modified from Mitchell [1982] and its issues of interest identified. The advantages and disadvantages of a long list of mutually supportive management strategies are examined, and examples of their use throughout the world are given. Of particular interest to this discussion is Sorensen's statement that small island nations, because of their perceived widespread problems and coastal orientation, are prime candidates for nationwide integrated coastal management. The role of modern resource management programs is to regulate the use of uncontrolled and new elements (increased material and energy flows) within a culture in order to minimize negative impacts and, at the same time, to evolve with that culture in a compatible and appropriate manner.

Environmental Management: Cultural Perspective

Prior to modern resource management, individual cultures provided for an appropriate level of management [Johannes, 1982] which evolved slowly with the culture and was incorporated into mythologies, mores, taboos, folklore, and religious practices [Campbell, 1969]. Examples abound in many cultures where the earth, forest, or reef is the provider to be protected by various means of limiting resource use. The traditional resource management practices were adequate to deal with the use of coasts as long as material and energy flows through the culture remained relatively low and steady (or only slowly changed). However, with increasing outside influence, the flow of energy and material accelerated. Population, cash flow, energy consumption, and consumer goods increased. These changes rendered traditional management practices inadequate. New resource management techniques and principles, or adaptations of traditional techniques, became essential if the indigenous resources were to be used to sustain the culture over long periods of time.

Energy utilization and consumption are particularly important to requirements for resource management. Every pre-fossil fuel culture had a natural energy flow composed of sunlight (direct and stored), food, and a harnessed energy source (*e.g.*, man's own strength and that of draft animals). This flow shapes the culture [Seagraves, 1974; Sahlins and Service, 1960]. As long as the amount of energy used and consumed by man remains small in relation to the total energy flow, he cannot significantly impact his environment; hence, the need for resource management (or "people management") is small. As Odum [1971] points out in discussing a Pacific coral atoll culture, "the energy resources available to man are insufficient for him to damage his supporting system." A man with a digging stick can inflict minimal impact, but the same is not true of a man controlling a bulldozer or a dredge. In American Samoa the changes included an expanding road and automobile system, rapidly expanding housing and population which usurped other land uses, filling of the reef, overfishing, cultivation and construction on steep hillsides, water quality problems due to domestic and industrial (tuna canneries) wastes, and litter and solid waste disposal problems.

The Setting

Tutuila is American Samoa's largest island (54 square miles) and is home to over 90% of the territory's approximately 32,000 people [U.S. Army Corps of Engineers, 1981]. The remainder of the population lives on five smaller islands. American Samoa is an unincorporated territory of the United States, and its citizens are U.S. nationals who may immigrate to the U.S. without passports and apply for full citizenship. American Samoans have an elected governor (since 1977) and a bicameral legislature, although they do not vote in U.S. national elections. The topography of the Samoan islands is characterized by their volcanic origin and shows the rugged relief common to Pacific volcanic islands. Tutuila is a mountainous spine created by five overlapping extinct volcanoes [Sterns, 1944], with tropical forested flanks dropping precipitously to the sea. Narrow palm-lined coral sand beaches fringe most of the island, and fringing coral reefs provide subsistence fishing grounds and wave protection. Villages are located in the small areas of flat land where the intermittent streams reach the coastline, although the advent of roads has encouraged nontraditional strip development.

The entire territory of American Samoa is in the coastal zone, with the exception of Rose Atoll which is a federal refuge managed by the U.S. Fish and Wildlife Service. Because the coastal zone designation is broad, planning efforts for the entire island are consistent with ICM guidelines, and coastal management efforts are generally not in conflict with programs intended to benefit only one section of the island. It is interesting that in American Samoa, each village controls lands from the top of the adjacent ridge to the edge of the coral reef, an ecological unit (watershed). Rarely do sociopolitical boundaries and ecological boundaries exist conterminously, and it is an advantage in environmental planning when they do so. The most distinctive geologic feature of Tutuila is its protected harbor where the center of government and commerce is located. The deep harbor is a drowned river valley fringed by steep slopes and is one of the best harbors in the South Pacific, which explains the U.S. Navy's interest in administering the islands and using the harbor as a coaling (or fueling) station from 1899 to 1951. Thus, U.S. influence in American Samoa began some 85 years ago. Pago Pago Harbor is the current center of commerce, industry, government, population, and tourism in the territory, although a population shift away from the crowded harbor area to the flatter plain of the western district has begun.

A major result of U.S. influence has been the importation of U.S. money, food, and lifestyle, with a concomitant, six-fold increase in population. Early records compiled by missionaries indicate that the American Samoan population was stable (4,000-6,000 people) during most of the 1800's with an annual growth rate of less than 1% (doubling time of 70 years). Since 1900, the growth rate has jumped to 3.9% in the 1950's (doubling time of 18 years) with an average 2.2% growth from 1900-1980. Out-migration has limited the resident population increase to 1.8% [U.S. Army Corps of Engineers, 1981]. In many ways, traditional Samoan culture is the dominant force on the island (fa'a Samoa or the Samoan way) because its authority structure (the matai system) remains intact. Land ownership is communal; everyone has access to family land, even if he lives off-island, and use is controlled by the matai. Everyone is a member of an aiga, a large extended family of blood and adopted relatives, which lends a security to the life of the Samoan that westerners may not have [Sutter, 1982]. (There are no orphans in the Samoan culture nor does anyone suffer from "poverty.") The official language is Samoan. Meetings of the legislature (Fono), the matai in village councils, and other official ceremonies usually are conducted in Samoan although English is also spoken.

Village life and the councils of the chiefs who run the villages are very important parts of Samoan society; it is at the village level that most people are allowed to participate in the decision-making process. Hence, public participation occurs best at this level. Finally, Samoans still hold many of their traditional beliefs which have served them over the millennia.

The developed side of American Samoa boasts a cash economy with substantial U.S. aid and investment, with ready access to U.S. goods and markets. Thus, Samoans have a relatively high per capita income when compared with other Pacific island nations and a high level of cultural exchange with the U.S. and other developed countries. Samoan association with the U.S. since 1900 has led to an American style state agency structure, including numerous single-purpose agencies that could be found in any mainland state: fish and game, public works, energy, planning and economic development, local government, and others.

ICM as Technology Transfer

While ICM is not generally thought of as technology in the sense of hardware (*e.g.*, manufacturing or processing plants), it is technology in that it possesses significant technical information which developing countries probably do not have. In this sense the lessons of past technology transfer efforts may be useful in the transfer of ICM to developing countries because coastal management transfer parallels technology transfer to developing countries. Pollnac [1978] proposes a model for intermediate food technology transfer and points out that any acceptable technological innovation must possess five major attributes:

1. Complexity: the complexity of the new technology must match that of the recipient country.
2. Compatibility: the new technology must fit the existing cultural practices of the recipient country.
3. Advantage: the benefits of the technology must be perceived as an advantage by the recipient society.
4. Trialability: the innovation must be available for trial by a cross-section of the socioeconomic structure.
5. Observability: the innovation's success is closely related to observation of its successful operation.

Dahl [1982] concurs with some of these attributes and recommends simplified administrative procedures, simplified data collection and planning techniques adapted to local manpower, and decision-making procedures suited to local cultural conditions, as well as appropriate forms for information presentation. Additionally, Dahl recommends that technology and equipment be appropriate to local conditions and maintenance capability, that outside expertise sensitive to local culture be available, that decision-making be decentralized, and that public education accompany technology transfer. By adopting these attributes and techniques in the program development phase, the resulting coastal management program will be sensitive to traditional culture and values. Also, the level of technology transfer will be appropriate for the target country.

Developing a Management Program

Environmental management in developing countries must find its niche in the blend of old and new cultures. There was little evidence in American Samoa of the U.S. coastal management experience by the late seventies.

The general approach used by the author in developing a plan for American Samoa consisted of these elements:

1. Problem identification.
2. Development of alternative solutions and policies.
3. Position papers.
4. Consensus building by public and agency participation.
5. Implementation.

This approach, similar to that of Sewell [1975], had sufficient flexibility to mesh American Samoa's unique situation with the requirements of the federal CZM Act. Figure 3 is a flow chart illustrating the general process of plan development, connections, and iteration loops.

Problem identification

The Developing Planning Office (DPO), the agency charged with the responsibility for developing a coastal management program in American Samoa, had already begun a survey of village needs in the territory prior to program development. That survey turned out to be important since it served as a primary focus for public participation. Meetings were conducted in the usual formal village council setting where most important decisions in American Samoa are first made. The assembled matai were asked questions about village problems, such as housing, unemployment, flooding, fishing, and erosion, and their responses were noted in a wish list manner to denote their priorities.

A review of available literature augmented the village surveys; this review included territorial agency and federal agency reports, *e.g.,* the Honolulu District of the U.S. Army Corps of Engineers, which has responsibilities on the island, had conducted

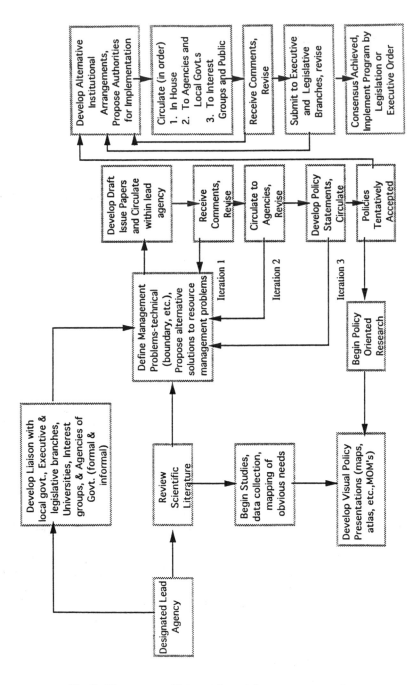

Fig. 3. The process of integrated coastal zone management.

several pertinent studies. Additionally, Samoans and palagi, both inside and outside of government, were interviewed to gain further verification and information concerning the problems identified at the village meetings.

Villagers and matais were skeptical about having their needs met, so follow-up efforts were critical for the success of the program. Funds available from other federal programs (primarily Housing and Urban Development) provided some of the items on the village wish list, and two results were evident. Funds were focused at a level that showed some immediate benefits to the village, and credibility for the developing management program was thus established at the local level. The additional funds from other U.S. federal programs were used in a "self help" manner: villagers provided labor, and the Development Planning Office, using federal funds allocated for that purpose, paid for supplies, such as cement and wood, to carry out the projects. The self help projects are an important part of village programs instituted by the territory's coastal management program. The local programs and planning were voluntary on the part of local governments but opened the way for them to participate in territorial programs and funding for which they otherwise would not have been eligible. The village plans were to include an inventory of existing conditions (the local surveys already discussed), an identification of expected land uses, and a land-use plan. The village plans also had to be consistent with the territorial-level policies and could become part of the American Samoa Coastal Management Program (ASCMP) after review by territorial agencies. If adopted, the villages could implement the plan using available territorial funding and agency assistance.

Contact was also established with government agencies, legislative and executive offices, and pertinent interest groups. Informal discussions elicited their problems and concerns and marked the beginning of working relationships. These discussions initially centered on technical or resource issues (*e.g.*, boundaries) and avoided politically charged questions of institutional arrangements or authorities. The latter issues can best be addressed after the technical issues are resolved and consensus building has had time to produce some results. The results of the informal discussions lead directly to the definition of management problems and suggested alternative solutions. By first focusing on technical concerns, alternative solutions are scientifically based and present technically viable solutions to the increasingly political review process as illustrated in Figure 1. From these combined efforts, a list identifying problems and needs was made. This list eventually became the basis for the goals and policies established by the coastal management program. Some of the problems were institutional in nature, some were cultural, but most were natural resource problems.

Alternative solutions and policies

The next step is to choose a range of alternative solutions to the problems and to develop goals and policy statements which present solutions. Goals are rather broad statements designed to express a direction or approach. More specific policies were enumerated as the means for accomplishing each goal. The draft policy solutions to the resource management problems were developed from prior studies, agency discussions, experience elsewhere, and office staff discussions. Policies for American Samoa fall into three broad categories: government processes, economic development, and resource conservation. Government process policies mandate sensitivity to *fa'a Samoa*, while the economic policies tend to promote and guide various kinds of development more western in scope. Resource policies attempt to protect such things as reefs, water and air quality, marine resources, drinking water quality, unique areas, and archeological, cultural, and historical resources.

Position papers

From the list of problems and policies, a series of draft position papers was prepared to explain the nature of each problem, what work or research had already been done, and propose alternative solutions. Since many of the problems fell under the jurisdiction of existing government agencies, the solutions, by necessity, involved those agencies. In most cases, the appropriate agencies were involved in the development of the alternative solutions, and their participation was encouraged by indicating to them the advantages of the program-increased staff and/or funding or the availability of outside help to assist with problem-solving. The draft position papers became the mechanism for sequential review by the lead agency, other agencies, local governments, interest groups, and to a lesser extent, the public through a series of feedback loop iterations, which incorporated revisions at each step. This process eventually led to acceptance by consensus.

Consensus building

In American Samoa, public participation occurs in the village councils. People discuss their problems with their matai, and each aiga arrives at a consensus. The matai then relates that consensus to the village council which also arrives at a consensus. Since this was the traditional way of reaching decisions, it was important to have agreement between the village councils and the government sector. There were few, if any, identifiable "third-party" individuals or groups in Samoa, and public participation took a somewhat different form than in the United States. Although it is a time-consuming process, it seemed best to involve the villagers in the traditional manner. Public meetings were held occasionally, but they were sparsely attended and served more as a forum for westerners and agency staff than for the Samoan public. Experience indicated that the traditional methods for venting grievances and solving problems were the most workable. Consensus building also occurred during the initial agency contacts and continued as working relationships developed during the position paper circulation and revision steps.

Implementation

American Samoa had a number of statutes (many passed in response to U.S. federal law), which, when taken as a whole, constituted sufficient authority for implementing a coastal area management program. No new legislation was required, and the program was implemented by executive order invoking existing legislation (networking). Timing was critical since American Samoa did not begin coastal management planning until program development eligibility was in its last years. A legislative approach providing the necessary legal authority to implement the program, while providing wider political representation, would have been much more time-consuming. The executive order which authorized the program required that government agencies incorporate the 16 program policies into their activities and carry out their responsibilities consistent with the coastal management program. A strong "top down" regulatory approach was not appropriate because of American Samoa's consensus tradition. The program attempts to achieve its goals by building consensus between villages, agencies, and territorial government, although there is a permitting process which seeks to enforce policies by the usual punitive means using existing permitting processes, such as building permits.

The Structure of the Program

The program which evolved from these efforts [U.S. Department of Commerce, 1983] is being implemented by the DPO, the agency designated by the governor. The choice of DPO as lead agency insures that planning will be coordinated and the policies of the coastal management program will be built into future planning efforts since DPO handles most on-island planning. The day-to-day management of the coastal area and enforcement of the policies are the responsibility of a number of agencies under a networking arrangement. The executive order which created the program requires that all agencies having jurisdiction over a permissible use incorporate the relevant policy considerations into their activities. In addition, any permit issued in American Samoa is required to be reviewed by DPO to ensure that policy considerations are implemented.

The three categories of policies, which provide guidance to territorial agencies, include specific areas of management: (1) government processes which include territorial administration and village development; (2) development policies which address shoreline development, coastal hazards, fisheries development, slope erosion, major facilities siting, and agricultural development; and (3) resource protection policies which include guidance concerning reef protection, recreation, water quality, water resources, drinking water quality, unique areas, archeological, cultural, and historic resources, and air quality.

Two areas of particular concern (special areas) were designated by the program: Pago Pago Harbor, a special area for development, and Pala Lagoon, a special area for restoration. Pago Pago Harbor is the economic center of American Samoa, and the objective of this designation is to develop the harbor in a way that promotes its value as a working port and safe harbor and protects its natural resources, including water quality. A special area plan was prepared, reviewed, and adopted by a policy-level committee composed of users to guide development in the harbor area and to accommodate water-dependent uses and activities.

Pala Lagoon is the largest semi-enclosed salt water lagoon on Tutuila and is partially fringed by mangrove swamps. It has high ecological productivity, but its water exchange has been reduced by the construction of an airport runway partly blocking its opening to the ocean. An outline for a special area plan was prepared during development of the program, and the government has begun to improve water quality by installing a sewer collection system with encouragement by the coastal management program. A detailed special area plan will be prepared to promote water exchange and to protect water quality and the unique ecological value of the lagoon. Since program implementation in 1980, designation of Fagatele Bay as a marine sanctuary has been proposed by the government of American Samoa, and an environmental impact statement [U.S. Department of Commerce, 1983] has been prepared. Fagatele Bay was approved for sanctuary status by the Department of Commerce in 1983.

Technical Assistance

Technical assistance was one of the more successful and useful applications of coastal management in American Samoa. A substantial amount of research and technical work had been done in the territory, primarily by territorial agencies, contractors, and U.S. agencies. However, much of it was difficult to locate and infrequently used. Experience and research [Dahl, 1982] indicate that information would be more useful if presented in a visual format (*e.g.*, maps). Therefore, the major technical contribution, in addition to a number of studies, was the development of the first atlas of American Samoa [American Samoa Government, 1981], incorporating all available existing and newly developed resource management information about the territory into a single, large, durable, and colorful document. In addition to traditional mapping parameters [*e.g.*, vegetation, soil, geology, and reefs] derivations such as slope, critical areas and areas of high erosion were mapped and cross referenced to the permit review form (Management Oriented Maps {MOMS}; Figure 1). For example, one policy covers building on slopes greater than 40%; the corresponding map plate presents slope percentages for all areas on the islands. The mapping process also pointed out gaps in existing knowledge and provided direction and focus for future research.

In evaluating a permit application, an analyst can note all the characteristics of the proposed site through use of the atlas and make permit modifications or conditions which mitigate the impacts of the proposed action or can reject the permit (See *Oceans*, January 1983, for a review of the atlas).

Recommendations for Integrated Coastal Management in Developing Countries

The U.S. model for ICM appears to be a valid coastal area management model for developing countries. It can be further fine-tuned to a particular situation by using the process shown in Figure 1 and applying the following recommendations:

1. Strong involvement with the local government through their traditional procedures is important in creating linkages, and the use of surveys, plans, and self-help projects for the at the local level gains their help and trust.
2. Public participation and consensus-building efforts should work at appropriate levels (in American Samoa this was the village level), although schools and television (public education) were later used.
3. Use of coastal management to reinforce territorial aspirations and goals ensures a high level of cooperation. Appropriate economic development can be promoted through coastal area management policies and objectives.
4. The use of traditional cultural experience, by appealing to local customs, ethics, and folk wisdom, reinforces aspects of the program, instills an ecological understanding, and adds traditional strengths to the program.
5. The use of visually formatted information (maps or atlas) to convey technical information to the villages, the public, agencies and governmental officials is effective.
6. Coastal management programs are useful in guiding technical assistance from outside the territory or country. A well-focused resource management program with clearly defined policies and goals can be a very useful conduit for guiding and coordinating outside assistance, both fiscal and personnel, into areas where help is needed while minimizing adverse impacts. This will help insure that territorial priorities are addressed and impacts are minimized.
7. The program must be flexible and operate at a level of bureaucratic and fiscal checks and balances which is compatible with territorial experience. For example, in a territory which values its extended family structure very highly and where families are large, nepotism is inevitable. While this is a problem to Americans, it may not be so for developing countries.
8. Experience with program implementation in American Samoa indicates that the U.S. system of checks and balances and resulting bureaucracies to control money is inappropriate in other countries. Traditional infrastructures generally cannot cope with extensive bureaucratic and fiscal procedures, and the creation of massive bureaucracies causes unnecessary and counterproductive burdens which will work against any program associated with them. A moderate system of checks and balances, tailored to developing island states, is more appropriate.

The Role of Science, and Particularly Oceanography, in ICM

Science and its findings were used extensively in developing the American Samoa program. To understand how this occurred it is helpful to divide things into process and substance. Science provides the substance which determines the content of the policy statements which guide government decision making. Scientific fact was incorporated into the position papers and the alternative solutions which led to policies so that decisions of the governmental structure would be based in scientific fact and knowledge. The process is the sequence of events that carries program development to a successful conclusion and is shown in Figure 1. During the process phase science meets politics and policy statements are the result. Process and substance are both important and essential if we are to succeed in achieving sustainable development. Since most scientists work only in the substance arena and rarely view the planning process as crucial, I would like to generally note some requirements that ICM planning puts on scientific results.

1. Scientific data must be in suitable formats. An example is the atlas of American Samoa; in general, spatially-oriented data is more useful to planners than lists, books, or other hard copies. Models are included under spatial data formats since their results are usually oriented in space. Simulation models are particularly useful since planners can then ask what-if questions and design more suitable policies to guide government.
2. Scientific findings must be timely and available. Decisions makers will continue to make decisions whether scientific data is available or not. For science to contribute to the decision it must be ready and useful. Many scientists wish to operate at the 95% confidence level at all times but that's not possible in government, and science must attempt to meet the decision makers' needs if it is to be pertinent. The joining of science and politics is an uneasy marriage but essential for sustainability. I suspect that scientists and policy makers will mutually benefit from the association; I know the public will be better served.
3. Science must fit the policy framework. It is often difficult, perhaps sometimes impossible, to fit complex scientific principles into a short policy statement which is pertinent to governance. Scientists should recognize that the details can be built into the implementing process elsewhere over time but that a short policy statement is necessary to put government on the right track. To a scientist policy statements seem too vague and general to be meaningful but they are, of necessity, compromises, something scientists don't like to do. The policy statement can be further defined as consensus is achieved and there is then support for the details.

In American Samoa oceanographic data was mapped into the atlas in the form of bottom contours, coral reef locations, coastal hydrography, ground water recharge to the fresh lens, fishing opportunities, water quality problem areas, and hazard areas including beach erosion and tsunami. In addition, studies were commissioned to determine current flow patterns and mixing times in Pago Pago Harbor and Pala Lagoon. These are the usual studies for coastal planning which involve oceanography, although there may be others. For further information I refer the reader to the Chapter 1 of this text where the science needs of ICM are discussed and correlated with the capabilities of oceanography.

Conclusion

It is becoming clear that with increasing population and other stresses measures must be taken to prevent the excessive depletion and despoliation of coastal resources and the resulting pollution as a nation makes the transition from undeveloped to developed status. The measures include a greater use of scientific information and appropriate technology to prevent one use of the coast from damaging another use. Science provides the substance and planning provides the processes to create management programs which merge into policy to be implemented by governmental authorities. The American Samoa lessons can be useful to countries which are new to these disciplines. The science of oceanography is particularly helpful on the "wet" side of coastal management which involves ocean processes. However, the knowledge gained from such sciences must be formatted to be compatible with the needs of those involved in developing and implementing integrated coastal management for developing island nations.

The desire for sustainable development is an especially worthy one since it requires the integration of environmental concerns into economic development decision making. It is not yet clear just what is entailed in such development but it is clear that science and planning are necessary. Developing nations must fashion policy to promote sustainability and those policies must maintain the quality of the environment so that it can continue contributing to the economic well-being of all citizens.

References

American Samoa Government, Development Planning Office, Atlas of American Samoa, Pago Pago, American Samoa, 1981.

ASCMP, NOAA and Development Planning Office, American Samoa coastal management program and final environmental impact statement, Department of Commerce, Washington, D.C. and Pago Pago, American Samoa, 1980.

Bank of America, Economics and business, Outlook, June/July, 1993.

Campbell, J., The Masks of God: Primitive Mythology, 504 pp., New York, Viking Press, 1969.

Dahl, A. L., Practical management of limited resources, Proceedings, Pacific Basin CZM Conference, Pago Pago, American Samoa, 1982.

Edgerton, S. A., K. R. Smith, R. A. Carpenter, A. S. Tougiq, S. G. Olive, C. P. B. Claudio, V. T. Covello, D. J. Fingleton, K. G. Kim, and B. A. Wilcox, Priority topics in the study of environmental risk in developing countries: Report on a workshop held at the East-West Center, August 1988, Risk Analysis, 10(2), 273-283, 1990.

Johannes, R. E., Traditional conservation methods of protected marine areas in Oceania, Ambio, 2, 258-261, 1982.

Miller, G. T., Environmental Science, 4th ed., pp. 134-138, Wadsworth Inc., Belmont, CA., 1993.

Mitchell, J. K., Coastal zone management: A comparative analysis of national programs, in Ocean 3 Yearbook, edited by Borgese, E. M and Grimsburg, N., pp. 258-319, University of Chicago Press, Chicago, 1982.

National Research Council, Coastal resource development and management needs of developing countries, National Academy Press, Washington, D.C., 1982.

Odum, H. T., Environment, Power and Society, 331 pp., Wiley-Interscience, New York, 1971.

Pollnac, R. B., Sociocultural factors influencing success of intermediate food technology, Food Tech., 89-92, 1978.

Seagraves, A. B., Ecological generalization and structural transformation of sociocultural systems, *American Anthropologist, 76*, 531-552, 1974.

Sewell, G. H., *Environmental Quality Management,* 299 pp., Prentice-Hall, Englewood Cliffs, New Jersey, 1975.

Sorensen, J. C., S. T. McCreary, and M. J. Hershman, Coasts: Institutional arrangements for management of coastal resources, Coastal Publication No. 1, Research Planning Institute, Inc., Columbia, S.C. for U.S. National Park Service and U.S. Agency for International Development, 1984.

Stearns, H. T., Geology of the Samoan Islands, *Bull. Geol. Soc. Amer., 55*, 1279-1324, 1944.

Sutter, F. K., Cultural obstacles and western management, Proceedings, Pacific Basin CZM Conference, Pago Pago, American Samoa, 1982.

Templet, P. H., American Samoa: Establishing a coastal area management model for developing countries, *Coastal Zone Manage. J., 13*(3/4), 241-264, 1986.

Templet, P. H., and S. Farber, The complementarity between environmental and economic risk: An empirical analysis, *Ecological Economics,* v. 9 (1994), pp. 153-165.

Towle, E. L., *The Island Microcosm in Coasts,* Coastal Resources Management: Development Case Studies, Coastal Publications No. 3, John R. Clark, Ed., pp. 589-738, Research Planning Inst. Inc., 1985.

U.S. Army Corps of Engineers, Honolulu District, American Samoa water resources study, 1981.

U.S. Department of Commerce, Draft environmental impact statement and management plan for the proposed Fagatele Bay National Marine Sanctuary, Washington, D.C. and Pago Pago, American Samoa, 1983.

23

Sustainable Development and Small Island States of the Caribbean

Erik Blommestein*, Barbara Boland*, Trevor Harker*, Swinburne Lestrade*, and Judith Towle*

Abstract

Small islands face many constraints which make them particularly vulnerable and dependent. One interpretation of smallness is that of limited development options. A narrow range of natural resources, restricted agricultural potential and limited opportunities for employment conspire to circumscribe the scope of small islands for active development policies. This is reflected in the sources of economic activity, human resources and environmental management.

Introduction

At the United Nations Conference on Environment and Development the special challenges to planning for and implementing sustainable development for small islands were recognized in a special programme area for the sustainable development of small islands.

Agenda 21 provides a planning framework for global, regional and national action towards sustainable development. Within Agenda 21 the programme areas, activities and means of implementation cover, by necessity, broad areas, which need to be made more specific at the national and regional level [United Nations, 1993].

Small Islands: Marine Science and Sustainable Development
Coastal and Estuarine Studies, Volume 51, Pages 385–419
Copyright 1996 by the American Geophysical Union

*Authors are employed by the United Nations and the Island Resources Foundation. The views are those of the authors and should not be attributed to their respective organizations.

In recognition of the two aforementioned points it was decided at UNCED to convene a Global Conference on the Sustainable Development of Small Island Developing States. This conference was held in Barbados in 1994 where a 'Programme of Action for the Development of Small Island Developing States' was adopted. Like Agenda 21 this Programme of Action provides a planning framework for the sustainable development of small islands [United Nations, 1994].

Somewhat arbitrarily we will define small island developing states (SIDS) as countries with a population size of 2 million or less, and very small developing states (VSIDS) as having a population of half a million or less. Some basic indicators of Caribbean SIDS are stated in table 1.

The Issue of Uniqueness

All developing countries share several problems. However, many of these problems are more pronounced in their impact on SIDS, particularly the very small ones (VSIDS). In summary, SIDS face a combination of economic and ecological constraints to attaining sustainable development. Small, fragile and ocean-land interface are the key words, whether these pertain to the numerous small and threatened ecosystems or to the equally small one or two dominant productive sectors. Vulnerable to outside influences, SIDS are restricted in their options and response mechanisms.

Much discussion has centered on the issue of the peculiarities or the "uniqueness" of island developing countries. Some commentators have justified the case for special and privileged treatment to small countries on the basis that these countries had certain unique characteristics that derived from their smallness. This contention has been rejected by certain other commentators. T.N. Srinivasan (1985), for example, in a thought-provoking World Bank discussion paper, concludes: "It would appear that many (though not all) of the alleged problems of small economies are either not peculiar to small economies or can be addressed through suitable policy measures". He continues "...causes of economic and social stagnation in some of these economies cannot be attributed to their smallness...".

As far as Farrel (1991) is concerned, it is more normal to be small than to be large; it is not true that small states are generally poor; it is not true that small States are not viable; and in specific reference to the Caribbean: "the essential problem of small states has little to do with their small size". Both commentators, however, do occasionally concede that one is not necessarily talking about absolutes, but that special problems of small sates (in particular, small island countries), may be more a matter of degree and of proportion. Srinivasan (1985) states: "... the particular vulnerability of small (island) economies is attributed to the disproportionate effect which natural disasters could have on them". While Farrel's (1991) essential problem with small states has little to do with their smallness, he admits that "this particular characteristic may exacerbate the problem and its effects".

TABLE 1. Selected Indicators (Latest available year)

	Land Area (km²)	Ocean EEZ (000 km²)	Coast-line (km)	Pop. (000's)	Pop. Density	GDP/CAP (U.S.$)	HDI	Openness
Anguilla					66			
Antigua Barbuda	440		153	62.9(m)	143	5636	0.781	221
Aruba				62.1(n)				
Bahamas	13942	759.2	3542	254.7(o)	18	11036	0.875	
Barbados	431	167.3	97	257.1(o)	597	5624	0.927	111
Br.Virgin Islands	150			16.6(m)	111	8313		
Cayman Islands	260		160	25.4(k)	98	23622		
Cuba	110860	362.8	3735	10574.9(n)	95		0.732	
Dominica	750	20.0	148	71.8(m)	96	2047	0.783	156
Dominican Republic	48730	268.8	1288	7169.8(p)	147	0954	0.595	61
Grenada	345	27.0	121	90.7(m)	263	1808	0.758	153
Guadeloupe	1780		306	385.5(o)	217			
Haiti	28000	160.5	1771	5939.0(n)	212	0154	0.276	
Jamaica	11424	297.6	1022	2248.2(m)	197	1559	0.722	
Martinique	1100		290	358.8(o)	326			
Montserrat	100			10.9(m)	109	6788		
Neth Antilles								
Puerto Rico	8900			3514.0(o)	395	0923		
St. Kitts Nevis	269			41.8(m)	155	3516	0.686	138
St. Lucia	616			133.3(m)	216	2550	0.712	187
St. Vincent/Gren	388			107.6(m)	277	1654	0.693	142
Trinidad Tobago	5128	76.8	362	1234.4(o)	241	3984	0.876	79
Turks and Caicos	430							
US Virgin Islands	340			101.7(o)	299	13706		
Cape Verde	4030			382.0	95		0.437	
Cyprus	9250			710.0	77			
Malta	316			357.0	1129		0.854	
São Tomé-Principe	960			124.0	129		0.374	

Notes: Population density: Number of people per square km.
HDI: Human Development Index
Openess: Index of the sum of the absolute values of exports and imports of goods and services divided by GDP
Sources: ECLAC on the basis of official data.

Other commentators, have emphasized the environmental and ecological fragility of small islands, because of the delicate balance between highly coupled terrestrial and marine ecosystems [Towle,1985]. In this context Brookfield (1990) states that "However the small size of islands does present a special case in considering the impact of modern changes....; the consequences may thus become much more visible in the short-term than in larger continental systems".

In an important sense this is all that needs to be established. Elsewhere, it has been suggested "much of this debate can be circumvented by not attempting to establish that small island countries are unique and special; or that they are characterized by circumstances which affect them and no other category of developing countries. It seems sufficient for present purposes to agree that small developing countries are afflicted by similar kinds of economic difficulties and development imperatives as in the case of developing countries generally, but that these problems seem more intractable in at least some small and particularly very small island countries such as those that are found in the Caribbean" [Lestrade, 1987]. So that the issue is not about smallness itself as a problem, but about sustainable development problems which may take on enlarged proportions in small, or very small island countries.

There is now an abundance of documentation on the characteristics of small island developing countries and on the implications of small size for their development potential. These characteristics are very well known, and only a summary listing of them is being provided here[1].

1. Small size of the economy, as manifested by small population, GDP and natural resource base leading to diseconomies of scale;
2. Concentration of human, social and ecological activities in a narrow coastal zone, resulting in conflicts between activities and frequent loss of coastal habitats;
3. High unit costs of transportation, communication and information reinforcing diseconomies of scale[2];
4. Vulnerability to natural disasters affecting the whole or the major part of SIDS[3];
5. Very open and dependent economies, as evidenced by the high ratio of external transactions to GDP[4], the concentration on a few primary export commodities or services; this makes SIDS highly vulnerable to changes in world demand and prices;
6. Weak indigenous technological capacity;
7. Above average dependency on fish and sea food for their calorie and protein supply[5];
8. Low bargaining capability in the private and public sector caused by human resource and financial constraints which often weaken the negotiating position of SIDS;
9. Limited public and private savings , a situation which requires significant inflows of concessional aid and foreign investment to finance capital formation. This greater dependence on external financial aid makes these countries susceptible to changes in the international environment for foreign aid, especially concessional aid;
10. Prevalence of monopolistic and oligarchic conditions in production, imports and exports, and consequent weak market forces to improve efficiency and productivity;

---

[1] Based in part on UNCTAD (1988), Lestrade (1987) and the ECLAC/CDCC Environment and Development Unit.

[2] In this context Briguglio (1993) states that the average index numbers for transportation and freight costs as a percentage of exports were 39.73 for SIDS (25 countries in the sample), 15.58 for non island developing countries (106 countries in sample) and 4.42 for developed countries (22 countries in the sample).

[3] Briguglio (1993) states that the average index number of disaster damage as a percentage of GDP (1970-1989) was 60.58 for SIDS (13 countries in sample), 32.31 for developing non island countries (61 countries in sample) and 5.35 for developed countries (4 countries in sample). Countries with zero incidence excluded from sample.

[4] Briguglio (1993) states that the average index number of trade dependence was 87.2 for SIDS, 76.7 for developing non island countries and 42.4 for developed countries.

[5] See table

11. High rates of migration with the result that in several SIDS populations remain stable or decline slightly and face skewed population distributions with under representation of those between 15 and 65.
12. Above average costs for tertiary and technical secondary education which reinforces below average enrollment ratio's and high losses of these trained people to emigration.
13. Existence of numerous small, and often fragile ecosystems which complicates management and sustainable development.
14. High degree of endemism and high rates of biodiversity (number of species) per square kilometer which limit development options if those species are to be protected.
15. Above average costs for the provision of non tradable goods and services.
16. Almost immediate repercussions on the coastal zone and marine environment of terrestrial events;
17. Circumferential coastlines and above average length of coastline and EEZ area per unit of land area.

The above factors apply to many developing countries, but it is the tendency of many or most of them to occur concurrently in SIDS which makes them particularly disadvantaged and results in relatively less resilience, and greater vulnerability and dependence.

The case has also been put in terms of limited development options: "One interpretation of smallness is that it is a situation of limited options; limited development options. Small countries are characterized by natural resource deficiencies, restricted agricultural potential, limited domestic market size which constrains the potential for non-export industry, and limited opportunities for the employment of both skilled and unskilled labor, a situation which results in brain- drain as well as significant levels of unemployment. These circumstances conspire to circumscribe the scope of the countries for active, not to mention, independent, development policies"[6].

Recent Economic Performance and Sources of Caribbean Economic Activity

The region remains dependent on the exploitation of its renewable and non-renewable resources for its economic and social development. Agriculture, tourism and mining remain crucial in terms of employment, foreign exchange earnings and contributions to government revenue. To maintain the productivity of the renewable resources on which agriculture and tourism depend is therefore of crucial importance to the long-term future of Caribbean economies.

---

[6]The foregoing is intended to be an illustrative listing of the characteristics or constraints of SIDS. Additional characteristics deal with maritime and archipelagic considerations, and implications for absorptive capacity of external aid, for example.

At various points in time each of these sectors had experienced export 'booms', the most recent example being tourism which is still a highly dynamic sector but for the most mature tourism destinations such as Barbados and the Bahamas. The question arises, however, whether small island states have been able to take advantage of export booms as an engine of growth, or whether the structural impediments inherent to small islands as referred to in section I, prevent this. Traditionally, export booms in the Caribbean tend to bias consumption and investment in favor of increased imports and to destroy the production capacity outside the booming sector[7]. Moreover, problems inherited from the 1950's remain- unemployment, lack of diversification and intersectoral linkages in the structure of production and a profound malaise in the agricultural sector, both for export and local food production.

In view of the high costs of transport and of non-traded goods and services, it is perhaps not surprising that in the last two decennia the region has focused on the accelerated development of high value but low weight exports such as tourism[8], financial services and illicit drugs.

Economic growth[9]

Perhaps in seeming contradiction to the earlier outlined constraints, the VSIDS of the Caribbean, have been foremost among the Caribbean countries in terms of rate of growth. As a group, the OECS countries recorded growth over 5 per cent per annum[10], as did, Cape Verde, Cyprus and Malta, during the 10 year period 1982-1991. The Bahamas averaged 4.4 per cent growth per annum from 1981-1990. With the exception of Cape Verde all of the small high-growth countries had vibrant and growing tourism industries. In the Caribbean the country with the smallest tourist industry, Dominica, grew the slowest in the group.

The next cluster of countries, the moderate-growth economies, had average growth rates of 2-3 per cent and comprised, Jamaica and Puerto Rico. These countries had relatively diversified economies and the performance of the various sectors was mixed. All had significant tourism sectors which recorded growth, particularly in Jamaica, making rapid strides after years of stagnation. Both suffered stagnant or contracting agricultural sectors, although agricultural export earnings increased significantly in Jamaica towards the end of the survey period. Both Jamaica and Puerto Rico showed increased manufacturing sectors, in the latter case the sector increasing from 36 to almost 39 per cent of GDP between 1982-1991. Export earnings were fastest growing in

---

[7] This occurs through changing the ratio of factor productivity between the booming sector and other productive sectors, thereby rendering production unremunerative in the latter. (St.Cyr, 1991)

[8] Srinivisan (1985) states that small countries face a cost disadvantage averaging 65% in building additional thermal capacity for power generation

[9] See ECLAC/CDCC (1993) and Harker (1993) for a more detailed exposure. The source of all data is ECLAC/CDCC based on national data.

[10] The average declined during 1992 and 1993, to 3.6 and 2.5 percent respectively. (Harker pers.com.)

categories such as garment manufacture and above all tourism in Jamaica and in chemicals, drugs and pharmaceutical in Puerto Rico. In Jamaica, the performance of the minerals sector was inconsistent and conditioned by external demand for aluminum and the price of petroleum, the major imported input into the industry. Jamaican growth performance was severely affected by the cost of servicing its national debt, since an average of almost 72 per cent of total fiscal revenues for each of the last five years had to be diverted to meet the costs of principal and interest. A large proportion of the earnings from exports of goods and services needed also to be diverted for servicing the external debt, 31 per cent in 1991, down from a peak of 46 per cent in 1986.

Barbados and the Dominican Republic fell into the category of low-growth economies, having an average growth rate of 0-2 per cent. In Barbados, growth was positive, though modest for the period 1980-1989. Severe contraction was, however experienced from 1989 onwards with a eleven percent fall in GDP during 1990-1992[11]. This was due to the decreasing competitiveness of that economy which made it unable to compete for market share in tourism, in its regional exports of manufactures and in its traditional agricultural exports. Contracting foreign earnings required concomitant contraction in domestic activities such as commerce, construction and government activities.

In the Dominican Republic a major factor limiting its growth was the cost of adjusting its economy from the traditional export activities -agriculture and mining- to new ones, mainly tourism and to a lesser extent activities in the export processing zones. Notable was the secular decline in the perennial mono-crop sugar. Export earnings from this product fell by half in value over the survey period and from 30 per cent of merchandise exports in 1980 to 20 per cent in 1991. Other factors limiting growth were large fiscal deficits incurred for non-liquidating activities in 1987-1988 and the consequences of this policy in subsequent years.

The last category comprises the contracting economies of Sao Tome, Haiti and Trinidad and Tobago, all of which experienced declines in GDP. Trinidad and Tobago, as an oil producer was affected by the declines in oil prices, but suffered volume declines in output as well. None of these countries benefited from the tourism boom experienced by other Caribbean countries, since their tourism sectors were small and in some cases earnings declined. Trinidad and Tobago also suffered decline in export earnings from sugar, while they had almost ceased in Haiti. Reduced export earnings redounded against domestic economic activities in both countries, since domestic consumption had to be sharply curtailed in sectors related to construction, distribution, finance, insurance and real estate. Stringent measures were also needed to curtail government activities, this being a pervasive trend for most countries in the region.

In Haiti, major declines were evident in manufacturing and commercial activities. While agriculture actually increased it share of GDP from 1980-1989, export earnings from the

---

[11] Modest growth was achieved in 1993. (Harker pers.com.)

sector fell steadily from 1983. Earnings from the free zones increased over the survey period from 25 to 50 per cent of exports between 1980-1988, but the rate of increase might have been faster and more sustained without the political disturbances endemic for the last years of the decade. This caused some firms to relocate elsewhere in the region and provided a setback for that country's efforts to diversify its economy.

Tourism

The tourism sector, by its steady growth, has provided most of the prosperity that has been experienced over the decade. All the countries in the Caribbean region either had a tourism sector or were trying to develop one. In consequence most Caribbean SIDS are vulnerable to tourism trends as shown in table 2. This table shows that while in the larger SIDS total tourist arrivals and earnings are higher, it are the VSIDS which, in relative terms, depend more on tourism.

For the period 1980-1990, the average rate of growth of tourist arrivals was approximately 4.3 percent, while tourism expenditures in current dollars averaged about 6.5 per cent growth. Above average growth was recorded by, Antigua and Barbuda, Dominica, the Dominican Republic, the British Virgin Islands, Saint Lucia, Saint Kitts and Saint Vincent and the Grenadines, or by established destinations recovering from a past decline, such as Cuba and Jamaica. Among the established destinations, growth was below average in the Bahamas, Barbados, Puerto Rico, and the Netherlands Antilles. Earnings contracted in Trinidad and Tobago.

Despite a picture of overall success, warning signals are being provided by declining productivity and profitability of many regional hotels. The spectra of declining competitiveness and productivity is appearing in some of the traditional destinations and there is a further danger of rigidity in adjusting to changing leisure patterns.

Blommestein (1993) notes that "there is the realization and acknowledgment (by the public and private sector) that adverse impacts, caused by the current level of tourism pressure, already threaten the tourism sector.[12] At the same time economic exigencies exert a strong pull on governments to rapidly expand tourism to satisfy employment and foreign exchange needs. This pull to expand is particularly strong in the face of limited alternative development options such as agriculture and manufacturing. This is the dilemma that governments of the region face in developing and managing tourism".

Blommestein (1993) continues; " If the Caribbean region is to maintain its advantage in tourism one pervasive myth needs to be dismantled. This is that environmental protection and the establishment and enforcing of standards and regulations, are cost increasing and hurt the tourism economy. While countries which are unwilling or

---

[12]For example Government of Jamaica, 1992

unable to become more environmentally efficient may and will suffer and experience losses of tourism earnings, those countries which comply will benefit greatly.

Over the last decade the demand for environment friendly tourism services has increased rapidly as a consequence of an increased environmental awareness by the tourist. Indeed market forces and global competition will force the region to become more environmentally conscious. However the adoption of a more environment friendly and sustainable tourism development model by then may be too late and the region would then have lost its competitive edge."

TABLE 2. Tourism Indicators

	Tourist Arrivals (ooo's)		Tourist Expenditures (US$mln)		Travel (net) as % of exports	Tourism Exp. as % of GDP
	1980	1990	1980	1990	1991	1990
Anguilla	5.7	31.2	1.3	29.3		67
Antigua Barbuda	86.6	205.7	42.0	298.2	302	87
Aruba	188.9	432.8	137.5	353.4	38	39
Bahamas	1181.3	1561.1	595.5	1333.0	319	47
Barbados		432.1	251.0	493.5	289	33
Belize	63.7	216.9	7.0	91.4	21	30
Br.Virgin Islands	97.0	160.6	42.3	132.1		96
Cayman Islands	120.2	253.2	44.6	326.1		54
Cuba	94.0	340.3	47.5	246.0		
Dominica	14.4	45.1	2.1	25.0	28	18
Dominican Republic	383.3	1533.2	167.9	899.5	109	
Grenada	29.4	82.0	14.8	66.6		43
Guadeloupe	156.5	288.4	110.9	230.7		
Guyana	40.0	64.2	...	30.0		9
Haiti	136.0	120.0	46.3	46.0	20	5
Jamaica	395.3	840.8	241.7	740.0	62	21
Martinique	158.5	281.5	74.6	240.0		
Montserrat	15.5	18.7	4.3	10.6		14
Neth Antilles	179.0	564.7	106.7	315.5	111	44
SMX	209.9	263.2	152.9	138.1		
Others						
Puerto Rico	1627.4	2645.4	594.7	1376.9		5
St. Kitts Nevis	32.8	75.7	13.4	63.4	219	48
St. Lucia	79.7	138.4	32.9	154.0	105	47
St. Vincent/Gren	50.4	53.9	13.7	46.0	38	28
Suriname		28.5	18.2	10.7		1
Trinidad Tobago	199.2	194.0	151.1	94.7		2
Turks and Caicos	11.9	41.9	4.2	31.0		44
US Virgin Islands	380.0	522.9	304.3	706.3		51

Notes:(a) 1989
Source: ECLAC/CDCC Sustainable Development Indicators Databank based on CTO and national data.
IMF for travel data

Apart from fisheries, tourism is the economic sector which depends most on the region's marine and coastal resources. Blommestein (1994) notes " The main driving factors are rest and relaxation, climate and beaches. This attention on climate and beaches has had as result that most tourist facilities are concentrated in the coastal zone." Jackson (1984) observes that the majority of Caribbean tourism facilities are concentrated within 800 meters of the high water mark. Often it is even more extreme. For example, in Jamaica one finds that 60 percent of the accommodation units are less than 15 meters from the high water mark. [Government of Jamaica, 1992]. Although there is this focus on the coastal zone, this does not imply that tourism as yet fully dispersed along these coasts. For instance in Jamaica 90 percent of the hotel rooms are to be found in four localities along the North Coast [Government of Jamaica, 1992]. Similar patterns can be observed in the other Caribbean islands.

Agriculture

Typically the agricultural sector comprises four components, *i.e.* export agriculture, domestic agriculture (which includes regional trade), fisheries and commercial forestry.

This sector has experienced weak or stagnant growth in most countries over the decade. This decline is due in part to policy, which has in many instances given priority to developing other sectors. As a result agriculture has become increasingly uncompetitive, not only accounting for a declining portion of GDP,(except in Haiti), but also experiencing absolute declines in total agricultural production in many of the VSIDS. While the contraction in the relative size of agriculture is in line with expected development trends, no viable commodity export earning activities have been identified, to supplant sugar and bananas, which continue to be of substantial importance and to exist because of protected markets.

The sugar industry continues on its secular decline. The industry is uncompetitive under current trade protection regimes, and survives only on quota systems provided by the European Community and the United States or, in the case of Cuba, special trading arrangements with Eastern Europe and the former Soviet Union. All these special arrangements are uncertain. In response to increased protected domestic production, sugar quotas from the United States were cut by 35 per cent in 1991 and arrangements between Cuba and former Council for Mutual Economic Assistance (CMEA) have been dismantled with the demise of that body, to be replaced by various bilateral arrangements which are themselves insecure.

The export earnings derived from sugar by the region,[13] declined over the survey period by about 36 per cent, representing a 40 per cent decline in the volume of sugar shipped.

---

[13] The aggregate analysis excludes Cuba and Haiti, due to inadequate data. The mean for the period 1980-1982 was compared with the mean for the period 1989-1991 to provide an indication of performance changes over the decade.

Declines were evident in all countries, with the exception of Jamaica which recorded growth in the last two years, as a result of vigorous field rehabilitation. For some countries, such as the Dominican Republic, the constraints on further expansion are related to the special arrangements to the United States market, for others, such as Barbados, the acreage under cultivation is being phased out and the efficiency of production has been declining. Output in Barbados in 1991 was the lowest in 60 years. For others, such as Jamaica the quest for foreign exchange has induced new measures to increase efficiency and output in the industry but any efforts will ultimately be limited by the size of the preferential quotas available to the region, since sale on the world market is currently uneconomic.

The banana producers benefited from improved banana prices but they maximized earnings by increasing output as well. Accordingly, earnings grew on average by about 16 per cent per annum between 1980-1991, volume increasing by about 11 per cent per year[14]. This volume increase was achieved by expansion of the banana cultivated area on marginal - often steeply sloped forested - lands, with the attended environmental impacts of increased erosion, deforestation and changes in the hydrological regimes.

Fairly widespread increases of over 10 per cent per year were recorded in export volumes in Dominica, Jamaica, St Lucia and St. Vincent. Yet, while the growth performance of the banana industry was one of the highlights of the period being surveyed, it is well to recall that for the three years prior to 1980 banana exports averaged 225,000 tonnes, a figure which was not surpassed by the region until 1986[15]. Especially rapid growth in export volumes was recorded in Jamaica between 1986 and 1991, although this fell short of 1973 production levels, by about 30 per cent. It is also noteworthy that banana production is not efficient enough to compete on the open market. The picture remains fairly good, as long as the United Kingdom market continues to be reserved for Caribbean producers, but this remains an uncertain prospect.

In many of the countries land tenure constitutes a significant constraint and there is at the same time an expansion of the agricultural frontier, as for example in Trinidad and Tobago by farmers who encroach on wetlands and forested areas, while at the same time arable land resources are under-utilized, or are being converted for housing, tourism and other development purposes [Blommestein 1993a].

Fisheries

Chakalall (1993) notes that the "coastal ecosystems of many island countries do not support highly productive fisheries. This is due to their limited or absent insular shelves

---

[14] Since 1991 banana production and banana exports have declines as a result in part by a depreciation of the UK pound sterling vis-a-vis the US dollar, in part by oversupply in the market and in part because of the new banana import arrangements set up by the European Union in the middle of 1993 (Harker Pers.com.)

[15] This was due to tropical storms and hurricanes

and the smaller quantity of nutrient run-off." This observation notwithstanding small-scale fisheries remains one of the most important socio-economic activities in SIDS.

Fisheries remains important for the region as fish and other seafood constitute a significant percentage in the daily protein and calorie supply especially in the VSIDS[16]. However, fishery resources in much of the insular Caribbean remain limited and there are anecdotal indications that the resources are over exploited. For example, in Barbados commercial fish landings declined by about 19 percent over the period 1980-1990. Likewise it is estimated that the insular shelf fish production in Jamaica declined from 10,500 mt in 1960 to 4,100 mt in 1992[17]. In the no- or negative growth countries additional stress is being placed on the resource by new entrants who are now engaged in fisheries because of a lack of alternative job opportunities.

Chakalall (1993) notes that many SIDS have not been able to take advantage of the resource potential of their Exclusive Economic Zones because of the capital and energy intensive nature of modern deep-water fisheries and the requirement of sophisticated skills. While some SIDS have licensed foreign fishing fleets to operate in their EEZ, DuBois (1985) notes that foreign fishing, especially the illegal one, represents a major obstacle to effective management of stocks.

Minerals

Minerals' producers have fared badly over the survey period As energy importers, the metals producers in the Dominican Republic, Haiti, and Jamaica were particularly hard-hit by energy price rises. Conversely, they were beneficiaries as energy prices fell and most[18] regional producers regained some of their competitiveness in the second half of the decade. As a result there was some resuscitation of the industry over the survey period although it was slow and halting. But in general the sector was buffeted by extreme uncertainty and consequent fluctuation in earnings.

While the metals producers benefited from the decline in oil prices, the oil producers were harmed. After 1982, prices fell as well as export volumes and consequently earnings. As a result, domestic activity in Trinidad and Tobago, the region's only oil exporter, was reduced drastically, with the economy contracting for 7 of the last 10 years.

For the future, the long-term prospects of Caribbean mineral producers will be shaped by the rate of growth in the industrial countries and consequently their demand for aluminum, the price of crude oil and the extent to which Caribbean operations can be

---

[16] See table 5.

[17] Griffith et. al. 1991 and Government of Jamaica, 1992.

[18] The bauxite industries in Haiti and the Dominican Republic did not survive the second energy shock.

reconfigured to increase their energy efficiency, since new efficient international production has been added to compete with regional producers in the last three years. In the final analysis, a development path cannot be predicated upon minerals exports alone. The terms of trade have been against minerals since 1950[19] and the amounts of minerals used to produce one unit of output have fallen steadily. Economies based on minerals have, therefore, grown much more slowly than, for instance, those specializing in services. Consequently, the expectations of those who supported the concept of raw materials-based development, fashionable in the 1960's and 1970's, have not been satisfied in the 1980's. Minerals' producers in the region will need to look at other resources to provide the main stimulus for growth in the future.

Mining in the very small Caribbean SIDS is largely limited to quarrying for aggregates and sand. The problem these islands face is that often the sand resources are limited and consequently beaches and dunes are mined. This invariably accelerates beach erosion, and diminishes the sustained earning potential of tourism.

Manufacturing

In view of transportation costs, diseconomies of scale, the wide existence of monopolies in commerce and manufacturing, it is not surprising that manufacturing in the Caribbean SIDS remains at a cross-roads[20] especially in the very small SIDS such as the OECS countries and the dependent countries, and, in general, has not kept pace with overall GDP over the survey period. Producers are being exhorted to reorient their production to global markets, yet, so far the means to do so seems to have eluded most of them. This failure is due, in part, to a protective regime within CARICOM which encourages high cost production for the regional market.

Over the survey period the sector increased its contribution to the GDP of Guyana, Jamaica, Puerto Rico and Trinidad and Tobago. Puerto Rico had the greatest growth in manufacturing and the greatest shift towards an economy based on manufactures. A similar shift also took place in Jamaica, mainly as a result of the increase in the garment assembly sector. In Trinidad and Tobago, the sector increased its relative contribution to GDP due only to a declining petroleum sector and contracting output. For the OECS countries, as a group, manufacturing contributed a relatively small percentage of GDP, at 7.6 per cent of GDP at the beginning of the decade, producing mainly light manufactures for domestic or regional consumption. By 1991 the sector had, however, decreased its contribution to GDP to less than 7 per cent.

The development and growth of the export processing zone (EPZ) or free zone, were noteworthy over the 10 year period, mainly because of its contribution to employment.

---

[19] It is estimated that in 1950, 100 units of minerals could buy 100 units of manufactures, but by 1986, 100 units of minerals could buy only 14 units of manufactures.

[20] These constraints are over and above those arising from protective regimes and other man-made obstacles.

Linkages to the rest of the economy remain few, and EPZs do not yet contribute significantly to GDP, since the value added by these activities is not high. They have been a source of much debate in the countries in which they have been introduced but are, nevertheless, expanding quickly in some of them. In general they seem, given the current level of manufacturing productivity and in the light of the prevailing unfavorable domestic policy framework, to be the only immediate viable option in this sector. However, unless workers and managers are able to upgrade their skills and proceed beyond low-wage, low-value added activities, it does not seem that EPZ can provide more than short term relief to the unemployment problem.

External sector

The fact is that all major merchandise export earning activities are in decline. Indeed, apart from the OECS banana producers, all countries of the region have shown an absolute decline in commodity exports (as measured in purchasing power) over the period between 1980 and 1991. The decline in the exports of agricultural products and minerals has not been compensated for by increased exports from domestic - and export processing zone manufacturing. Table 3 shows selected components of the balance of payments.

Sugar is inexorably being phased out in some countries and is only profitable in protected markets. Bananas have staged a comeback in output, but still depend on the preferential market in the United Kingdom. Since this market will not afford the same levels of protection in the 1990's the industry will need to contract by abandoning marginal lands and develop the means to stand on its own internationally. Nevertheless banana producing countries will need to pursue alternatives to bananas[21]. Minerals are of declining global importance, have unstable earning capabilities and in the long run cannot be relied upon to provide the main development thrust. As a consequence of weak sectoral performance, merchandise earnings have not been able to keep pace with expenditures so that merchandise trade deficits are the norm.

Imports continued to grow in all countries, except for those most affected by low-growth or contracting economies such as the Dominican Republic, Haiti and Trinidad and Tobago. The growth of imports was most pronounced in the high growth VSIDS depending on tourism. This, in part, reflects the dependency of this sector on imported goods and services and the inability to create linkages between tourism and other sectors.

Unrequited transfers from migrant workers remain important for most SIDS. The reliance of migrant remittances is most pronounced for Cape Verde where in 1991 net transfers were seven and half times as large as exports. The exceptions are small island

---

[21] As in St. Lucia, for example, where travel earnings are already higher than commodity exports.

TABLE 3. Balance of Payments
US $ million

	1981								1991							
	X exports	M imports	Trade Balance (X-M)	Current Account	Travel (net)	Private unrequited transfers	Travel as a % of X	Transfers as a % of X	X exports	M imports	Trade Balance (X-M)	Current Account	Travel (net)	Private unrequited transfers	Travel as a % of X	Transfers as a % of X
Antigua Barbuda	60	-115	-55	-19	40	7	67	12	32	-318	-286	-35	296	16	302	50
Aruba (1)									902	-1399	-497	-208	341	0	38	0
Bahamas (2)	200	-801	-601	-17	525	-20.0	263	-	320	-1154	-834	-180	1022	-2	319	-
Barbados	121	-481	-300	-26	233	21	193	17	144	-618	-474	-30	417	32	289	22
Belize	115	-146	-31	-10	2	20	2	17	120	-224	-104	-49	25	15	21	13
Dominica	10	-48	-38	-14	2	6	20	60	60	-104	-44	-26	17	13	28	22
Dominican Republic	962	-1520	-558	-720	7	724	1	75	658	-1729	-1071	-58	723	330	109	50
Grenada	17	-49	-31	0	2	12	12	71	27	-106	-80	-28	...	17	-	62
Haiti	215	-318	-103	-104	36	52	17	24	163	-300	-138	-11	33	86	20	53
Jamaica	963	-1038	-75	-166	229	82	24	9	1145	-1551	-406	-198	710	168	62	15
Neth. Antilles									254	-1018	-764	28	282	-47	111	-
St. Kitts-Nevis	24	-41	-16	-2	7	8	29	33	21	-98	-76	-50	46	6	219	28
St. Lucia	46	-113	-67	-33	34	11	74	24	105	-267	-162	-81	111	17	105	16
St. Vincent/Gren.	21	-52	-31	-9	12	12	57	57	66	-111	-45	-9	25	25	38	38
Suriname (1990)	514	-454	61	16	-14	7	-	1	466	-374	92	32	-12	-8	-	-
Trinidad & Tobago	2542	-1789	753	335	13	-42	1	-	1751	-1210	541	-17	-8	-15	-	-
Cape Verde									6	-123	-117	-40	0	45	-	750
Cyprus	489	-1079	-590	-242	147	33	30	7	875	-2363	-1488	-179	746	21	85	2
Guinea-Bissau (1990)	14	-73	-59	-41	-1	-17	-	-	19	-68	-49	-35	...	-2	n.a	-
Malta (1990)	510	-285	-375	45	275	33	54	6	1154	-1753	-599	-56	358	32	31	3
São Tomé-Principe(1990)	17	15	2	1	0	1	-	6	4	-13	-9	-14	0	0	-	-

Notes:(1) Includes oil not involving a change in ownership
(2) Excludes oil not involving a change in ownership
Source: IMF Balance of Payments Yearbook

countries such as Aruba, the Bahamas, the Netherlands Antilles and both Virgin Islands with an extensive tourism and off shore banking sectors and large migrant populations[22].

Nevertheless, current account deficits have been a chronic feature in the decade and have been financed by incurring foreign debt. For some countries this is no longer a viable option, however, since the cost of debt servicing now constitutes an impediment to continued growth.

On the domestic front, policy makers have increasingly come to realize that in small, open economies domestic economic space is defined by external performance. The assumption that external and domestic accounts could be insulated from each other, so that government action might be used to stimulate growth, has proven to be false. Accordingly, economic management over the decade has been characterized by a continuing quest to achieve balance in the fiscal and foreign accounts and to provide the policy measures which will stimulate the expansion of foreign exchange generating activities. While some success is being achieved in attaining the fiscal and external balances, the search for new sources of income has been less successful.

Human Resources

Population

There are wide divergences among the rates of population change in the Caribbean region which vary from -0.6 percent in Montserrat to +4.2 percent in the British Virgin islands (the latter is the result of immigration). The average rate of growth during the eighties stands now at 1.34 percent per year for the whole region. Although the majority of countries register growth rates, many VSIDS experience either close to zero growth, which is a growth rate of below 1.0 percent per year, or negative growth. These rates reflect, of course, different mixes of natural increase and migration rates (see table 4).

For the 1985-1989 period death rates vary among countries from about 5 to 11 per thousand. With the exception of Haiti, the observed differences are more the result of age structure differences than of mortality levels. Indeed all countries of the region, again Haiti excepted, now have life expectancy at birth of around 70 years or higher.

Fertility levels in the region have declined with many of the countries having experienced a nearly 50 percent drop in the total fertility rate levels since 1960, representing one of the more outstanding demographic transition phases on record.

---

[22] Negative net private transfers in Trinidad and Tobago may reflect past migration patterns.

Some countries, such as Aruba, Barbados and Cuba, have reached fertility levels close to or below replacement levels in contrast to other countries, such as Haiti, with fertility rates in the range of 5.0 to 6.0 per woman.

TABLE 4. Selected Demographic Indicators

	1990/91 Population ('oo's)	Total Fertility Rate	Crude Rates per Thousand Birth	Death	Natural Increase	Ave. Annual Rate of Growth 1980-1990
Anguilla						
Antigua Barbuda	62.9(m)	...	17(r)	6(r)	11(r)	...
Aruba	62.1(n)	1.8(g)	16(r)	6(r)	10(r)	0.2(j)
Bahamas	254.7(o)	2.1(a)	20(b)	6(b)	14(b)	2.0
Barbados	257.1(o)	1.6(c)	15(c)	9(c)	6(c)	0.6
Belize	190.8(m)	5.0(c)	37(b)	5(b)	32(b)	2.5(i)
Br.Virgin Islands	16.6(m)	...	19(c)	6(c)	13(c)	4.2(i)
Cayman Islands	25.4(k)	...	16(r)	5(r)	11(r)	4.1(j)
Cuba	10574.9(n)	1.9(g)	18(r)	6(r)	12(r)	1.0(j)
Dominica	71.8(m)	...	18(c)	5(c)	13(c)	-0.3(d)
Dominican Republic	7169.8(p)	2.8(g)	28(r)	7(r)	21(r)	2.7(q)
Grenada	90.7(m)	4.5(a)	33(a)	8(a)	25(a)	0.2(i)
Guadeloupe	385.5(o)	2.2(g)	19(r)	6(r)	13(r)	1.7(j)
Guyana	794.2(n)	2.8(b)	25(b)	6(b)	19(b)	0.5
Haiti	5939.0(n)	6.4(l)	47(r)	16(r)	31(r)	1.6
Jamaica	2248.2(m)	2.9(b)	25(b)	6(b)	19(b)	1.2(j)
Martinique	358.8(o)	2.1(g)	18(r)	6(r)	12(r)	1.0(j)
Montserrat	10.9(m)	2.3(e)	17(r)	11(r)	6(r)	-0.6(i)
Neth. Antilles						
Puerto Rico	3514.0(o)	2.3(g)	19(r)	7(r)	12(r)	1.0(j)
St. Kitts-Nevis	41.8(m)	2.8(c)	21(c)	10(c)	11(c)	-0.4(i)
St. Lucia	133.3(m)	3.4(b)	25(b)	6(b)	19(b)	0.8(i)
St. Vincent/Gren.	107.6(m)	3.1(a)	24(a)	6(a)	18(a)	0.9(i)
Suriname	402.5(n)	3.6(a)	26(r)	7(r)	19(r)	1.2(j)
Trinidad&Tobago	1234.4(o)	2.5(b)	21(b)	7(b)	14(b)	1.3
Turks and Caicos						
US Virgin Islands	101.7(o)	2.8(g)	23(r)	5(r)	18(r)	0.6(j)
Cape Verde	382.0	...	36	8	26	2.3
Cyprus	710.0	...	19	8	10	1.1
Guinea-Bissay	984.0	...	43	23	30	2.1
Malta	357.0	...	15	8	7	1.0
São Tomé-Principe	124.0	...	35	10	25	2.3

Notes: Caribbean data are census data. Others midyear estimates by UN Population Division.
(a)1987 (b)1989 (c)1988 (d)1981-1991 (e)1985 (f)1985-1990 (g)1990 (h)1991 (i)1980-1991 (j)1980-1989 (k)1989 census (l)1985-1987 (m)1991 census (n)January 1990 estimate (o)1990 census (p)1990 CELADE projection (q)1981-1990 (r)1985-1989 average
Sources: Boland, Barbara. 1993. Population Dynamics and Development in the Caribbean, ECLAC Population Unit.

Of concern remain the high incidence of teenage pregnancy and the low and stagnating contraceptive prevalence. Teenage births are associated with low levels of education and unemployment. Given the close association of these factors with poverty, it is likely that adolescent childbearing constitutes both a consequence and a cause of poverty. It also represents an obstacle to sustainable development in that young women who give birth in their teens, not only compromise their own educational, economic and personal development, but also jeopardize their children's life opportunities.

Infant mortality has in many Caribbean countries declined to less than one third of the immediate post war levels. Currently the rates are between 20 and 30 in most countries, with the exception of Haiti and the Dominican Republic (100 and 65 deaths per 1000 live births respectively). Advances in sanitation and public water supply and public health programs have been primary determinants of this reduction.

Caribbean populations remain young with over 50 percent under 25, although the proportion is slowly declining. With continued fertility declines it is expected that this trend will continue. In contrast and reflecting past population growth, massive increases are being experienced in the labor force age group (15-64), which is expected to grow even further by the year 2000. Percentage increases over the past two decades range from 85 percent for the Bahamas to a low of 11 percent for the Dominican Republic. The implications for education, training, employment and sustainable development policies are enormous.

Migration

The most recent data available for the 1990/1991 censuses suggest that, during the eighties, the Caribbean islands lost about 1.2 million people through emigration. Together with the migration losses during the fifties, sixties and seventies, the combined total is about 5.2 million people. This amounts to approximately 16 percent of the 1990 population of the total region[23].

In terms of absolute numbers the heaviest losses were experienced by the larger, most populous islands (Cuba, Dominican Republic, Haiti, Puerto Rico and Jamaica). As proportion of the population, however, the percentage was higher for the small Eastern Caribbean islands. For instance, Grenada recorded losses amounting to 56 percent of its 1980 population, while Montserrat and St. Kitts-Nevis recorded 65 and 70 percent respectively. The result was that for several of these countries, their population remained stable or even declined between 1980 and 1990/1991[24].

---

[23] Guengant (1992) quoted in Boland 1992.

[24] Simmons and Guengant, Caribbean Exodus: Explaining Country Variation in Net-migration Balance. 1990. Quoted in Boland (1992).

Migration appears to have been prompted by multiple causes; political crises; economic forces (poverty and unemployment); changes in immigration policies in potential destination countries; "culture of migration traditions"; and international linkages to cultural and kinship networks.

Caribbean migration to North America and Canada appears to have become female dominated[25], selective of the region's youth under the age of 29, and tend to be predominantly selective of highly skilled individuals[26], and other categories of human capital perceived to be especially scarce.

The impacts and effects of migration on development are, again, most pronounced in the VSIDS. For instance, the smallest of the region's islands will soon be the first to experience the dual treat of both "depopulation" and aging as a consequence of continuous migration. Another effect is the loss of investments in education[27] and the productive outputs of migrants in their most economically active years, including those who are highly educated and skilled. These effects are considered major obstacles to the development efforts of the Caribbean region. To a certain extent this is compensated by positive influences on development such as a reduction in unemployment or the receipt of remittances from abroad[28].

Education and human resource development

The West Indian Commission reports a skewed pattern of educational attainment in the English Speaking Caribbean. "All West Indian children have a chance of entering primary school, and a very high percentage of them stand a chance of completing. Nonetheless, other data suggest that large numbers of students graduate from primary school without the numerical skills needed to secure a job in the modern sector of the economy.

The probability of moving further up the educational ladder drops sharply at the secondary level, and even more sharply at the tertiary and university stages[29]."

The limited human resources are exacerbated by these low tertiary enrollment ratios which are particularly pronounced for the VSIDS[30]. While it is acknowledged that there

---

[25] This in contrast to earlier migrations which were highly male selective.

[26] For instance 45 percent of the university graduates in St. Lucia migrated over the 1975-1980 period. In another example the West Indian Commission reports that while there are about 4000 tertiary graduates per annum in the Caricom, region, about 1600, or 40 percent, emigrate.

[27] For example, it is estimated that in Jamaica the loss experienced in the 1950-1980 period was equivalent to sixty percent of the country's stock of graduates during the 1977-1980 period. This loss in investment was estimated at US $194 million.

[28] See also table 3, the columns of private unrequited transfers and transfers as a percentage of exports

[29] Report of the West Indian Commission, 1992.

[30] For example the West Indian Committee quotes a World Bank report which shows that the probability of entering Caribbean university is 0.25 percent, and 3 percent for entering other tertiary in Dominica.

are other factors which also contribute to and compound this low level of graduates, there can be little doubt that, because of transportation costs and diseconomies of scale, the very small SIDS are subjected to higher than average unit costs of producing a tertiary graduate, or of a secondary level technician. As the West Indian Commission states; "It is a hard reality that West Indian Governments are already devoting a share of their national budgets, and of their GNP's, which compare well with those of many other developing countries, and even of developed countries, "and further on" ...economic growth is unlikely to provide very substantial increments to the resources available to education." and " At the same time, there is only limited scope for increasing the share of education in national budgets by cutting back on other sectors [West Indian Commission, 1992]."

The Caribbean Environment Programme [UNEP, 1992] notes that " In most of the Caribbean countries the general public still lacks an adequate understanding of the linkages between development and environmental protection, and of the short- and long-term benefits and disadvantages of economic (including fiscal) and environmental protection," and notes further on that "investment in education and public awareness building is relatively small when compared with investment in development, and is the most cost-effective contribution to sustainable development". This lack of education and public awareness negates the one advantage of small islands, that is their social cohesion and the easy access by the general population to politicians and other decision makers. In view of the conclusions of the previous paragraph it may be very difficult to achieve an increase in investment in sustainable development education and public awareness. Nevertheless, this will be essential if Caribbean enterprises are to remain competitive.

Poverty

The analysis of economic indicators provides a useful picture of economic trends but is incomplete since it says little about the distribution of the costs and benefits of such economic performance. To the varying fortunes of the economies need to be added differing rates of population increase from country to country and varying impacts of economic contraction, especially of the reduction of government services, on varying groups within each country.

It is, nevertheless, possible to conclude that people in four Caribbean islands - the Dominican Republic, Haiti, Jamaica, and Trinidad and Tobago, as well as Sao Tome and Principe - have become poorer in the past five years. The decline in personal income, especially for the poorest, is made more onerous by the reduction in government services in all of these countries, given the high levels of public expenditures which prevailed and the need in some of them to earmark large portions of public resources for the repayment of the debt.

The decline in economic activity has affected the lives of people in these countries in a number of ways. Jobs have been lost and even for those fortunate enough to retain their employment, standards of living have fallen. The most readily available symptom of such decline has been the steady depreciation of the currency, which in small, open economies has a much greater impact on all sectors than in bigger countries having a large reservoir of domestic production[31].

While personal standards of living were falling in these countries, the capacity of governments to provide a social safety- net for the poorest was also diminishing, due to a contracting revenue base. In order to reduce growing fiscal deficits and the accumulating debt burden, public expenditures had to be reduced, since revenues could not easily be raised. The reduction in expenditures sometimes resulted in a reduction in the delivery of social services, in areas such as health, education, housing and, in some cases, nutrition.

Any attempt at evaluating the social impact of expenditure cuts has, nevertheless, to be treated with care. While the quantum of funds available for social services has in some cases been reduced, the proportions allocated for personnel emoluments and materials have also been skewed in favor of the former, further reducing efficiency. At the same time, new means might be used to deliver traditional services in a more efficient manner or to a more precisely defined target group so that a mere evaluation of expenditure might not signify reduced delivery. Finally, the backlog of social services might not be immediately quantifiable as the deficiencies in health or education might not become observable until after a large lag; yet by then rectification, if still possible, might be protracted. Attempts had, therefore, to be made to discern social trends despite the lack of precise data to measure them.

Bearing in mind the caveats noted above, it is, nevertheless, possible to arrive at a few tentative conclusions. In some countries the incidence of poverty has increased significantly. In Jamaica[32] the incidence of individuals living below the poverty line increased from 66 per cent of all employed workers in 1977 to 76.5 per cent in 1989. The rate of growth of the hard-core poor was even faster, since those earning 75 per cent or less of the poverty line income increased from 51 per cent in 1977 to 68.6 per cent in 1989. The segment showing the highest incidence of poverty remained peasant agriculture, although even here the percentage of the segment living in poverty increased from 83 per cent in 1977 to 91 per cent in 1989.

In Trinidad and Tobago[33] the level of poverty is estimated to have tripled between the years 1981-1989, with between 11-15 per cent of households, depending on the criteria used in the definition, living in poverty in the latter year. People of African descent showed the highest level of poverty, as did female headed households. There was a strong link between poverty and unemployment, and this corroborates the estimate of

---

[31] In Jamaica for instance, domestic food production meets only an estimated 10 percent of domestic food needs.

[32] Taylor, LeRoy (1992).

[33] Henry 1992.

growing poverty over the decade since unemployment increased from 10 per cent in the early 1980's to 22 per cent in 1989. Where actual poverty studies are not available inferences can be drawn from the steep economic decline which has been experienced in various countries.

Anecdotal evidence suggests deterioration in the social indicators in other countries, such as, for example, the Dominican Republic and Haiti. Among the other countries covered in this survey Cape Verde, Sao Tome and Principe, Haiti and Guinea Bissau were all categorized as countries having a low level of human development, with low levels of income and social services and fairly low social indicators, as a consequence. The economic difficulties experienced by these countries over the last decade suggest a decline in the quality of life of these people as well.

In all the cases studied the most vulnerable groups among the poorest were pregnant and lactating mothers, children and the aged. Rural groups tended to be poorer than urban groups and the largest families were in all cases the poorest.   Where economic contraction has taken place for a period longer than the decade, signs of social erosion are discernible in health, education and nutrition, so that relative standards have fallen as compared with competing countries having recorded more balanced growth over the decade. But absolute standards have also fallen in some instances, implying declining indicators in the same country over time. The significance of this erosion for future development prospects should not be underestimated, especially as this will be predicated on human resourcefulness.

For some countries the backlog in housing is quite considerable, leading to urban slums and often illegal squatter settlements. While high levels of unemployment have been endemic, the growth of the informal sector is a manifestation of the effort of the unemployed to devise and carry out their own survival strategies. Yet the concentration on low productivity activities, such as itinerant petty peddling and the provision of simple services, attests to the low levels of skills possessed by the hard-core unemployed.

Environmental Management Issues

Quite apart from the earlier referred to economic and human resource constraints, the region is also faced with global, regional and national environmental problems which cast doubt on the long-term capability of the region's renewable resources to serve as economic resources in the future. Increased acknowledgment of these problems and their potential negative implications for the region's development potential has resulted in a marked increase in environmental awareness at all levels. In the region this has been expressed in strong support for the Caribbean Environment Programme and for the Caricom states, for the Port of Spain Accord on the Management and Conservation of the Caribbean Environment. In recent years and increasingly through the Association of Small Island Developing States (AOSIS), SIDS have been active participants in the negotiations for UNCED and the SIDS Conference and global treaties such as the Convention on Biological Diversity, the Framework Convention on Climate Change or the Montreal Protocol.

Impacts arising from global issues such as climate change and sealevel rise, or depletion of the ozone layer may negatively impact on the region's resources and hence may further constrain the already limited development options. Impacts of global issues have in common that the people of the region bear little responsibility on their occurrence, have no management control on their causes, but bear potentially high risks and costs. The increase in flood risk, for example, is larger than average for small islands, while for SIDS the annualized costs for protection or adoption to sea level rise as percentage of GNP could be as much as 10 to 300 times higher than for continental countries in North America or North and Western Europe[34]. Current levels of the GDP of the SIDS and support by the international community are unlikely to generate sufficient funds to successfully implement mitigation policies and programs. These costs do not take into account future potential losses in tourism earnings arising from climate change impacts and the depletion of the ozone layer, which would further depress the GDP of SIDS. Such losses could arise from a loss of beaches and from an increase in malignant skin neoplasms in the tourist markets which could reduce the demand for sun, sand and sea holidays. There remains, however, a need to a better understanding of the range of issues involved in the implications for SIDS of global issues.

The impacts of regional problems, such as, the transboundary trade and transhipment of hazardous, toxic and nuclear wastes, the dumping of solid wastes by cruiseships, the management of transboundary fishstocks or the transhipment of illicit drugs, exacerbate the impacts of global environmental issues. Again the region has little management control, although at times, as in the case of cruiseships, for example, they bear somewhat more of the responsibility[35].

Here the management problem is one of enforcing national and international legislation of the ocean areas under national jurisdiction, which, for small islands, are several times larger than their land mass. Again, expressed either as a percentage of GDP, or in terms of human resource requirements, higher than average costs are likely to inhibit effective management of the ocean resources.

Superimposed over global and regional issues are national issues, over which SIDS have, at least in theory, management control. Here the dilemma of SIDS, like other developing countries, is the need for further and accelerated economic growth, particularly in the poorest SIDS. Poverty levels cause resource management problems such as deforestation and overfishing or environmental health problems such as the

---

[34] See Intergovernmental Panel on Climate Change (1992), table 5.

[35] However the proposed OECS port reception facilities project for Annex 5 wastes of Marpol 73/78 does not address waste minimization or reduction on board of cruiseships. Hence, implicitly, it assigns the sole responsibility of reducing solid waste dumping at sea to governments.

disposal of liquid wastes. At the same time, the more modern sectors, such as mining or tourism or manufacturing ,cause problems which are beyond the regulatory and management capabilities of the SIDS. In this the SIDS are not unique. However what makes the problem a more challenging one for SIDS is that island habitats tend to be small or very small with a high degree of endemism in relation to their size. For example, the International Council for Bird preservation (ICBP) indicates that "the Caribbean supports over 130 restricted range species, with nearly every island in the region included within or forming an endemic bird area[36]".

Although there has been progress during the 1980's, it appears that the Caribbean remains faced with three types of problems with respect to sustainable development. The first one relates to the planning, allocation and use of renewable resources and non-renewable resources, the second is linked to problems of resource management, while the last pertains to environmental degradation and pollution[37].

TABLE 5. Per Capita Protein Supply (grams/day)

	1969-1971			1986-1988		
	Grand Total	Total Animal	Fish and Seafood	Grand Total	Total Animal	Fish and Seafood
Antigua Barbuda	60.0	36.9	10.1	67.3	41.9	7.7
Bahamas	79.4	48.6	4.3	76.5	47.9	2.7
Barbados	79.5	43.9	7.9	99.1	61.7	9.4
Cuba	67.9	33.9	4.3	77.4	37.3	6.0
Dominica	53.9	24.3	7.0	70.2	31.4	4.7
Dominican Republic	43.8	16.4	2.4	48.7	17.3	2.0
Grenada	63.0	30.1	11.5	82.3	46.4	11.2
Guadeloupe	66.9	30.0	9.8	89.2	52.7	14.8
Haiti	45.0	6.7	0.4	49.4	9.7	1.6
Jamaica	65.9	30.3	9.2	63.5	26.1	4.4
Martinique	69.5	35.7	12.2	84.7	45.2	15.5
Neth. Antilles	69.4	43.6	5.3	88.9	58.2	8.9
St. Kitts Nevis	47.4	22.0	...	76.6	41.4	7.8
St. Vincent/Gren.	51.5	21.3	8.2	66.9	31.6	6.0
Trinidad & Tobago	65.1	26.9	3.9	76.2	34.9	4.3
Cape Verde	45.0	7.5	3.9	64.6	15.5	6.9
Malta	96.8	39.0	3.0	97.9	51.7	4.5
São Tomé-Principe	47.6	10.2	5.3	47.3	13.2	7.5
World	64.8	21.7	3.3	70.4	24.5	3.8
Developed Countries	95.9	51.5	6.4	102.9	59.4	7.6
Developing Countries	52.0	9.5	2.0	59.8	13.2	2.6

Note: Other countries: no data

Source: ECLAC/CDCC Sustainable Development Statistics Databank, based on FAO Food Balance Sheets, FOA (1991)

---

[36] ICBP (1992).

[37] The identification of the problem areas is based on Field et al. 1988.

The problems of resource use allocation and planning relate essentially to policy matters and tend to affect several economic and social sectors and by implication, several government ministries and agencies. It is evident that responsibility and inter-ministerial co-ordination and co-operation are often enigmatical, resulting either in inaction or reacting to crises situations only. Furthermore it is not certain whether development in the region is carried out within a framework of an environmental policy and it is often alleged that decisions on both renewable and non-renewable resource use are taken without considering environmental consequences.

The problems of resource management tend to pertain to the management of single sectors and areas of responsibility are usually more defined. These problems stem partly from the uncertain broad policies as referred to above, from unclear or conflicting sectoral goals and objectives and are exacerbated by financial and manpower constraints, weak management and the existence of many small ecosystems and high rates of bio diversity. In this respect, specific areas of concern are, for example, forests, biodiversity and marine resources, water, beaches and soil erosion.

As far as environmental degradation and pollution is concerned, among the results of policies to achieve economic growth, are pollution of fresh and coastal waters, resource depletion and habitat destruction. The exact dimension of pollution and resource deterioration is difficult to determine as systematic monitoring and data- analysis are still in a rather embryonic phase. There are, however, sufficient data from ad hoc studies to warrant the conclusion that the disposal of industrial and agricultural effluent and of domestic waste is less than adequate.

The continued existence of the problems stem in part from past approaches towards environmental management and sustainable development, which have tended to:

-focus on the public sector and on physical planning in particular,
-exclude the impacts which environmental degradation may have on production, in particular on agriculture, fisheries and tourism,
-rely on physical planning guidelines and legal instruments, both of which face constraints in enforcement,
-neglect of the use of economic and financial incentives and disincentives, and
-fail to show that incorporating the environment pays.

A summary of the key environmental issues is contained in the *"Environmental Agenda for the 1990's*[38]*"* which was specifically prepared for the VSIDS of the Eastern Caribbean. This summary is reproduced below:

---

[38] Towle, Judith (ed), 1991.

Key Environmental Issues

Forests

*   Economic pressures to increase export crop production which have resulted in accelerated clearing of forested land and the establishment of agricultural plots on ever steeper hillsides - areas highly susceptible to erosion.
*   The absence of controls to protect against soil erosion, resulting in an increased risk of landslides, flooding, and excessive downstream siltation when forested lands on steep slopes and unsuitable soils are cleared for agriculture, production forestry, fuelwood, road building and other construction activities.
*   The failure of Eastern Caribbean Governments to consistently defend critical water catchments against non-forested land use activities, thus placing at risk the continued availability of reliable sources of domestic water.
*   The lack of land use regulations designed to prevent agricultural and residential encroachment in designated "protected" forest areas.
*   Uncontrolled livestock grazing, contributing to land deterioration, deforestation, erosion, and general denudation of the landscape.
*   The absence of incentives to increase agroforestry and plantation forestry on private lands or to compensate landowners for improved landscape/forest management practices such as terracing or reforestation.
*   The need to improve the understanding of political decision-makers about the value of forests, particularly with regard to soil and water conservation.

Watershed Management and Protection of the Water Supply

*   Increased demand for a reliable source of safe drinking water, particularly near urbanized areas and clustered tourism facilities, at the same time that pressures to expand non-compatible land uses to critical water catchment areas have also increased.
*   The failure of legislative restraints and existing land use control measures to fully protect upper water catchments and forest reserves against deforestation and encroachment.
*   Limited expansion of protected watersheds by planned programs of land acquisition.
*   The relative ineffectiveness of land use controls in placing restrictions on the use of private lands within catchment areas, a particularly critical issue since a significant proportion of the catchment area needed for the maintenance of water systems is under private ownership.
*   The difficulty of finding solutions to water-related problems when the issues involved are political and social - as well as technical.
*   Failure to tie metered water income to revenue requirements for watershed management and acreage expansion.
*   The need for more comprehensive water policies and the integration of water supply requirements into national development planning and decision-making.

Coastal/Marine Sector

*   Lack of detailed information on critical marine habitats for use in decision-making.
*   The need for a better balance between the pressure to develop the fishing industry and the necessity for improved management of the resources upon which the industry is based.
*   Accelerated "piecemeal" development of the coastal zone with minimal consideration of the cumulative impacts of development activities and projects.
*   Significant loss of critical coastal habitats such as mangroves and coralline structures that serve as important controls of shoreline erosion.
*   Increased water quality degradation, associated with accelerated urbanization and tourism development in the coastal zone.
*   Unregulated removal of sand and vegetation in the coastal zone, resulting in increased rates of coastal erosion.
*   Increased risk of coastal flooding in the absence of enforced coastal set-back requirements.
*   Absence of a standardized requirement for environmental impact assessments of all large coastal development projects.
*   Lack of comprehensive development control guidelines and policies committed to maintaining the quality of coastal resources.

The Agricultural Sector

*   Economic pressures to expand banana production acreage and inputs without sufficient regard for the environmental consequences of unregulated banana cultivation on the natural resource base.
*   The need to improve the institutional structures, managerial performances, and technical expertise of the small farm sector, following the decline of the previously dominant large estate system.
*   The lack of sufficient incentives, extension services, soil conservation investment subsidies, and marketing assistance to further diversify the agricultural base away from its current emphasis on annual subsistence and semi-perennial export crops.
*   The prevalence of land tenure insecurity among small farmers who - in the absence of other incentives - are unwilling to pursue costly land conservation strategies, the benefits of which might not accrue to them in the future.
*   The inadequacy of quantitative data on agrochemicals (importation, use, impacts) upon which to base informed decisions.
*   The general failure of OECS countries to effectively implement extant pesticide legislation or to provide up-to-date pesticide control regulations and monitoring procedures.
*   The lack of adequate land use planning or zoning restrictions in the agricultural sector to ensure the continued availability of environmentally-suitable and economically-productive lands for cultivation.

Tourism Sector

*   Unquantified and unresolved linkages between growth in the sector and associated environmental and social issues.
*   Inadequate strategic planning in the sector, resulting in the emergence of tourism "styles" more often by chance than by deliberate choice as part of a planning process which has assessed the social and environmental implications of each alternative approach.
*   Failure of OECS Governments to give sufficient attention to an integration of economic planning and land use planning so that tourism development reflects carefully analyzed carrying capacity considerations.
*   Tendency to increase the importation of goods and services to support tourism as the industry expands.
*   The lack of comprehensive coastal zone management programs to reduce the negative environmental impacts of tourism infrastructure and activities in coastal areas.
*   Existing problems with basic services and infrastructure exacerbated by rapid expansion of tourism.
*   The need for visitor impact mitigation and management strategies.
*   The need for better public/private sector coordination in the identification, development and management of natural and historical attractions and amenities.
*   The need to diversify the tourism base by more aggressive marketing of nature-based tourism, along with a concurrent development of site management plans for targeted natural areas and the identification of sufficient resources to manage each site.

Pollution Control

*   Low level of awareness among decision-makers, businesses, and the general population about pollution issues and their costs to the community and the economy over time.
*   Lack of strategic planning to develop comprehensive national plans for limiting waste generation and for putting in place policies for waste management, pollution control and recycling.
*   Seemingly costly solutions to environmental pollution problems without proper assessment of innovative means for raising needed revenues by establishing licensing and discharge permit fees for waste disposal services.
*   Failure to provide for contingency planning at the national level for oil and other hazardous material spills on land, in ports, harbors and marinas, and in coastal waters.
*   Insufficient documentation on the quantitative and systemic aspects of environmental pollution to permit easy development of remedial or regulatory measures.

* Out-dated public health and water legislation, lacking regulations, national standards and modern criteria for water quality, pollution control, and waste management.

Parks and Other Protected Areas

* Relative lack of understanding and appreciation for the economic and social benefits to be derived from the protection of indigenous natural and cultural resources, often resulting in government policies of benign neglect.
* Lack of effective protection for scarce or threatened resources unless they are placed under the management control of a specific authority or happen to fall within the boundaries of a designated protected area.
* Fragmented institutional responsibilities for the protection, development, and management of critical natural areas and historical resources, including shipwrecks and archaeological sites.
* The absence of environmental impact assessment procedures for major development projects, including an evaluation of natural and cultural resources prior to project approval or commencement of construction.
* Failure to provide an integrated management framework for the development and use of outstanding natural and historical resources and scenic areas.
* Failure to provide a national system for establishing priorities and criteria for protected area designation, for phasing designated areas into the system, and for managing classes or categories of protected areas within the system.
* The need for better linkages between park planning and other development sectors, especially tourism, water supply, recreation, education, and fisheries.

Environmental Management

* Environmental decision-making on an ad hoc basis, focusing on short-term or interim rather than long-term policy objectives for the environment.
* Fragmentation of environmental management functions within OECS Governments due to the diffusion of these responsibilities among a number of departments representing several ministries and statutory bodies.
* Ill-defined lines of institutional authority with regard to the management of land, water and cultural resources.
* Generally weak coordination among government agencies and statutory bodies with related environmental management responsibilities.
* Lack of an officially approved, up-to-date land use planning frame work in all OECS countries, thereby reducing the overall effectiveness of development control procedures.
* The need to strengthen the planning process and to upgrade the responsibilities and capabilities of physical planning units, including the environmental expertise of planning staffs.

*   The need for environmental impact assessments, as one means of forcing a systematic examination of environmental and social issues at an early stage in the planning process.
*   Shortage of trained and experienced technical personnel for environmental planning, monitoring and enforcement activities.
*   Outdated environmental legislation or lack of supporting regulations to much of the extant body of environmental law, thus diminishing the overall effectiveness of the legal base for resource protection.
*   Failure of OECS Governments to take full advantage of the creative energies, resources and expertise of NGO's with environmental interests and agendas.

For a region that still remains dependent on the exploitation of its natural resources a sustainable approach to development is crucial. For the next decade or so it is anticipated that tourism will remain the most dynamic sector in all but a few Caribbean countries. During the decade of the eighties environmental issues have come to the foreground, but have not as yet resulted in significant policy changes. While SIDS and VSIDS may face particular constraints in their development options it does not reduce their sensitivity to environmental degradation. Indeed one can argue that SIDS are more dependent, given their focus on tourism and agriculture, than larger countries on integrating environment and development. Not all the characteristics of SIDS are negative, however. For example, their small size can be an advantage in that community participation may be easier to achieve. Also, again because of their small size, experiments in sustainable development, may be easier to evaluate, especially the effects of environmentally sound development on economic variables such as GDP, trade in goods and services, employment and government revenue.

Towards a Sustainable Development Strategy

ECLAC/CDCC (1992) has argued that there may be need for a greater degree of precision as to what the issues are as they pertain to SIDS and precisely in which areas or for what kinds of programs these countries would wish support from the international community. These countries might wish to be careful not to argue on the basis of the supposed uniqueness of their situations, nor on the basis that they are among the neediest countries of the world community. Many of the recommendations for dealing with SIDS too closely resemble general development prescriptions and do not often enough appear to hinge on particular characteristics of these countries. Uniqueness may not so much be inherent, but may consist of specific features such as the magnified effect of certain natural and economic occurrences, the dependence on a very few (usually tropical agricultural) commodities, the lesser resilience of these economies, and the high per capita cost of public administration, physical infrastructure and overseas representation. On the positive side certain SIDS characteristics may render them particularly deserving of international support in certain areas, such as the development of tourism potential, and off-shore financial services.

The question arises as to precisely what it is that the SIDS are seeking from the international community. Clearly they would wish the international financial institutions and bilateral donor countries to have regard to their situations in negotiating financing arrangements. These terms should be as concessional as possible and for as long a term as possible. This may appear to be justified by the strains on the budgetary and fiscal situations of these countries deriving from the inevitable smallness of the tax net, and the high per capita cost of public administration and infrastructure of all kinds. This is a small country consideration. They would wish arrangements to be put in place which seek to address particular identifiable vulnerabilities. Examples of such arrangements might be export compensation arrangements, disaster relief and rehabilitation programs, and measures to deal with global warming and sea level rise. Problems in these areas are felt particularly severely in island developing countries; and constitute justifiable areas of need from this standpoint.

From a more all-encompassing standpoint, they require special programs of economic diversification, either within the agricultural sector, or away from the agricultural sector depending on the situation of particular countries. This exaggerated dependence, which tends to afflict island economies, especially the smallest among them, is an important aspect of the vulnerability of these countries. Focused and sustained programs for economic diversification would be particularly desirable for purposes of assisting the countries to reduce dependence, increase the possibilities of sustained economic growth and also deal with a pervasive aspect of smallness. It may be that the scope for diversification is itself constrained by the small size of SIDS, for which reason it has been argued that "flexible specialization" is the approach that small countries should adopt towards decision-making on the allocation of resources to productive activities[39].

Regional cooperation arrangements, possibly including regional economic integration schemes, aimed at combining the scarce resources and limited potential of individual island countries, commend themselves to the international community. It may perhaps be difficult to make a case for support on the basis of some of the island (especially, small island) characteristics, while these countries continue their hesitation in taking certain obvious steps towards further cooperation in areas such as joint overseas representation. Another possibility for addressing the high level manpower resource shortages of SIDS which takes into account the openness and dependence of these countries, is for United Nations and other bodies to assist in identifying of the implications of certain global developments for SIDS.

In addition to arguing the case, which has been done in many a forum, even now in more focused terms along the lines indicated above, island developing countries must

---

[39] Flexible specialization refers to an approach to the organization of production which has been recommended for SIDS and has been adopted by a number of them. Fiji calls it a "niche-filling export strategy". It involves competition on the basis of products rather than price, by creating niches in a variety of small-scale, high value added industries, in which design would be a major component and in which all aspects of production from management, design, distribution, and sourcing to strategic thinking would be flexible. UNCTAD 1992.

now seek to intensify their efforts to put specific proposals to the international community.

Quite apart from compensating for disadvantages, project proposals with an SIDS perspective may be geared to taking advantage of potential advantages which these islands possess. Pursuit of such project possibilities would be part of the "normal" pursuit of development possibilities, and exploitation of development potential in which all countries engage. To say this, in no way minimizes the necessity, with donor support where appropriate, of identifying this potential and taking every possible opportunity of developing it.

Following the recognition of Agenda 21, the SIDS Programme of Action[40] goes beyond the generalities in identifying - albeit largely environmental - a more sector-focused approach. It remains generic in the sense that the Programme of Action aims to cover all islands. Hence, there still remains, a need for programs and projects to be formulated for presentation to the national populace and its political leadership and, where appropriate, for presentation to donors. At the national level this is largely the responsibility of the countries themselves.

It must be borne in mind that Barbados is but one culmination in a long term process. It facilitates the international recognition of problems we - the islands - have been facing for a long time but never put together in a clear comprehensible package. This has now been achieved and it is largely up to us to determine the way we follow through.

At this interim stage, conclusions remain tentative, and perhaps providing a guideline for future thought. The complexities and interrelationships of causes and effects between environment, economic growth and development pose a formidable challenge to both decision makers and environmentalists. This challenge is not just intellectual, but is even more fundamental, since ultimately it affects the quality of life of the region's inhabitants. This challenge is also posed within a framework of rapid political and economic change in the rest of the world. The small states of the region will have to face continuing and possibly accelerating adjustments to changes in comparative advantage. Perhaps the real dilemma of sustainable development within the Caribbean region lies in the development of the capacity to react to short and medium-term social and economic changes without losing sight of long term goals [ECLAC/CDCC, 1994].

If the region gets caught in a vicious cycle of stagnant or deteriorating economic and social conditions, then improving natural resource use will be difficult, if not, outright impossible. If, however, the region can achieve positive economic and social changes

---

[40] The SIDS Programme of Action has taken a three prong approach: (a) The management of environmental problems (climatic change and sealevel rise, natural and environmental disasters and management of wastes); (b) The management of resources of SIDS (such as land, coastal and marine, fresh water, energy, tourism, biological diversity) and, (c) Building the capacity to promote sustainable development (national and administrative capacity, regional institutions and technical cooperation, transport and communications, science and technology and human resource development including education, training, health and population; transfer of technology and finance).

then there is the chance that sustainable development can be achieved at higher levels of production and quality of life [ECLAC/CDCC, 1992].

Assessing the consequences of attempting to achieve sustainable development will imply an evaluation of the economic, social and natural systems of the region and of their individual components. But it is more than that, because we need to carry out such an analysis guided by the formulation of appropriate questions and assumptions. In its most simple format, we could perhaps derive a triad of questions relating to the main developments which can be expected within the region over the short, medium and long terms; expectations regarding resource use and the production of wastes and physical changes; and the impacts which these changes will have on the societies of the region. Such questions need to be answered within a framework of suitable and variable assumptions with respect, for example, to economic growth, technology and social cohesion. The assessment then, could include an analysis of the causes of the current problems; the factors which will influence the future, past, current and future options for actions and policies; and the derivation of strategies which could enhance positive impacts and reduce negative effects. The outcome of such an evaluation would not be a forecast or plan since planning under the long-term horizons implicit in sustainable development is impossible. What it would provide, though, is some indication of where interventions may be meaningful and feasible. Perhaps more important is that it would force decision makers to question the validity of existing development paradigms and dogmas, thereby initiating a process of change [ECLAC/CDCC, 1991].

References

Blommestein, E., Environment and Agricultural Credit, Inter American Development Bank, Washington, D.C., 1993(a).

Blommestein, E., Critical Issues in Sustainable Tourism and Environmental Health: Proposals for Action, WP/93/6, ECLAC, Port of Spain, 1993(b).

Blommestein, E., Towards Sustainable Tourism in the Caribbean, (forthcoming), ECLAC/CDCC, Port of Spain, Trinidad and Tobago, 1994.

Boland, B., *Population Dynamics and Development in the Caribbean*, ECLAC, Port of Spain, 1992.

Briguglio, L., Les Vulnerabilites Economiques de Petits-Iles En Voie De Developpement, Fondation d'Etudes Internationales de l'Universite de Malte, Valetta, Malta, 1993.

Brookfield, H.C., An Approach to Islands, *In Sustainable Development and Environmental Management of Small Islands*, edited by W. Beller *et. al.*, pp. 23-34, UNESCO, 1990.

Chakalall, B., Fish Processing Technology at the Microenterprise Level in Small Island Countries, FAO Regional Office for Latin America and the Caribbean, Santiago, Chile,1993.

DuBois, R., Coastal Fisheries Management. Lessons Learned From the Caribbean, *Coastal Resources Management: Development Case Studies*, Renewable Resources Information Series, Coastal Management Publication No.3, edited by J.R. Clark, pp. 20-32, National Park Service/Agency for International Development. Washington, D.C., 1985.

ECLAC/CDCC, Sustainable Development in the Caribbean, LC/CAR/G.307, p. 14, ECLAC, Port of Spain, Trinidad and Tobago, 1991.

ECLAC/CDCC, The Issue of Special Status for Island Developing Countries, LC/CAR/G.374, p. 26, ECLAC, Port of Spain, Trinidad and Tobago, 1992.

ECLAC/CDCC, Overview of Economic Activities in Caribbean Countries -1992, LC/CAR/G.398, ECLAC, p. 49, Port of Spain, Trinidad and Tobago, 1993(b).

ECLAC/CDCC, ECLAC/CDCC and the Global Conference on the Sustainable Development of Small Island Developing States, LC/CAR/G.418., p. 24, ECLAC, Port of Spain, Trindad and Tobago, 1994.

FAO, *Food Balance Sheets*, p. 384, FAO, Rome, Italy, 1991.

Farrel, T.M.A., The Concept of Small States: Current Problems and Future Prospects with Special Reference to the Caribbean, p.30, Institute of Social and Economic Studies, Cave Hill, Barbados, 1991.

Field, R. M., A. Goodridge, H. McShine, R. Ramdial, and R. Warren, An Assessment of Options for Environmental Management in Trinidad and Tobago, p. 42, UNEP/ROLAC, Mexico City, Mexico, 1988.

Griffith, M., V. Inniss, J. Wilson, and I. King, *Barbados National Report to UNCED 1992*, Ministry of Labor, Consumer Affairs and the Environment, Bridgetown, Barbados, 1991.

Government of Jamaica, Jamaica National Report on the Environment and Development, Government of Jamaica, Kingston, Jamaica, 1992.

Guengant, J-P., Current Demographic Trends and Issues, UNFPA Symposium on Population and Development, UNFPA, New York, 1992.

Harker, T., Towards Sustainable Development Policies, p.26, ECLAC, Port of Spain, Trinidad and Tobago, 1993(a).

Henry, R., Regional Conference on Poverty: The Trinidad and Tobago Case, LC/CAR/G.405, p. 50, ECLAC/CDCC, Port of Spain, Trinidad and Tobago, 1992.

Intergovernmental Panel on Climate Change, Global Climate Change and the Rising Challenge of the Sea, p. 35, UNEP/WMO, Geneva, Switzerland, 1992.

Jackson, I., Enhancing the Positive Impact of Tourism on the Built and Natural Environment, Reference Guidelines for Enhancing the Positive Socio-cultural and Environmental Impact of Tourism Vol.5, p. 42, Organization of American States, Washington, D.C., 1984.

Lestrade, S., Economic Issues Affecting the Development of Small Island States: The Case of the Caribbean, *The Courier*, No.104, July-August, 1987.

Simmons, A., and J-P, Guengant, Caribbean Exodus: Explaining Country Variation in Net-migration Balance. Mimeo, ORSTAM, Guadeloupe, 1990.

Srinivasan, T.N., The Costs and Benefits of Being a Small, Remote, Island, Landlocked or Mini-state Economy, World Bank Discussion Paper, pp. 1-31, March 1985, World Bank, Washington, D.C., 1985.

St. Cyr., E.B.A., Some Fundamental Propositions in the Theory of Caribbean Economy, *Social and Economic Studies*, Vol.40, No.2, pp. 137-152, Institute of Social and Economic Research, University of the West Indies, Kingston, Jamaica, 1991.

Taylor, L., Poverty in Jamaica: A Review of Recent Evidence, CONS/92/14, p. 22, ECLAC/CDCC, Port of Spain, Trinidad and Tobago, 1992.

Towle E., The Island Microcosm, in *Coastal Resources Management: Development Case Studies*, Renewable Resources Information Series, Coastal Management Publication No.3, edited by J.R. Clark, pp. 52-74, National Park Service/Agency for International Development, Washington, D.C., 1985.

Towle, J., (ed.), Environmental Agenda for the 1990's, A Synthesis of the Eastern Caribbean Country Environmental Profile Series, p. 71, Caribbean Conservation Association and Island Resources Foundation, Bridgetown, Barbados/ Washington, D.C., 1991.

UNCTAD, Report of the Meeting of the Group of Experts on Island Developing Countries, UNCTAD, Geneva, Switzerland, 1988.

UNCTAD, Specific Problems and Needs of Island Developing Countries: Report of the Secretary General of UNCTAD to the General Assembly, UNCTAD, Geneva, Switzerland, 1992.

UNEP, Environmental Problems Affecting the Marine and Coastal Environment in the Wider Caribbean Region, UNEP (OCA)/CAR/IG.9/Inf., p. 47, UNEP Caribbean Environment Programme, Kingston, Jamaica, 1992.

United Nations, Report of the United Nations Conference on Environment and Development, Vol. I, Annex II, Agenda 21, A/CONF.151/26/Rev1 (Vol. 1), p. 486, United Nations, New York, 1993.

United Nations, Programme of Action for the Sustainable Development of Small Island Developing States, (advanced unedited text), United Nations, New York, 1994.

West Indian Commission, Time for Action: The Report of the West Indian Commission, The West Indian Commission, Black Rock, Barbados, 1992.

24

Socio-Economic Databases in the Caribbean: Status and Desiderata

Lancelot A. Busby

Abstract

This paper examines the question of data availability to assist the planning process. It evaluates the holdings of information sets that can effectively guide governments to making development plans that place importance on sustainable development. The paper examines data sets that exist at present and attempts to explain their existence and incorporation as desirable information for planners through effective communication to governments of the importance of the phenomena dealt with in these data sets. The importance of the environment is made and the desirability to have it enter into the consciousness of the planners is registered. The paper identifies effective communication and persistence as the two pillars on which success of the move to gain more prominence for the environment as a planning concern must rest.

Introduction

Any discussion on sources of information within the frame of reference as indicated by the title of the paper must address a wide spectrum of statistics. It must also examine their usefulness in the data set that must inform the planner, the entrepreneur interested in marketing, or indeed any other aspect of business. Within the context of a discussion on the environment, one might wish to focus on the effect of styles of primary production, agro-processing and industrial activities, and their effect on the environment as they contribute to the social and economic progress of our Caribbean societies.

Small Islands: Marine Science and Sustainable Development
Coastal and Estuarine Studies, Volume 51, Pages 420–436
Copyright 1996 by the American Geophysical Union

At the outset I wish to make the point that information cannot be compartmentalized and used to the maximum in a partial manner (*i.e.,* divorced from other statistics). The entrepreneur operating in the agricultural sector or, indeed, any other sector, would do well to make use of a data set wider than statistics of agricultural production and daily retail prices of selected agricultural products or gross domestic product (GDP). The decision to purchase any good or service is made against the backdrop of a complex of variables considered and is usually explained through the instrument of a demand function. Other items such as industrial production statistics, imports, demographic variables, and a host of imponderables influence the decision to purchase or not to purchase. This, in turn, will exert some effect on the decision to invest. The investor must, therefore, be in tune with the determinants of demand.

The planner, on the other hand, must be aware of the implications of the proposed investment, not only on the labor market but also on the entire society through the impact of the investment on the country's natural resources and other elements of the environment. The need for a wide range of statistics is, therefore, well understood. In addition, the need to make country comparisons makes the case for a regional database of socio-economic statistics available for wide access and use.

The Nature of the Database

A database is in its simplest form any set of information that can be used for any purpose. In the context of this discussion, the use referred to is planning. In a slightly more technical sense, a database should possess a number of qualities, among them being the following: its records can be edited, updated, or deleted; new records can be added; its records can be searched through the use of Boolean logic; and a report can be generated. The preparation of a regional database introduces a complication. To the extent that the countries use different currencies or other units of measurement, across-country searches and computations are impaired in the absence of a common "intervention" currency or other unit of measurement. Size difference and reporting differences in terms of unit of measurement used will, in addition, compromise the usefulness of the search and filtration process across countries. This difficulty explains the fact that many regional databases are really collections of country databases. They are more often than not limited in their ability to present normalized data across countries.

A major explanation of the conceptual and methodological diversity can be cited in the sovereignty claimed and exercised by countries in the preparation of their statistical series. A modality of forging a common approach to the preparation of statistics will be outlined in this paper. Whereas to date there has not appeared a statistical database that complies with the requirements as outlined above for an ideal database on the Caribbean, several Caribbean data sets have been put together and offered for use. These data sets are discussed below. The sources of information are presented with a view toward demonstrating the origins of the data and the difficulties involved in their collection, processing, and production.

Databases at the National Level

Agricultural Statistics - Major Sources

The major generator of information on any country in the field of agricultural production, prices, or practices must, of necessity, be that country. Statistics as prepared by subregional or international agencies must be seen as secondary to the national effort. This does not deny the usefulness of the subregional or international statistics, as these can present an overview of the phenomenon being studied in a manner that is not addressed at the national level. This paper acknowledges the need to consider as of greatest importance the national data sets.

At the national level, agricultural production is measured either by the Ministry of Agriculture or the Central Statistical Office, often acting in close collaboration with the Ministry of Agriculture. Farmers' registers held by the Ministry of Agriculture are used to assist in the building of sampling frames. These registers assist in the selection of the sample and are crucial in making national level inferences from the results of the sample survey. The register can produce disaggregations of activity groupings of farmers/holdings.

The example of Trinidad and Tobago may be cited here. There has for many years been a close working relationship between the Ministry of Agriculture and the Central Statistical Office. This has fostered many surveys and censuses in the agricultural sector and given the field staff of the Central Statistical Office an ability to make eye estimates of agricultural production. One may look at the present activities of the Central Statistical Office as an indication of the data availability on the agricultural sector. What follows is a listing, as of September 1991, of the survey program of the Central Statistical Office in Trinidad and Tobago.

1. Quarterly survey of pig farmers.
2. Bi-monthly broiler farmers survey.
3. Bi-monthly survey of table egg farmers.
4. Bi-monthly survey of hatcheries.
5. Bi-monthly survey of broiler processors.
6. Bi-monthly survey of food crop farmers.
7. Monthly survey of milk production.
8. Monthly survey of public abattoirs.
9. Monthly survey of private abattoirs.
10. Quarterly survey of livestock feed millers.

Throughout the English-speaking Caribbean, the relationship between the Ministry of Agriculture and the Central Statistical Office varies and this affects the quantity and quality of statistical coverage and output. In cases where the link is strong, the nation benefits from a greater coverage of agricultural statistics. It may thus be stated that the organizational structure of the information-generating institutions is an important determinant of the amount and quality of the information that is made available. It is clear that the linkages between public sector information-generating departments must be strengthened to provide maximum information for planning at all levels.

Secondary-source data are also collected where it is felt that these data are for the most part accurate or at least show a reliable trend. In the case of Trinidad and Tobago, included in this category are data such as sugar and agro-chemicals from the Central Bank, cocoa and coffee from produce marketers, citrus from the cooperative Citrus Growers' Association, copra from the Coconut Growers' Association, and tobacco from the West Indian Tobacco Company.

Client-Oriented Sources of Agricultural Information

The type of service offered by the Ministry of Agriculture or the Central Statistical Office very often does not satisfy the businessman who wishes access to a data set that is current. One notes the emergence of databases that are "client-oriented" and deal with a specific set of commodities. These emerging databases are characterized by their aim to collect information on as current a basis as possible. In Trinidad and Tobago, as well as in a number of other Caribbean countries, one notes the role in this area of private sector organizations which seek to supplement the work done by the Central Statistical Office. Their orientation is towards the collection of price and quantity data, on as timely a basis as possible.

These special-interest organizations enjoy a flexibility of operation which derives from a clear focus and an administrative structure that is lighter than that of the public service. In many cases, the emphasis on currency of data runs the risk of compromising the quality of the collections. What is left for these organizations to do is to ensure optimum data collection, dissemination, and interpretation modalities.

Whereas the present paper presents an overview of the sources of statistical and related information in the Caribbean, it is at best only a partial inventory of the data sources in the region. By and large, the public sector information capability is designed along similar lines throughout the English-speaking Caribbean countries. Differences in quantity and quality of output in the various countries are a function of the official recognition of the value of information and the exercise of political will to bring about an information facility for the benefit of decision making.

Business Statistics

Establishment surveys (or surveys of business establishments) are conducted for the sake of collecting and presenting information on the structure, size, and other characteristics of business. Survey techniques are used in the selection of the establishments, and a methodology for inputting for non-response and for raising the sample results to national estimates is used. The information collected is readily utilized in the national accounts compilation exercise. The data are extremely useful for the decomposition of current valued estimates into constant price estimates in which the price effect is eliminated, yielding a picture of real output. The concept of real output is the one that guides the analyst in his or her assessment of the undisguised performance of the economy. These establishment statistics provide information on a host of areas of interest such as employment in the establishment/subsector sector, hours worked, cost of production structure, profile of goods and services produced, analysis of inventories, analysis of sales, including exports, and analysis of sources of financing.

At the national level, the major public sector sources of establishment-type statistics are the Central Statistical Office and the Industrial Development Corporation or its equivalent. In the private sector information is obtained from the establishments, the Chamber of Commerce, manufacturers' associations, and small business associations. In a number of Caribbean countries, there is an important role in this area for the national development foundations (NDF's or NRDF's). The level of sophistication of these agencies is being improved on. This process can be accelerated by means of dialogue and interaction with the local statistical office. The usefulness of establishment statistics is determined by the rate and quality of the responses provided by the establishments in the sector. There still exists the problem of a relatively high non-response rate from establishment surveys. In addition, the long period of time taken to collect the returns renders the survey results a bit limited in usefulness to the researcher interested in current figures.

The United Nations' International Standard Industry Code (ISIC) is used as the basis for classifying establishments into branches of economic activity. The adoption by the Caribbean countries of this classification affords a measure of comparability in measurement from country to country.

Population Statistics

One of the most critical pieces of information in any planning exercise is the population size. The socio-economic or economic-demographic characteristics of the target population are main determinants of response to official or private sector economic or social stimuli. In at least three English-speaking Caribbean countries a continuous sample survey of the population is conducted. These surveys yield inter-census estimates of the population of the country but by far the most authoritative of the data sets on population are those yielded by the decennial census.

The Latin American Demographic Center (CELADE), in its response to the clamor for small-area demographic information, put in place REDATAM (retrieval of small area DATA on microcomputers) to store and retrieve population census and similar type data. It is now possible to request searches of this large database and produce hardcopy of the results on a microcomputer in a fraction of the time (and at a fraction of the cost) that it would take for the same job to be done on a mainframe computer. The first sets of data to be treated by this software were the 1980 census data for many Caribbean and Latin American countries. After all of the countries have processed their 1990 round of censuses, REDATAM will be used to analyze the results by small areas, in accordance with requests, and the resulting tabulations saved into a database using the technique developed by the Economic Commission for Latin America and the Caribbean (ECLAC) in its work on selected statistical indicators and agricultural statistics. REDATAM is able to describe exactly and access exclusively the enumeration districts that must be studied. It is a major planning tool, as it can help in the determination of the best sites for the location of major elements of public or other works such as schools, hospitals, or business establishments. The Demography Unit of ECLAC/CELADE in Port of Spain collects and publishes time series of demographic statistics. These are obtainable upon request to that unit.

External Trade Statistics

External trade statistics are of prime interest to the investor. The need to know the extent of the domestic market for any commodity that is at present imported cannot be overstated. Data on the quantity and value of goods imported will indicate the extent of local demand and indicate a market size. This information, coupled with price and cost of production data, will indicate the viability of the venture. Other information afforded by the trade statistics include country of origin of imports and country of destination of exports. This indicates the direction of trade. Export statistics coupled with country of destination information may lead the exporter to intensify his marketing in one or more countries or shift to other countries that may prove to be better markets for his or her product.

External trade statistics are one of the few statistical series compiled by every country in the English-speaking Caribbean. They are, in fact, by-products of the customs procedures in the countries. The Customs Departments are revenue-gatherers, so the trade statistics may be considered to be secondary statistics. Most of the trade statistics published are based on the classifications elaborated by the United Nations. Within more recent years, much debate and work has been devoted to establishing another trade classification scheme to supersede or co-exist with the present Standard Industrial Trade Classification (SITC) which itself has gone through three revisions. The SITC Rev. 3 provides a broad analysis of exports and imports into one of 10 sections, each of which is further subdivided into divisions (2-digit level), groups (3-digit level), sub-groups (4-digit level), and basic headings (5-digit level).

More recent efforts have sought to have all English-speaking Caribbean countries move to the harmonized system (HS) of recording trade. This system classifies products by the materials used to produce them. The HS has replaced the Brussels Tariff Nomenclature (BTN) and the Customs Cooperation Council Nomenclature (CCCN).

The HS is a six-digit numerical code. It has 21 sections, 97 two-digit chapters, 1,241 four-digit headings, and 5,019 six-digit subheadings. Many countries have added one or two more digits to the HS to highlight products of particular national interest. Some of the Caribbean countries are yet to present their trade statistics in HS format.

There are merits for the use of the HS and for the use of the SITC. The latter facilitates economic analysis and provides information on the stage of processing of the products and their economic use. As an example, the HS classifies both wood charcoal and wooden statuettes in the same category, whereas under the SITC wood charcoal falls under one category and wooden statuettes fall under another. Some countries are aiming at producing their trade statistics using both the HS and the SITC Rev. 3 through a table of equivalencies. In the case of countries that produce their trade statistics using the HS, conversion to the SITC Rev. 3 is possible. Conversion from SITC Rev. 3 to HS is not possible. In earlier nomenclatures, CCCN and SITC Rev. 2, there exists the possibility of bi-directional conversion, *i.e.*, conversion from CCCN to SITC Rev. 2 and from SITC Rev. 2 to CCCN. The earlier BTN and the SITC Rev. 1 classifications are also capable of bi-directional conversion.

The United Nations has developed a broad economic category (BEC) classification in order to provide a general analysis of international trade trends. This classification distinguishes between food products, industrial supplies, fuels, capital goods, transport equipment, and consumer goods. All Caribbean countries do not themselves produce analyses of the trade by BEC, but figures are published in the *International Trade Statistics Yearbook*, published by the UN Statistical Office. Whereas the data may appear in UN publications, it is in the interest of the individual countries to embark on their own exercises of analyzing their trade statistics by BEC, for the most authentic figures on any country must be the ones compiled by that country.

Trade statistics are not perfect. As with any other type of information collected and presented, they are subject to error. The market researcher must bear in mind that the figures reported are most likely an underestimate of actuality in that several transactions may have gone unrecorded. Within the last 10 years the Caribbean has seen an upsurge in the incidence of "suitcase trade" in which people mix vacations with purchasing for resale, or selling commodities, admitted into the country as personal effects. The extent to which this is prevalent will determine the extent to which reliance can be placed on the trade figures describing the items affected. Other inaccuracies in trade statistics may derive from faulty classification or recording of the goods, values or quantities, and discrepancies in the declarations of destination and origin by the exporter and the importer.

By and large, trade statistics tend to be more reliable than many other types of statistics and indeed, in the case of the national accounts exercise, are used as a fallback position when direct measures are not possible or of good enough quality. A greater degree of comparability in the processing of trade statistics is being put into place at present through the implementation of ASYCUDA, a customs warrant processing system introduced into the region by UNCTAD, together with its companion statistics-generating module called EUROTRACE.

In the Caribbean, the CARICOM Secretariat is putting into place CARTIS, a trade information system that seeks to establish a regional trade information network to support the development of intra- and extra-regional trade. The main users are expected to be exporters and exporting organizations, trade promotion agencies, importers/investors, researchers, and ministries of trade/commerce. The delivery service that is proposed includes the dissemination of trade information from databases covering company profiles, market opportunities, suppliers' profiles, trade documents, and trade statistics.

For access to a data set that would encompass a large number of countries at the level of the item traded, the researcher can turn to the UN Statistical Office. Information is available from this source in SITC Revs. 1, 2, and 3 format. Data from this source are available in hardcopy, microfiche, diskette, and through online searches. The Organization for Economic Cooperation and Development (OECD) covers the countries in that Organization[1] and processes trade statistics in SITC Rev. 3 format. Information can be received in hardcopy, microfiche, or diskette. Other databases exist and may be sourced through inquiries to the UN Statistical Office.

General Regional Statistics

Sources of General Regional Statistics

Whereas it is important that countries prepare their own data sets, there is usually a role that can be performed by an organization operating at the supra-national level. To this extent the subregional and international agencies that operate in the Caribbean can present data that add value to the several data series as published nationally. The re-packaging of information as is currently done adds the element of comprehensiveness and cogency to the otherwise separately obtained data. The publications of the CARICOM Secretariat and ECLAC are examples of integrative work that brings together information on a number of countries. In the case of ECLAC, the appearance of the Selected Statistical Indicators Report is annual, while the Agricultural Statistics Report is biennial in periodicity. The occasional delays in publication are almost invariably the result of the inability of the countries to publish their statistics on time. Both ECLAC statistical publications have the added advantage of having been converted to computer online databases. This means that the data can be accessed on a computer. The packages are self contained and do not need any other software for them to run.

---

[1] Australia, Austria, Belgium, Canada, Denmark, Finland, France, Germany, Greece, Iceland, Ireland, Italy, Japan, Luxembourg, the Netherlands, New Zealand, Norway, Portugal, Spain, Sweden, Switzerland, Turkey, the United Kingdom, and the United States.

This means that a researcher can have at his or her fingertips a vast set of information and can easily repackage some of it for his or her own specific purposes. The Selected Statistical Indicators Report contains macro information on the following: national accounts; balance of payments; trade statistics at SITC section levels; trade balances; monetary statistics; banking statistics; central government revenues and expenditures; retail price index statistics; tourism indicators, as well as charts that describe the series of major interest.

The package can run on a network and can thus be used simultaneously by several people. It can also be remotely accessed through the use of appropriate communication packages. The number of countries covered by the ECLAC/CDCC document is greater than the CARICOM coverage. The ECLAC/CDCC publication entitled *Agricultural Statistics* is scheduled for publication every two years. Extreme difficulties in collecting the data have at times caused setbacks in the appearance of the document. ECLAC is contemplating the discontinuation of the series, largely as a result of the data supply problem. The document reports on population, rainfall, sectoral gross domestic product in current and constant prices, land tenure, agricultural production, and stocks. As in the case of the Selected Statistical Indicators Report, hardcopy is supplemented by a diskette version that is itself an online database with a design similar to that of the Selected Statistical Indicators Report.

Supporting both major statistical series is a document produced and published by ECLAC. This document, entitled *Major Statistical Publications-Abstracts,* presents to the researcher a first idea of where to look for the statistics required. This document describes the major statistical publications and their contents. In addition, it provides information on the publishing agency and indicates the holdings of these documents by the Documentation Center of the ECLAC/CDCC. This document is the output of a regional organization and addresses the major statistical publications of the Caribbean countries that make up the Caribbean Development and Cooperation Committee (CDCC). This document is essential to providing an orientation to the researcher as to what is available and where.

The Caribbean Development Bank (CDB) produces a document that pulls together economic and social statistics of the Caribbean countries. There is an obvious overlap in the CDB and ECLAC data sets. Any discrepancies in the numbers would almost certainly be due to differences in timing of the data collections, as there are at times significant differences between preliminary and revised estimates.

Caribbean-Wide Meta-Data Sets

A number of databases are hosted on the computers at ECLAC in its Documentation Center or on other computers. The Documentation Center uses among other hardware, one HP3000 minicomputer with approximately 1 gigabyte of storage space. This computer can be accessed remotely and bibliographic searches can be made of the databases that are described below.

CARISPLAN is a bibliographic database of publications relating to social and economic planning and development in the Caribbean. It is produced by ECLAC and covers areas such as economic and social policies, development planning, public finance and investment, human resource development, education policies, environment, demography, technology transfer, industrial development, trade, and statistics. Detailed abstracts are included. Some 25,000 records are included in this database.

CARPAT is a bibliographic database of patent applications received by CARICOM countries. It covers many areas of technology, *e.g.*, food technology, mineral mining, oil production, pharmaceuticals, and agricultural chemicals. There are about 2,300 records in this database which is produced by ECLAC.

CAGRIS, produced by the University of the West Indies, is a bibliographic database (with abstracts) of Caribbean agricultural literature on plant protection, post-harvest technology, animal production, fisheries and aquaculture, plant production, water resources, irrigation and drainage, forestry, soil science and management, education and extension, pollution, agricultural economics, and other related topics. The size of this database is of the order of 12,000 records.

LABORDOC, produced by the International Labor Office in Geneva, is a bibliographic database on employment promotion, training, development of cooperatives, social security, occupational safety and health, workers' education, and industrial relations. The size of this database is of the order of 40,000 records. Among the in-house search capabilities at ECLAC are databases on CD-ROM. These include the following.

1. NTIS, which is produced by the U.S. Department of Commerce. This database contains summaries of completed and ongoing U.S. and non-U.S. government-sponsored research and development and engineering activities and business information.
2. Communications Infodisk, produced by Faulkner Information Services, is a hypertext database of in-depth technology overviews, product reports, comparison charts and summaries covering communications hardware, software, services, and emerging technologies.
3. Among the diskette-based online databases held at ECLAC are the Selected Statistical Indicators Report and the Agricultural Statistics Report databases. These are available upon request and the sending of three formatted double density 3.5 inch diskettes.
4. The Energy-Economic Information System (SIEE) presents energy information on the member states of the Latin American Energy Organization (OLADE). The database takes the form of a compiled dBASE application and allows access to prices of imports and exports of oil and other forms of energy. Among the information presented are data sets on the following: prices, reserves and energy

resource inventories, supply-demand, energy balance, information on energy installations, economy and finance of the energy sector, and overall economic indicators.

The data are updated on a quarterly basis. ECLAC is about to place this database onto a computer that will be dedicated to answering queries from remote users as soon as the technical difficulties and hardware acquisition problems and remote interfacing with the package are overcome.

Strengths and Limitations of Caribbean Information Sources

An evaluation of Caribbean sources of information on trade reveals the major part being played by the public sector in that the basic data is obtained from the administrative process. The statistics produced by the statistical offices are for many people the only data sets to be consulted. The packaging of this type of information is, however, not always in the form desired by the user, hence the appearance of other databases. One major strength of the public sector originating data is that it is revised so that errors and late documents are addressed in subsequent publications. One major weakness is that the time lag before publication is often too great for the entrepreneur who must act immediately to take advantage of a present situation that may not last for six months. Whereas national data sets coming out of the Central Statistical Office are updated, the tendency is for Caribbean data sets prepared by entities outside the Caribbean not to be too prompt in reflecting these updates. It is possible, on the other hand, for database constructors in the private sector to be too solicitous of timeliness and thus forget to keep an eye open for updated information, thereby adversely affecting the quality of their holdings.

Bibliographic databases are perhaps the best maintained sources of online information in the Caribbean today, although in many cases the software used is not very user-friendly. A reasonable infrastructure for bibliographic information interchange has been laid down and remote accessing is already in place and working. No completely satisfactory statistical database software has yet emerged, although many numerical databases use "off-the-shelf" software. This represents a fair start, but the future of numerical databases must lie in the free form structure as opposed to the rigid field definitions that characterize many commercial packages if the integrity of the national data sets is to be maintained. A few numerical databases have attempted to structure themselves along lines that are somewhat parallel to the bibliographic databases, and possess a distinct hypertext flavor.

One area of strength of Caribbean statistics is that the figures presented bear units of measurement that reflect our production levels and show more effectively the movement in the figures than other data sources that monitor global trends and, therefore, report in higher units of measurement. The integrative work of regional and international agencies is of great importance because they summarize information that surrounds several national (and in many cases, uncoordinated) efforts into a study that places the national situations into a regional perspective.

A major limitation is that many databases are now in existence but do not reach out to the prospective clients because of inadequate communications. In many cases the information is obtainable at a price, and this may deter some users from taking advantage of the information available.

The price to be paid for access to databases may now pose a problem to some users, but there is no escaping the fact that information is collected, packaged, or re-packaged at a cost, and someone, most logically the user, must bear that cost. As time passes, the benefits of being connected to an information network will become so evident that the problem of payment will disappear.

Why these Databases and not Others?

One notes within the data sets available varying degrees of reliability and availability. To the extent that these data serve primarily a planning or a plan-monitoring function, to that extent are they viewed as being necessary. One can trace the evolution of the areas of concern of the planners and note the amplification of the data sets in accordance with the widened view of the world as perceived by the planner. Statistics on international trade have been collected for very many years and continue to be the best statistical series in most countries of the Caribbean. This is the direct result of the need as seen from colonial times to measure the trade in exports, especially from the colonies. The exports provided a good indication of the effect of economic activity on the estates. It can be argued that the colonies had no national income for obvious reasons. Beginning with the narrow concern for growth in GDP or GNP, the economic statistics and national accounts received the attention of the planners and statisticians who collected and processed the information required. At a later stage, questions of distribution were impressed on the minds of the planners when growth was recorded with no real impact on the level of well being of the population because the wealth was concentrated in the hands of a relatively small proportion of the population. The questions of income distribution and quality of life first made their impression on the international financial institutions and then received attention by the planners who then amplified their area of interest to include income distribution, poverty, and quality of life (living standards) concerns. In both stages described, the planners responded to a signal of the importance of the new concerns. The communication of the new concerns was reinforced by technical seminars that addressed the measurement of these indicators. These seminars addressed planners and statisticians alike and were successful in conveying the message that there were concerns other than aggregate measures of growth that were important. What is the use of boasting of a 6% rate of growth of GDP while it does not translate into any increase in well being for the majority of the population? Today, what was formerly economic planning has evolved into social and economic planning as a direct result of the permeation into the consciousness of the planners of the importance of the "people" aspect of development.

There has been within recent times some concern for the effect of development on the environment. The concern has been expressed that a major challenge is to achieve a development style that is in harmony with the environment that does not degrade the environment or upset the ecological balance. Every planner would state that environmental concerns are taken on board when major project implementation takes place. They point to the now mandatory environmental impact assessments that must be made before any major project is approved. Surely, this means that planners are sympathetic to the issue of sustainable development.

The short answer to the question that captions this section (why these databases and not others?) lies in the ability of some concerns to be made clear and to have struck a resonant chord with the planning apparatus. One is at the same time mindful that even if a case is made and conveyed, it may not strike the same chord as other cases drawing attention to other concerns. This may be due to the low ranking of that concern in the entire set of competing concerns that confront the planner. A key to official acceptance and incorporation of registered concerns must, therefore, be the extent to which the particular concern can be taken on board to co-exist with other policies.   After consideration of a representation, it is the prerogative of the administration to make a judgment based on cost-benefit analysis. The decision may very well be to regard the downside of a particular policy as a cost and continue to enact it with the confidence that the positive benefits will outweigh the negative.

The Case For a Database on the Environment

By and large, Caribbean countries are island countries. Land mass is extremely finite and further effectively limited in many cases by a (usually central) mountain range. This leaves a relatively narrow coastal plain to support the greater portion of the population and its economic activity.   The marine environment that surrounds them should logically loom large in importance in terms of sustaining the peoples in that it is a major source of protein, iodine, and other nutrients that support human well being.   The effluents from industry that we allow to enter into the sea, as well as the soil eroded by the indiscriminate denudation of the hillsides in the name of peasant farming or land development, are a direct threat to the marine life of our shelf areas. The use of the sea for the disposal of raw sewage, especially without regard to the need for an intimate knowledge of ocean current patterns, has been a major cause of the pollution of our beaches. When we base much of our development thrust on beach tourism, we begin to appreciate the absolute need for a close association between tourism development and waste disposal. Tourism is a very volatile product and will evaporate at the first sign of a health hazard associated with the pollution of beaches and other tourist centers. Land disposal in a number of cases has affected adversely the quality of ground water and, by extension, the supply of potable water. The problem of land disposal of liquid and solid wastes in the Bahamas has been discussed and is an apt example of human action impacting negatively on the environment.

The development alternatives are relatively few in the Caribbean. This stems from undiversified structures of production and to some extent rigidities in the conception of development paradigms. The present importance of the services sector has seen the rise to prominence of tourism as a main determinant of revenue and a major engine of economic growth. The activities associated with tourism tend to be coastal in nature and location. Because of the large numbers of people involved in tourism, additional strain is placed both on the infrastructure and on the environment. The infrastructure is stressed to the extent of increased demand on water, electricity, vehicular traffic, noxious emissions from internal combustion engines, and the need for increased medical support. The environment, on the other hand, is stressed to the extent of aqua sporting activities which upset the ecological balance near the coastline, the problem of waste disposal, and the increased destruction of the fauna, both terrestrial and marine.

The conservationists, among whose ranks we can include the physical oceanographers, have made their pleas for greater environmental concern and sensitivity when the styles of development are being designed. Whereas they can from time to time describe the environment in terms of a number of parameters such as sea level rise, degree of pollution of the marine environment, loss of coral reefs and coastal erosion, they have not been able to package their concerns with the effectiveness required to impact forcibly on the planning mechanism. This need is as urgent in regard to land resources as it is with respect to marine resources.

The finite nature of our land resources should urge extreme caution and optimum usage. Not many countries have charted out their land resources through the conduct of a land capability survey in which land is characterized by soil type, mineral content, gradient, and number of distinguishing features. This should form the basis of the decision as to how land use consistent with sustainable development should be determined. Marine resources should similarly be coded and care taken to avoid depletion of species. Such an inventory should document the resources available from the judicious farming of the sea.

The Move Towards the Incorporation of these Concerns

There is no short cut to the process of convincing planners of the importance of the environmental dimension in planning. A communication strategy must be devised and aimed at both the administration and the populace at the same time. The effective use of audio-visual material or visual material with voice can go a long way towards sensitizing the public to the consequences of practices that violate the environment and upset the ecological balance. Presentations must include references to actual works and their results, as well as the discussion of several indicators of a deteriorated situation. Indicators can take the form of determinants and consequences. This approach begins to suggest a number of indicators that should be included in a database that monitors the environment. Among the series to be included are the following:

1. Population by age group, gender, social group, *etc.*
2. Births, deaths, migration.
3. Employment, wages, economic activity.
4. Household income and expenditure data.
5. Exports by economic sector (and trends).
6. Land productivity.
7. Land capability.
8. Rainfall and temperature.
9. Storm frequency and path data.
10. External demand for tourism and agricultural production.
11. Proportion of beach tourism to total tourism.
12. Water quality and sources.
13. Infrastructure.
14. Indicators of coastal erosion.
15. Indicators of beach quality (water quality).

The above listing represents some of the data sets that are relatively easy to obtain and are available in varying degrees throughout the Caribbean. It will require effort to unearth some of the above data from their current sources because so much of this type of information has been collected as part of the administrative process, most probably is not analyzed, and does not at present find itself into the data set referred to for many activities. In constructing time series for the above indicators, care must be taken to make the data as homogeneously collected over time as possible. The data must, therefore, be collected with a certain known periodicity and with consistency in the manner of collection. The purpose of collecting the data sets suggested above is not merely to chronicle the degradation of the environment, but to take corrective (and repair) action. For example, in Cuba, an exhaustive study of coastal erosion in the northwestern portion of the island has led to successful beach reclamation and repair activities. Far in excess of merely reclaiming some beach area, this activity means preservation of a major projected earner of foreign exchange and engine of growth. ECLAC has committed itself to the development of an environmental database for the Caribbean countries.

Apart from the above type of data, there is information of a scientific nature that is either bibliographic in nature or measured on a rather infrequent basis. Much of this scientific data cannot easily be translated to useful information to be used by our planners. In such a situation the Delphi method of consulting a group of experts who can brainstorm and do some "crystal balling" on the scientific data, relating it to its impact on our societies some time in the future, can be used to bridge the information gap between scientists, society at large, and the planners, in particular.

Emerging from the "crystal balling" will be an information set that describes the state-of- the-art concerning the area of interest. The nature of this information set is bibliographic. It will contain references to the facets of the problem, *e.g.*, in the case of the environment, the question of sea level rise or the disappearance of species. Following on this, inventories should be conducted that attempt to quantify the problem. These inventories can be taken from time to time in order to assess direction of movement and present the policy makers with an idea of the dynamics of the problem. The phenomenon can be associated with human or natural activity in an attempt to provide further explanation of the causes and determinants of environmental degradation. At present, at least one country is attempting to establish an Environmental Management Agency that will coordinate all environmental matters and ensure a regime of planning operations that is free from internal inconsistency.

Communications - The Key

The foregoing has served to highlight the role of effective communication in making known the crucial role of the environment in the sustainable development consideration. The approach to communications must take the form of an outreach that breaks the restricted intra-disciplinary conferences that scientists and other special interest groups tend to convene. It is only by bombardment with information that other circles of influence will become aware, then sensitized, to the need to be solicitous of the environment in all its aspects. The economist may well look at the degradation of the environment as a cost incurred in economic development, but this view may be considered a short-term perspective.   Development policies should take into consideration the longer term and seek to leave for the generations to come a world that is no less accommodating to life than it is at present.

The Challenge Facing the Physical Oceanographer

The task of communicating the effect of human activity on the environment is sufficiently serious as to warrant concerted and well-planned action on the part of the scientific community concerned with sustainable development. It is all the more critical in the case of small island states. The objective is to ensure that the environmental concerns are taken into account in planning. The tactics that can be employed to achieve this objective should combine the education of the public at large with aggressive moves to insert the viewpoint of the environmentalist into fora of planners and generators of public opinion. For example, seminars to students at varying levels in the school system can be designed to bring to the general public the effects of current practices on the ecological balance of the country. Public debate can be stimulated by the periodic promotion of conservation activities and their coverage in the media. At another level, alternatives to development paths may be suggested in the case of activities at present seen as desirable though disruptive of the terrestrial or marine environment. In this context, every entry into the environmental database should be designed to support the need for achieving development that is sustainable.

The physical oceanographer or related scientist must assume the role of change agent and continuously seek to utilize any forum for conveying the message of sustainable development. Communication is an area that requires expertise. To the extent that the scientist in most cases is not a proven communicator, to that extent should marketing tactics be employed in contracting the services of the communication specialists to prepare material for dissemination. Effective communication requires patience and the willingness to try again until the objective of making an impression on the mind of the decision-makers is achieved.

25

Numerical Modeling of Small Island Socio-Economics to Achieve Sustainable Development

Guy Engelen, Roger White, Inge Uljee, and Serge Wargnies

Abstract

This paper develops a modeling framework for the exploration of policies aiming at sustainable development for small islands. The essential requirements for realistic modeling of small island socio-economics are first presented briefly: the approach must be integrated, dynamic, and spatial in nature. Next, the modeling framework itself is developed. It consists of a number of coupled modules that represent interaction mechanisms at different geographical scales. This approach allows for a tight linkage of environmental qualities with the socio-economic characteristics of small islands at very detailed scales. Finally, we elaborate on a number of model extensions considered to be useful in "measuring" sustainability and in displaying its intrinsic uncertainty. The paper constitutes an extension of our earlier work on modeling and decision support tools for analyzing the impacts of climate change on socio-economic systems.

Introduction

Small islands exemplify rather explicitly the fact that man must survive in an environment with limited physical dimensions and non-renewable resources, and has a limited set of technological means for exploiting it. Small islands with their dispersion, their isolation and their limited population, economy and geographical extent, are very open systems (Marshall, 1982) directly and strongly influenced by distant international market, world prices and transportation costs (Jacome, 1992). From a purely geometrical point of view, a large proportion of their territory is classified as coastal zone. Furthermore, this is the area where an important part of the socio-economic activities

Small Islands: Marine Science and Sustainable Development
Coastal and Estuarine Studies, Volume 51, Pages 437–463
Copyright 1996 by the American Geophysical Union

are usually concentrated, where there is competition for space and where pollution and waste generation are high (Blommestein, 1993). It is also the area where unique but fragile ecosystems, both aquatic and terrestrial, are typically present in more than average proportions and where the effects of sea level rise and climate change would be felt first and most dramatically. The great value of small islands in terms of their biodiversity, as well as their strong dependency on the natural environment for food production, tourism and agricultural exports, demands a carefully managed interaction between man and the environment. Their generally disadvantaged position makes this task both more difficult and more urgent.

The expression "management" of natural resources is chosen here rather than "preservation" because it is a pragmatic reality that for the development and the survival of mankind a certain degree of exploitation of non-renewable natural resources be allowed (The World Bank, 1992). The definition of sustainable development (WCED, 1987) accommodates this differentiation: "development meeting the needs of the present generation without compromising the ability of future generations to meet their own." It has been argued that this definition, as well as the general concept of sustainability, is not precise at all and that it is difficult to work with on the operational level:

> "It is intrinsically inexact. It is not something that can be measured out in coffee spoons. It is not something that you could be numerically accurate about. It is, at best, a general guide to policies that have to do with investment, conservation, and resource use. And we shouldn't pretend that it is anything other than that" (Solow, 1991).

Perhaps we should end this chapter now, concluding that the prospect of modeling sustainable development by means of numerical models is an illusion. But on the other hand, is Solow's point not true for most socio-economic concepts and measures? Are social scientists not always trying to work with approximate measures and partial representations? "In order to even think about a situation we must invent words and concepts with which to discuss it, symbols which are simpler than what they represent" (Allen and Lesser, 1991). Are social scientists not always faced with subjectivity in their choice of variables and mechanisms with which to describe socio-economic systems? "Social science is speculative in nature. In contrast to the natural sciences..., social science does not easily succumb to one or even a few theoretical foundations" (Dendrinos, 1994). Consequently, models used by social scientists and the results which they produce will remain to a certain level subject to interpretation and debate.

In this chapter we will not succeed in proposing the ultimate model and measures for fully capturing the concept of sustainable development. Rather, we will try to move in that direction, and propose an integrated modeling approach thought to be useful in the analysis of options for sustainable development in its broadest sense. Part of this discussion will be conceptual and part of it will be based on the experience we have built up in modeling the socio-economic implications of climate change and sea level rise.

Modeling Sustainability

The WCED's definition of sustainability implies knowledge of the status of the systems affected by human activities, not only now but even more so in the future. Indeed, inherent in the definition is the hypothesis that it is known what the legitimate actual needs for the use of natural resources are "to meet the needs of present generations." And so is the hypothesis that future resource utilization requirements are known, that will permit "future generations to meet their own." Hence, in order to successfully direct man's activities towards sustainability, it is essential to have a good understanding of the current status and the carrying capacity of the systems of which man presently is a part, and equally to be well informed on how these will evolve in the future. However essential such knowledge might be in formulating sustainable development schemes, we have to admit that it is very difficult to come by and extremely complex. Part of the complexity resides in the integrated nature of the human-environment system. Thus, we require not only (1) a very good understanding of the functioning of natural environmental systems and the ways in which they evolve, grow and regenerate, as well as (2) a good understanding of man's social and economic needs and activities, including the changing technologies, consumption patterns and cultural values, but also (3) knowledge on the linkages and interactions between the natural and the human systems; the way in which technologies, exploitation schemes, pollution, and waste disposal affect natural systems, how much human-induced stress natural systems can withstand, and to what extent they can recover from it.

Only modeling of some sort, qualitative or quantitative is likely to provide the required understanding of the behavior of these complex integrated systems. But, despite the fact that important research efforts are underway, much additional work will be required because our understanding of both natural environmental and socio-economic systems is only partial and often descriptive in nature. Furthermore, only a very limited effort has gone into understanding the interplay of environmental and socio-economic systems. Only few models and decision methodologies or tools deal with the man-environment interface in a comprehensive, integrated manner. Traditional divisions in scientific disciplines have led to research efforts that treat human and natural systems as separate issues. Socio-economic models will typically ignore environmental factors and environmental models will rarely include the impacts caused by man. "Although there is now a growing body of work analyzing the physical dependence of the economic process on ecological relationships and the environmental costs of economic activity, these two aspects have not so far been well integrated into a comprehensive model of economic-environmental interaction" (Barbier, 1989).

In the next section we elaborate on what we believe to be essential requirements for a modeling framework for it to be useful in the context of sustainable development. We will elaborate on its integrated structure, on its dynamic, evolutionary quality and on its spatial, geographical nature.

Hybrid Modeling Framework

No economic or social system today can be studied as a closed system. To some degree all such systems are open to exchanges of both energy and material with the outside world. For small islands, however, the open character of the socio-economic system is very explicit. Due to their small population size and their isolated location far from markets, they lack economies of scale and have limited opportunities for local production. Moreover, for their foreign exchange they traditionally rely heavily on the export of a few, mostly agricultural, products. Unlike closed systems for which the final state of thermodynamic equilibrium can be predicted, open systems are continuously in a transition phase, evolving from one state of organization into another. Hence, small islands are highly dynamic evolving systems strongly interacting with the outside world and highly subject to all sorts of influences, good or bad.

A good example of the perils of openness is provided by the banana debate in the Commission of the European Union (EU) (De Meyer, 1991). With the continuing unification of the EU after 1992, the question has risen as to whether the imports of Caribbean bananas should be favored throughout the entire EU. Because of the post-colonial links with the United Kingdom and France, Caribbean bananas have a guaranteed market in some EU countries (specifically the U.K., France, Italy, Greece, Spain, and Portugal). The possibility exists that due to the "disappearing" borders between the European countries, these markets will be lost to the cheaper and better looking "dollar-bananas" imported by U.S. fruit multinationals from Central and South America. The dollar-bananas enter the EU mostly via Germany and the Benelux countries and can now be shipped and sold freely throughout the EU.

This banana debate is typical of the discussions that take place thousands of kilometers away from the Caribbean islands, involving stakeholders of all sorts weighing one agreement against another, which decide to a very large extent the future well-being of the populations of a number of the Windward Islands.

On the islands themselves, activities are frequently concentrated in a narrow coastal zone. This results in conflicts of interest and competition for space and causes irrecoverable stress on the unique but fragile terrestrial and marine ecosystems of the coastal zone. So, even if we were to know from the islands' position in the international market that a particular economic sector, such as bananas, would expand, that would not tell us where on the island such growth would take place, nor what sort of conflicts it would create with other activities already established. Clearly, the knowledge as to whether the additional activity will concentrate in one location or another, or will spread out over large parts of the island is of great importance in estimating the sustainability of the development.

Hence, in order to explore the options for sustainable development we need models that capture realistically the phenomena driving the dynamics of small islands that operate at very different geographical and temporal scales. To be useful, the models should:

1. Be capable of describing the main natural, environmental, and socio-economic characteristics of the individual island, and the linkages of these systems within and across the island boundaries. Hence, we need *integrated models* that represent the islands as open systems.

2. Contain the main mechanisms of change of the natural, environmental, and socio-economic systems, both those operating on the short term as well as those active on the time horizon of a generation or more, to project forward in time the possible combined effects of growth or change of the socio-economic and natural systems. Hence, we need *dynamic models* that cover a sufficiently long time interval but that step through time in sufficiently detailed a manner to capture short term, rapid changes too.

3. Provide an estimate of both the direct and indirect effects of events on the local, regional, and world scales that will affect the sustainability of the island systems and their development. Hence, we need *spatial models* that can represent spatial interactions at different scales: the socio-economic interaction of the island with the outside world as well as the interaction of spatial actors on the island itself.

Apart from these three requirements we could mention others that are of a more practical nature. Certainly data availability will very often limit the use of certain types of models. For example, to apply dynamic models data will need to be available as time series. For small island countries, certainly the developing small islands, data availability can pose a serious limit on any numerical modeling attempt. In such cases models should be selected that require very limited data and the available indigenous knowledge should be used to a maximum extent in alternative, more qualitative modeling attempts. Next to data availability, the costs of collecting the data and developing and running the models can pose important limitations. With the growing capabilities of personal computers (PCs) in terms of calculation power and storage capacity, this is gradually getting to be less of a problem. Finally, the usability of a model, its user friendliness, its calculation speed, the intake of the inputs and presentation of its outputs, will determine to a large extent its success in practical applications. To some of these points we will return later.

In our search for adequate models to assess the vulnerability of low-lying coastal areas and small islands to climate change and sea level rise (Engelen *et al.*, 1993a; Engelen *et al.*, 1993b), we have mentioned these points as essential requirements. In that context we have made the comment that "medium term or long term models of socio-economic systems rarely cover periods longer than 10 to 20 years in the future." This is because the level of uncertainty in the model increases dramatically for longer periods. For the study of sustainable development involving processes that change over generations, we do not have instruments with real predictive capabilities in the sense used by Simmonds (1986): "prediction refers to the process of making (definite) statements about the future. This is more specific than the use of the word in statistics, where prediction is taken to mean a calculated rather than an observed value." The models that we will present are at best exploratory models, instruments that allow "a structured exercise of the imagination" (Batty, 1976) and that permit us "to discover other realities that may be logically possible" (Enchinique, 1972). When using the models for decision-making purposes, it is important to remember that the numbers which they produce are laden with uncertainty and that they should be treated with caution. We will come back to this later in this chapter.

The above requirements are, generally speaking, not fully met by any particular type of model in isolation. Typically, *macro-economic models*, including those developed in relation to sustainable development (Barbier, 1989; van den Bergh and Nijkamp, 1991) treat the problem in a non-spatial manner. *Spatial interaction models* are very powerful instruments for describing the dynamic and spatial behavior of socio-economic agents (White, 1977; Allen and Sanglier, 1979; Engelen, 1988). They are typically written as sets of linked differential equations and hence become computationally impractical if a detailed spatial representation is required. They also suffer of the so-called "scale problem" (Huggett, 1980) since they represent interaction mechanisms that are only operational at one single spatial scale. *GIS* (Geographical Information System) *models* allow us to describe geographical systems at a very high degree of spatial resolution. They lack, however, the capacity to represent the spatial and temporal dynamics (Marble and Amundson, 1988; Brimicombe, 1992) underlying the spatial structures which they display so beautifully. *Spatial allocation models* typically allow a detailed spatial representation and do consider local spatial interaction to some degree, but are mostly of a non-dynamic nature (Wright, 1989).

None of the types of models mentioned allow us to model spatial interaction processes at a detailed geographical scale. In search of such a method, we have recently developed a cellular modeling methodology which is based on the principles of *cellular automata* (White and Engelen, 1993; White and Engelen, 1994; Engelen *et al.*, 1993a; Engelen *et al.*, 1993b). This methodology, although it does well what it has been developed for, is incapable of dealing with long-range spatial interactions and long term time scales.

From the above overview we conclude that no one type of model used in isolation is capable of modeling the spatial and temporal dynamics of both long- and short-range interaction in systems with a detailed spatial resolution. Rather, a number of complementary techniques must be used in combination. These should operate in such a manner that the change due to long range interaction mechanisms can be allocated in a spatially detailed manner on the basis of local characteristics as well as short range interaction mechanisms (Figure 1). Thus, applied to small island states, the model could show where change predicted at the national level, whether due to endogenous factors or change in relations with the rest of the world, would become visible. Conversely, it could show what the combined effect of very local phenomena would be on the state of the island as a whole as well as on the interaction of the island with the rest of the world.

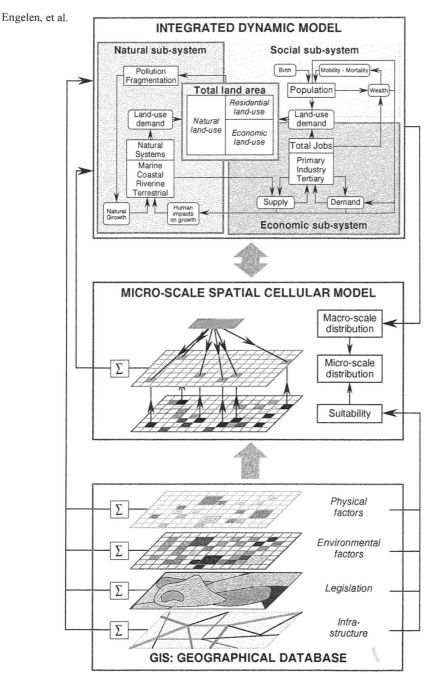

Fig. 1. A two-level model. On the macro-level, long range interactions are modeled by means of an integrated model. The macro growth coefficients are fed into a micro-level cellular model to perform the detailed allocation based on short range mechanisms. Both levels of the model will retrieve data from the same GIS.

We have demonstrated that such coupled models successfully reproduce the dynamics of urban land uses, and have used the approach to model the dynamics of the population, the economy, and the resulting land coverage on a prototypical Caribbean island (Engelen *et al.*, 1993b).

The ISLAND Model: A Hybrid Model to Assess the Vulnerability of Low-Lying Coastal Areas and Small Islands to Climate Change and Sea Level Rise

To demonstrate the applicability and use of coupled models, we present briefly the model which we developed in the context of a study carried out for UNEP. In this model, called the ISLAND model, we distinguish between the macro-level, describing the long-range and long-term interactions of the island with the external world, and the micro-level, describing the short-range, short-term interactions between the spatial, economic, and demographic factors internal to the island. Each level is described by means of a model capable of representing the relevant spatial dynamics adequately. Both levels have a number of variables in common and exchange information with one another continuously. They both get their level-specific data from the same GIS database. Generally speaking, the approach is fully regionalized on both levels but, in the application of the ISLAND model, the macro-level is modeled as a single region, as one point in space interacting with the external world. Hence, a single set of linked dynamic equations describes the evolution in time of the integrated socio-economic and natural systems. For small islands, this simplification is legitimate as long as we may assume that due to the very small size of the island, exchanges with the outside world dominate the long range interactions internal to the island, and that interactions with the external world are of an island-wide character, rather than at the level of some of its parts. At the micro-level, the region is subdivided in small geographical units, called cells, representing areas of a rather homogenic nature in terms of overall socio-economic, land use, and physical-environmental characteristics. The model is solved by simulating it forward in time for a maximum period of 40 years. At each time step, the growth coefficients calculated at the macro-level for each state variable are fed into the model at the micro-level. The latter distributes and allocates the growth onto the individual cells (a grid of 70 by 80 cells, each covering an area of 500 m on the side). The distribution is based on micro-level spatial interactions specified in cellular automata rules and on physical suitability measures retrieved from the GIS. The cellular model returns its results to the macro-level, thus guaranteeing the full coupling of the macro and micro scales.

The Macro-Level Dynamics

At the macro-level essentially three coupled subsystems are modeled: the natural, social, and economic subsystems. Generally speaking, the modeling framework allows for different types of mathematical representations of each of these subsystems, thus permitting a more or less detailed description of certain aspects of these systems. For the ISLAND model, we have selected model representations that fit the application of the modeling framework to a hypothetical small island in a Caribbean setting, referred to as ISLAND. It reflects the archetype nature of the work and leaves room for a number of improvements.

The *economic subsystem* is modeled by means of a highly aggregated input-output model (Leontief, 1966), solved at each iteration, in which the economy is aggregated into five major sectors: subsistence agriculture, export agriculture, tourism, construction, and city and other activities. The economic subsystem is coupled with the demographic model (through household demand), with the natural system (climate changes influence export demands), and with the micro-level via a land productivity expression (activities require space). The input-output model will calculate for each of the five economic sectors the inputs required, in terms of jobs and imports, to meet the changing demands as generated by households, the export sector or the other final demand sectors of the economy. We have chosen to model the economic subsystem by means of an input-output model because of its representation of the interdependencies between economic sectors, and because of the way it integrates these with labor requirements, imports, and final demands. Thus, it shows the propagation throughout the economy of the changes in any one sector. The technical coefficients of our model do not change over time. This implies that the structure of the economy is not changing in the period modeled. This is a very weak hypothesis, certainly for long-term modeling and for economies which are highly technical and where technological innovation and substitution are frequent. In principle, the technical coefficients of the model will have to be recalculated every time substitution takes place (see, *e.g.*, Richardson, 1978). Despite these deficiencies, input-output models are in widespread use among planning and statistics agencies. Hence, very often data for the model will be readily available, even in small developing islands.

In the *social subsystem* the demography is modeled as a single population group growing exponentially as the result of births, deaths, and migration. The birth rate is specified to follow a long term, externally provided, structural trend, while mortality and migration rates depend on both forecasted long term structural trends, and on the economic well-being of the island population as indicated by the employment participation rate. The structural components represent the long term changes occurring on the ISLAND or in the larger region of which it is part. Such changes are the expression of long term effects set in motion in the past. They are the result of cultural values, education, family planning, and health care programs. They reflect improving (or worsening) living conditions that gradually penetrate the area, affecting mortality and traditional migration trends, as well as the overall evolution of well-being in migrant receiving regions (Marshall, 1982). Grafted on this, there is a socio-economic component expressing the immediate effect of changes in economic prosperity on the ISLAND itself.

The *natural subsystem* consists of a set of linked relations representing the expected change in time of sea level and temperature, together with the consequent effects in terms of the loss of land and changes in precipitation, storm frequency, and external demands for services and products. Knowledge concerning such relations is still limited. Despite the fact that the relations are expressed as numerical functions, this part of the model is kept semi-qualitative in nature, with relations representing expert knowledge or hypotheses rather than empirical laws. Part of the natural subsystem concerns a measure of the physical suitability or aptness of each micro-level region for receiving each of the activities modeled. The definition of suitability measure is much as it is used in GIS research (see *e.g.,* Burrough, 1989; Berry, 1993). It concerns a composite measure—a combination, weighted sum or product—of a series of characteristics of each cell (Wright, 1989). These characteristics may include topography, soil quality, water availability, road infrastructure, or planning regulations. Most advanced GIS packages have overlay analysis or cartographic modeling techniques available to perform the calculations required to produce suitability maps. The individual suitability measure of the cells will play an important role in the allocation of activities in the micro-level model, while the suitability values summed over the ISLAND are used to calculate land cost and land marginality in the land productivity expression.

At the macro level, the model calculates the growth of the population, the changes in employment per sector, and the amount of imports required. New people and new jobs will require space to deploy their residential and economic activities. In the model the translation of economic and residential activity into demand for space is performed by means of the *land productivity* expression. Land productivity is a function of the total amount of land available, the suitability of that land for the activity carried out, and the land (and its suitability) already occupied. Generally speaking, land productivity will increase if land gets scarce, hence more expensive, and activities will concentrate more on the same amount of land. It will decrease if the land occupied gets more marginal, as marginal land is less productive. The total amount of land required by each activity is used as an input in the micro-level part of the model.

Micro-Scale Dynamics

More than in the macro-scale representation, the main mechanism underlying the micro-scale dynamics is the fact that socio-economic activities will interact with one another in geographical space. It is a well documented phenomenon in regional economics and it is implicit to concepts such as competition for space, distance decay effects, agglomeration effects, or intervening opportunities. (see, for example, Richardson, 1978). It translates the fact that activity X does not flourish in the neighborhood of activity or feature Y, while Z does very well near Y. Push and pull forces of this nature are represented in the cellular model by means of distance decay functions (expressing how strongly one activity attracts or repels another if they are separated by a given distance) for each pair of activities or land uses represented in the model. These functions are key elements in the *transition rules* of the Cellular Automaton model (see, for example, Gardner, 1970; Tobler, 1979; Couclelis, 1985; Langton, 1986; Couclelis, 1988; Gutowitz, 1991; White and Engelen, 1993).

It is by means of such Cellular Automata transition rules that we calculate the changing land use on the micro-level. These rules can be either qualitative or quantitative in nature, or can be a combination of both. Rules are applied to a defined *neighborhood* which in our model is circular in shape and consists of the 113 cells nearest a given cell. Rules are applied according to their priority level. The highest priority is given to user interventions, such as the user-imposed allocation of an activity to a cell. The next highest priority concerns the elevation of a cell relative to the sea level. For example, a "land" cell that floods due to sea level rise changes to a "sea" cell. The lowest priority goes to rules that calculate the likelihood for a cell to receive a specific activity. As explained earlier, this likelihood is strongly determined by the mix of activities present in the neighborhood of the cell but, in addition, the intrinsic suitability of the cell itself is taken into account. For each of the cells in the grid the potential for transition to each of the land uses modeled is calculated. Time progresses discretely and each of the cells change state simultaneously. In principle, cells will change to the state, the land use, for which their transition potential is highest, until the total amount of land required for the activity is met. The total amount is calculated by the macro-level model and is imposed on the cellular level. At each iteration the micro-level model will produce a new land use map that will display the new location of the different socio-economic activities. Next to the land taken in by the economic activities (subsistence agriculture, export agriculture, tourism, construction, and city and other) and the residential activities, a number of land uses are represented on the micro-level that do not have an equivalent on the macro-level. These concern natural land states such as sea, beach, coral reef, mangrove, as well as economic land uses such as an airport. For these latter the macro-level model does not calculate and impose the overall growth coefficients. Nevertheless, these land uses have a direct effect on the distribution of activities on the micro-level, and thus influence somewhat more indirectly the dynamics at the macro level.

Application of the Model

For the ISLAND model, no calibration as such has been performed since no particular real island is being modeled; rather general principles concerning socio-economic implications of climatological changes and the resulting spatial re-organization have been explored. However, the fictitious ISLAND has been conceived with the concern of representing an authentic Caribbean island having most of the characteristics typical for the region: (1) a relief combining low lying areas and highlands; (2) an economy primarily based on export agriculture, tourism and, to a lesser extent, tertiary or industrial activities; (3) a subsistence agriculture sector producing some of the elementary foods; (4) major imports of all kinds of consumer goods; and (5) a rapidly growing population. The suitability map for each land use has been drawn so as to be realistic in relation to the relief of the ISLAND, the access to the road infrastructure, and the prevailing (eastern) winds, and consistent with the suitabilities for other land uses on the ISLAND. The positioning of the activities for the initial condition has been done on the basis of the suitability maps (for the economic and residential activities), on the basis of the relief (for the natural land uses) and on the basis of examples taken from actual islands. The initial values of the state variables and parameters describing the economy and demography are based on statistical data from the region (The Diagram Group, 1985; The World Bank, 1992; ECLAC, 1991). We have designed an island with a population of 24,500 and a total employment of 11,000.

Because of the theoretical nature of the model, we will not dwell on a more exact description of the ISLAND, nor will we discuss any scenario in particular. The interested reader is referred to Engelen *et al.* (1993b). Rather, we show in Figure 2 six maps as they result from a typical scenario in which temperature and sea level are rising by 2°C and 20 cm over the simulated period of 40 years. On the maps it is clear how the loss and partial relocation of mangroves and beaches affects the demand for and location of tourism activities. Residential activities are expanding due to the population growth (in 40 years the population grows by some 56%). The net increase in population, as well as the relative decline in the employment in the export oriented economic sectors, leads to an important increase in the subsistence agriculture sector. The expansion of all activities in the coastal area, as well as the competition for space in the coastal area, drives the subsistence agriculture into the marginal, steep, lands in the center of the island.

Although we are presenting no real island in particular, we believe we demonstrate in the example a possible evolution that some of the small Caribbean islands could experience under conditions of population growth, relative decline of export demands, and climate change. Clearly, for small islands the consideration of climate change in any of the scenarios related to their future development and its sustainability is of prime concern because "they are considered extremely vulnerable to global warming and sea level rise, with certain small low-lying islands facing the increasing threat of the loss of their entire national territories" (Agenda 21, paragraph 17.127).

However, in order to use the modeling framework in the broad context of sustainability, some changes are desirable. These will largely relate to the strengthening of the feedback loops representing man's impacts on the natural systems. Indeed, the human impacts on the environment are discussed largely in terms of the amount of "natural" land lost and the suitability, productivity, and marginality of the land used for specific purposes. The model could be improved by incorporating explicitly the dynamics of natural systems such as mangroves, coral reefs and beaches, so that it would describe how these grow and regenerate, and how they are being affected by different sorts of human influences. Among others we need to look at impacts such as the mining or damaging of beaches (lowering the beach, altering its sensitivity to erosion, building of

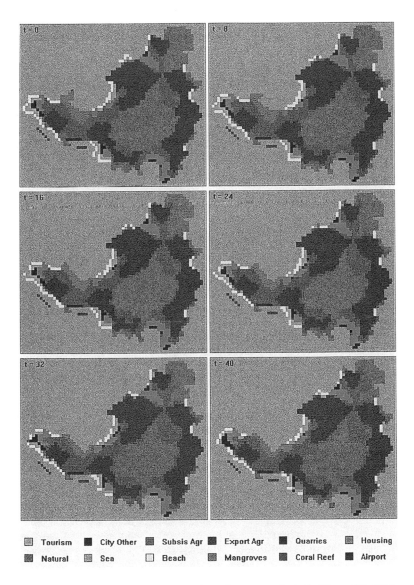

Fig. 2. Six land use maps showing the evolution of a prototypical Caribbean ISLAND under a RISE scenario: sea level rise of 25 cm and temperature rise of 2°C in 40 years.

defense structures); the mining, damaging or polluting of coral reefs (influencing their attractiveness for the tourist industry or reducing their role in beach protection); the eutrophication of rivers, deltas and coastal waters; the pollution of beaches and coastal waters; the loss of mangrove stands (used as dump sites and as sources of wood); and deforestation (causing erosion and depletion of aquifers, *etc.*) To this end, and ideally, for each natural system, a separate sub-model should be created with its own dynamics and linked to the other natural and socio-economic components of the model. Thus, it would be possible for the user to run sustainability scenarios that show not only the results in terms of socio-economic indicators, but in terms of biological, ecological, and environmental variables equally well.

For the representation of the economy, the input-output algorithm should be changed, or replaced, in order to take into consideration more accurately the long term dynamics, technological changes and effects of product substitution in the economy. However, we should be careful not to overload the model with detail and sophistication. The beauty of models resides in the fact that they are simplified representations of reality, easy to manipulate, cheap to build and use, and great to learn from.

Decision Support Systems and Sustainable Development

Thus far our discussion has focused on the general characteristics a modeling framework must have if it is to be useful in the search for sustainable development schemes. We have hardly touched upon its use or users. But in practice, a model is not very useful in solving the problems faced by small islands unless it is made available to the end-user in a transparent and responsible manner (Holtzman, 1989). One way of doing this is to embed the model in a decision support system (DSS).

Decision support systems are computer-based information systems that are built to help decision makers address semi-structured problems by allowing them to access and use data and analytic models (El-Najdawi and Stylianou, 1993). Their emphasis on semi-structured (ill-defined) problems means that DSS are relevant instruments in relation to sustainable development. Indeed, if we study sustainability in its broad sense, rather than in the narrow economic sense of "optimal resource and environmental management over time, requiring maximizing the net benefits of economic development, subject to maintaining the services and quality of natural resources" (Barbier, 1989), then it is not defined in terms of a single goal, but rather as a set of different goals that all should be met where and when possible. This involves "satisfying" rather than "optimizing," as there is not a "unique" or "optimal" answer to the questions raised; rather different alternative solutions coexist and need to be evaluated.

In the given circumstances, simulation is a very useful research technique (Boersma and Hoenderkamp, 1985). It aims at describing and explaining the behavior of systems through time. It provides the means for constructing theories or hypotheses to elucidate the observed behavior and for using these theories to explore the possible futures of the system. Simulation is very much an experimental technique and has the great advantage of allowing the decision maker to deal with problems in a very plastic manner. For example, it will show him the results of a "what-if" analysis on sustainable development as a set of land use maps that are displayed chronologically on the computer screen. In a very tangible manner he will see how specific situations come about. He can compare model outputs with preferential or ideal development schemes, experiment with alternatives and draw conclusions concerning the policies or decisions likely to result in the desired outcomes or, on the contrary, those that are to be avoided.

For small island users, a good DSS should not only be relevant, reliable, user friendly, flexible and fast, but it should also be cost effective. Often only limited financial resources are available to develop, use and maintain it. Small islands are limited in their human resources too and decision making is performed by generalists rather than specialists. In such circumstances DSS can make available the required specialized knowledge at a fraction of the cost, providing that it does so in a way that is straightforward to the generalist, that is transparent in terms of input required and actions to take, and available on a computer platform that is both affordable and familiar to the user.

A Prototype DSS for the Sustainable Development of Small Islands

At this moment the ISLAND model discussed in "The ISLAND Model" section has been embedded in the model base of a DSS. Thus, it is complemented with additional models and analytic methods, as well as a number of editors, input, presentation and communication tools, and data translators, all of which can be manipulated directly or indirectly by the user. Each of these is of a generic nature, and hence can be applied to any island or country. These tools allow the user to develop interactively his own application. He decides on the area to be modeled, its spatial extent and resolution, as well as on the type and number of socio-economic activities and land uses to be included. To start a new application, he will either import a DTM (Digital Terrain Model) and a land use map from an existing application (*e.g.*, a GIS), or enter it through mouse-clicking by means of the built-in "land use" and "elevation" editors. Similarly, the transition rules, the model variables, and the parameters will be entered or changed by means of graphical editors.

The user has access to the macro-level model via a graphical representation of the model which displays the principle subsystems, their variables (as boxes), and the most important feedback loops (as arrows) of the model (Figure 3). This scheme not only serves in understanding the architecture of the model but is also a graphical interface to the different components of the model. If the appropriate model-box is clicked, a window opens, thus allowing access to the relevant variables and parameters for that

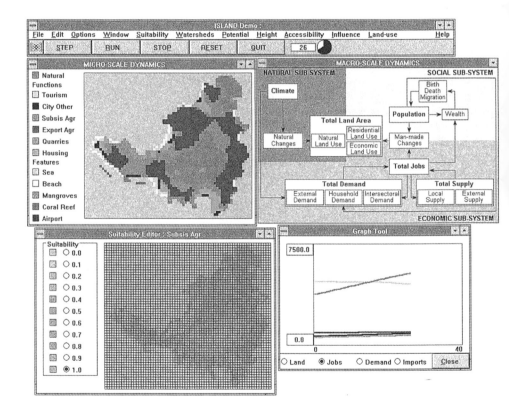

Fig. 3. A view of the graphical user interface of the prototype decision support system. Menus allow the user to manipulate simulation files and to use the different tools of the system. Once a simulation file is opened, the micro- and macro-level models will appear in separate windows. A control panel allows for the control of the simulation.

part of the model. While the simulation is running, the different windows are being updated to display the current values of the model variables and parameters.

The DSS has been implemented in MS-Windows 3 on top range IBM-PC compatible computers, equipped with a 80386 or 80486 processor with a minimum of 4 Mb of RAM and preferentially a Super VGA graphics board and screen. This platform has the great advantage of being widely available, of being cost effective and of being sufficiently powerful to perform the calculations required within reasonable time limits. Our system is making extensive use of the Multi-Tasking and Inter Program Communication facilities offered by Windows. This allows for an easy updating and high level of reusability of the components of the system and the exchange of data at run-time with other applications (*e.g.*, the Microsoft Excel 4.0 spreadsheet package). To compare the different "what-if" scenarios carried out by the user, a comparison tool has been included. It does a cell by cell comparison of two simulation outcomes and displays a map with the differences. Such comparisons are important when a very detailed analysis of a specific area is required. In the analysis of general trends and long-term changes in land use and land-coverage, more general similarities need to be studied, and measures are required that can detect and recognize patterns, shapes and contiguity. An example of such a measure is the fractal dimensionality of the patterns on land use maps (White and Engelen, 1993).

Towards a Generic Decision Support System for the Sustainable Development of Small Islands

Each small island is unique, but small islands share sufficient similar characteristics for them to learn from one another's experiences[1]. This would be made possible if groups of small islands were to adopt a common methodology and supporting software to report on their development initiatives and experiences, in order to share and exchange them with other islands. Sharing would be easier if a regional central site were selected, where data bases, case bases, knowledge bases, and model bases could be stored. This regional information center would update its material on a continuous basis as new data became available from sources such as a census, remote sensing, research, or field work. The approach would involve the development of a DSS based on a set of generic and robust models covering the range of topics and problems typical for the region as well as models to deal with problems that are specific for the individual islands. For each island a simplified and dedicated version of the DSS would be assembled. Relevant models

---

[1]These ideas have been presented in the project proposal "Planning for Sustainable Development: A Decision Support System for Small Islands," initiated by Earthwatch UNEP and developed by Delft Hydraulics (The Netherlands), the Research Institute for Knowledge Systems (The Netherlands), and The World Conservation Monitoring Center (U.K.) and submitted to the Conference of Small Island States in Barbados, April, 1994.

would be selected from the model base and relevant parts of the central data and knowledge bases would be incorporated. The island version of the DSS should be a stand alone application, having full autonomy, so that island users can solve their planning and policy problems without having to connect to the central site. Further, it should allow island users to monitor changes that are occurring on their island and to update their DSS accordingly. This new information and knowledge would then be shared with the regional center to update the data, knowledge, and model bases and to be exchanged with the other islands in the region. This sort of approach would to a maximum extent make use of the indigenous capacities and knowledge in the region and allow specialist know-how to be available without needing the expert in person.

Additional Modeling Tools to "Measure" Sustainability

We have argued that our two-level modeling framework developed as part of a decision support system can be instrumental in studying and "measuring" sustainability, because it allows study of the temporal and spatial dynamics of the demography, the economy and the natural system in a truly integrated fashion. It enables one to learn what overall changes are to be expected, when these will occur, and also where in space these are likely to take place. For effective policy making knowledge of the timing, magnitude and location of changes is essential as it allows development alternatives to be interpreted in terms of natural area lost, biotopes endangered, stress on natural systems and marginal lands, living standards and job opportunities for the population, *etc.* To ease the comparison of policy alternatives, a number of specific evaluation modules are being developed and made available as part of the DSS. In this section we present three modules, each of which is used in combination with the two-level model presented. The intention is to provide the user with important indicators on the sustainability of his or her policies. We will briefly discuss each of them, namely the "transition-potentials," the "pollution-flows," and the "probability-of-impact." The three measures have an explicit geographical dimension and are presented as maps to the user.

Transition-Potentials

"Small islands have ... cultures with special adaptations to island environments and knowledge of the sound management of island resources" (Agenda 21, paragraph 17.125). Modern development, however necessary for some of the islands, is often driven by exogenous economic powers that go against these indigenous, sustainable activities that have developed over time. Blommestein (1993) reports on such evolution "... for example in Trinidad and Tobago ... farmers ... encroach on wetlands and forested areas while at the same time arable land resources are under-utilized, or are being converted for housing, tourism, and other development purposes." The sort of development sketched by Blommestein shows clearly the inequality of economic powers at work, leading to speculation on the land, to shifting land uses, to invasion of valuable natural systems. But, when the best land is taken in by an economically stronger activity, the weaker activity will be pushed into marginal, often natural land, thus damaging natural and ecosystems. Deforestation, erosion, and depletion of drinking water reserves will be the result. Biodiversity will be affected, because of the decreased spatial extension and fragmentation of habitats. Not only plant and animal life is affected, but also the well-being of groups of the socially weak is affected as more human input is required to produce the same amount on marginal lands and in otherwise difficult situations. The result is social stress and even violent conflicts (Homer-Dixon *et al.*, 1993).

If economically weak activities are being pushed away from the most appropriate lands, thus endangering sustainability in the sense of weakened self-reliance and self-sufficiency, policy makers have to intervene and seek to protect these activities, through zoning and other regulations. In this sense, a good indicator of sustainability consists in comparing the actual location of an activity with the location of the most appropriate land for it. The *transition-potential* indicator aims at visualizing these discrepancies. In the paragraph on the micro-scale dynamics, we have explained that the cellular model calculates for each cell a transition potential expressing the likeliness for it to change from its actual land use to any other land use. Generally speaking, a cell will take on the land use for which its transition potential is highest. Transition potentials are calculated on the basis of the physical, environmental, and institutional suitability of the cell and its functional suitability. The latter is calculated by means of distance decay functions which express the push and pull forces that socio-economic activities exert on each other within a fixed radius. The amplitude of such forces differs considerably among activities and is a function of the economic bargaining power, expressed in terms of productivity and output per unit area, of each activity. Hence, in the calculation of the transition potentials, activities such as subsistence agriculture, for which output per unit area is low, is getting less weight than, say, commercial activities for which the value added per unit area can be very high. Thus, although a cell can figure among the highest transition potentials for subsistence agriculture, it could have at the same time a much higher transition potential for commercial activity and change to a commercial cell in the next simulation step.

Take the example of an island that is expanding its tourism sector. Due to the value added generated in this sector, almost any piece of land that is well located for tourism activities and wanted for expansion, will not be able to withstand the pressure for conversion to that use, irrespective of its value for, say, subsistence agriculture. This sort of expansion works fine as long as demand for tourism exists. From the economic point of view building hotels is a good and durable investment. The wealth generated in the tourism sector and its spin-offs might increase considerably the well-being of the island population generally or even of the poor in particular. If the environmental impacts of the additional tourists is taken care of, the expansion could be called a good step towards sustainability. However, what happens if the demand for tourism diminishes? At that time the sort of expansion described has to be evaluated in terms of valuable agricultural land being lost irreversibly; a non-renewable resource is gone forever, as is the knowledge of how to use it for food production. Hence, in terms of sustainability, the allocation of the land to the expanding tourism sector has to be accounted for negatively.

To demonstrate the use of the transition-potential map, we show in Figure 4, at the top, a map with the transition-potentials for subsistence agriculture for each cell. At the bottom is the actual land use map, calculated by the model, hence taking into consideration the competition for space among the different activities. It is very visible from the map, especially on the western coast of the island, how the best pieces of land are not taken in by subsistence agriculture, rather that subsistence agriculture has been pushed away by housing, tourism and city functions into the marginal lands (low transition-potential) of the central highlands. The same conclusion can be drawn from the frequency table displaying the distribution of the transition-potentials for subsistence agriculture. Superimposed on this is the distribution of the transition-potentials for subsistence agriculture of the cells actually taken in by the activity. It is clear that the tail of the distribution, representing the most appropriate cells, is not taken in by the activity. In comparison, the frequency table for stronger activities, such as tourism, shows that the most appropriate cells are taken in by the activity.

Pollution-Flows

In the previous sections we have mentioned the importance of having a detailed representation of space in order to understand the impacts of man's activities on the environment. We have discussed the fact that such spatial knowledge has its importance in terms of understanding the marginality of the land and the amount of stress that such land can withstand, as well as the fragmentation of natural areas and the effect this has on biodiversity. Another very important aspect of man's impact on the environment, however, is the fact that each human activity brings with it some sort of waste production and pollution. These side products find their way, in a more or less controlled manner, into the natural environment. Non-spatial models of sustainable development, such as the ones put forward by Barbier (1989) or van den Bergh and Nijkamp (1991), consider the environmental quality as being represented by one macro index, reflecting a sort of "average" or "total" environmental quality of the system studied. However, pollution, in any of its states, gaseous, liquid, or solid, is not spread in an average or equal way over an island or country. Clearly, pollution has an apparent source and moves along specific channels towards a distinct sink. In the remainder of this paragraph we will concentrate the discussion on water pollution.

Small islands, due to their size, their geographical location, and relief may often suffer drinking water shortages, so the pollution of any of the reserves has an immediate effect. And, human activity almost inevitably results in water contamination. If this pollution reaches the coastal waters it may get trapped in the near-shore currents and affect coral reefs, sea grasses, mangroves, fisheries, and beaches.

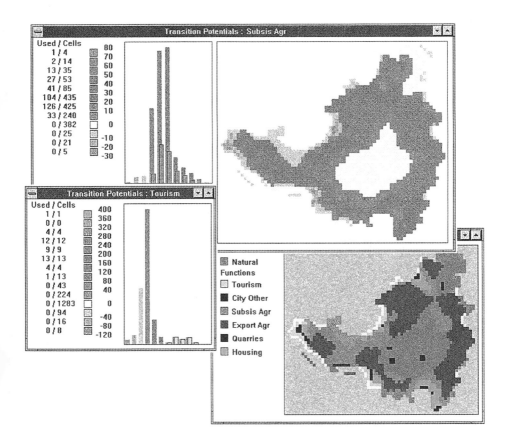

Fig. 4. A transition-potentials map for subsistence agriculture and the corresponding land use map. Notice that subsistence agriculture is located in sub-optimal areas.

In our pollution-flow module an algorithm calculates how in normal circumstances (following the laws of gravity, hence excluding engineering solutions) water carries pollution from source to sink. These algorithms, which are standard in GIS, get an added value in our modeling framework, because they show how the changing location of activities will result in changing concentrations and types of pollution that get channelled into the coastal waters at specific outlet points. Hence, they show how the growth and spread of human activities creates problems of pollution downstream from where the activity takes place. Such knowledge would be valuable in safeguarding sensitive and valuable coastal ecosystems—marine, but also terrestrial. The combination of the above module with a hydrological model describing the near-shore currents could enhance the detailed understanding of the transport and dispersion of the pollution in the coastal waters. Figure 5 shows the most important drainage basins of the island (area exceeding 15 km^2) and shows the outlet points of each. For each basin, an overview table can be opened that shows, while the simulation is running, the changing land uses and linked pollution generated in the basin.

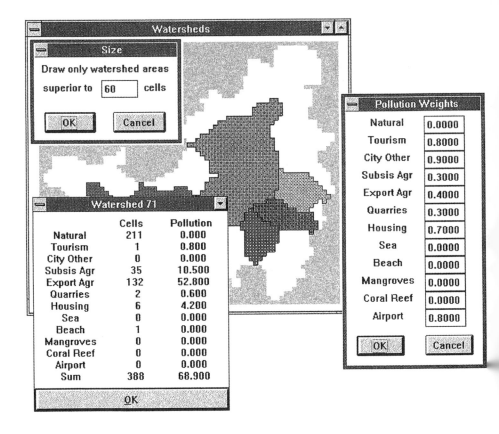

Fig. 5. Main drainage basins of the ISLAND. The pop-up window displays the land uses and their associated contribution to pollution in the largest basin of the ISLAND.

Dealing with Uncertainty

We have mentioned in our introductory paragraphs that socio-economic systems are very complex, that they display an ever changing behavior and that as a consequence they are largely unpredictable. Models and tools made available to policy analysts can only partially capture this complexity; hence, any suggested solutions or development alternatives must be considered approximate. Since models are incapable of telling what will happen exactly, they are often used to find out what not to do in forecasting and planning exercises. The model is run under different scenarios in order to corner the conditions under which the system would be directed towards undesirable states. In this context, a systematic study of the possible but unpredictable factors on the system is of key importance. Best and worst estimates should be tested to see how the system responds to these and to anticipate the different behaviors that are to be expected. Another, somewhat similar exercise involves running the model repeatedly with slightly different but realistic parameter values in order to produce a probabilistic synthesis of the different model outcomes (see, for a similar exercise, Allen and McGlade, 1987). To illustrate this we have run our ISLAND model 100 times, each time with a stochastic perturbation of the parameter expressing the effect of well-being on the mobility of the island population. In each run of the model this fluctuation is introduced as a random perturbation of the best estimate we have for the parameter. This perturbation has a uniform distribution and it causes the parameter to vary between 50% and 150% of its estimated value. Thus, we express the amplitude of the uncertainty. For each land use a land use probability map is generated, which shows for each cell the percentage of the runs in which the cell gets the particular land use, given the variance in the value of this one model parameter. Figure 6 shows such a map for natural land use.

The practical value of such probability maps as indicators for sustainability is that they show the decision maker where in space certain risks might appear in a more or less consistent manner and independent of the uncertainty in one or more of the parameters of the model. If activities appear in an area where they are a threat in terms of biodiversity, pollution or environmental stress, then the decision maker, faced with limited resources of all sorts, should be able to prioritize his interventions. Although the probability maps will warn the user of possible dangers, they do not tell him what specific parameter values give rise to the high probability scores in the sensitive areas. To get more insight into these factors and thus to be able to design planning interventions that could significantly reduce the likelihood of adverse impacts, additional research and systematic what-if and sensitivity analysis is necessary.

Conclusions

We have adopted in this chapter a broad definition of sustainable development and have stressed the vagueness of the concept: sustainability does not succumb easily to normalization, optimization or to modeling in general. We have argued that to study it

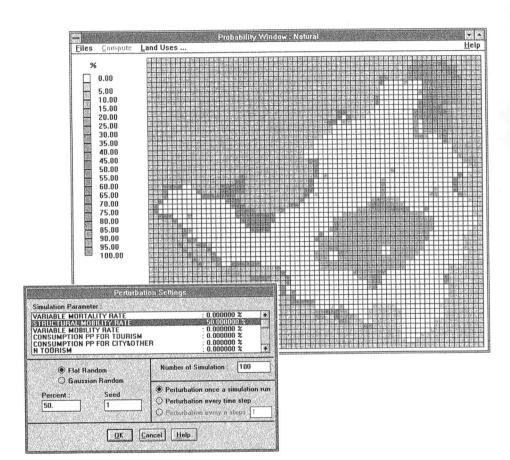

Fig. 6. Land use probability map showing the probability of having natural land-cover on each cell after 100 runs of the model with random fluctuations in a migration parameter.

or to design successful policies for sustainability, dynamic models representing complex reality in an integrated manner and dealing with geographical characteristics on a detailed scale are essential. Although this requirement is generally well understood and accepted on a conceptual level, fulfilling it presents a challenging task. Scientists still lack sufficient insight into the essential mechanisms and interactions that cause natural and socio-economic systems to change over time and thus are not as yet well equipped to capture the behavior of these systems in their models. However, new scientific paradigms and the associated modeling methodologies will permit scientists to better capture the complexity and uncertainty that is inherent in the dynamics of socio-economic and socio-environmental systems. In addition, the rapid evolution of the information sciences and technologies will facilitate a better understanding and management of these systems (Han and Kim, 1990).

The rapid expansion of GIS technology and its use by many planning agencies is a sign of the growing contribution that the information sciences are making to policy formulation and implementation. But information technology should not replace human intelligence and liberty of action. Nor should it trap the users into a never-ending race for more and more detailed data. Nor should it confuse the users by leading them to believe that the beautiful visualization of very detailed data sets automatically represents a trustworthy or useful image of reality. Rather, the real needs for support in the decision making process should get more attention in the development of new information systems.

Decision making related to sustainable development means deciding on the future of complex systems. It requires a high level of knowledge. Consequently, the knowledge and intelligence content of information systems has to be strengthened to assist policy makers in the urgent and difficult task of bringing the systems they manage onto a more sustainable path. Policy makers must be kept aware of the uncertainty that is inherent to any run of a model and any scenario analysis. Information systems should provide adequate support in the design and implementation of robust policies that leave as many options as possible open for future generations, since, in an uncertain world, adaptability is an important characteristic in determining the level of sustainability.

References

Allen, P. M., and M. Lesser, Evolutionary human systems: Learning, ignorance and subjectivity, in *Evolutionary Theories of Economic and Technological Change: Present Status and Future Prospects*, edited by P. Saviotti and S. Metcalfe, pp. 160-171, Harwood Academic Publishers, Reading, U.K., 1991.

Allen, P. M., and J. M. McGlade, Modeling complex human systems: A fisheries example, *European J. Oper. Res., 30*, 147-167, 1987.

Allen, P. M., and M. Sanglier, A dynamical model of growth in a central place system, *Geograph. Anal., 11*, 256-272, 1979.

Barbier, E. B., *Economics, Natural-Resource Scarcity and Development, Conventional and Alternative Views*, 223 pp., Earthscan Publications Ltd., London, 1989.

Batty, M., *Urban Modeling: Algorithms, Calibrations, Predictions*, Cambridge University Press, London, 1976.

Berry, J. K., *Beyond Mapping: Concepts, Algorithms, and Issues in GIS*, 246 pp., GIS World Inc., Forth Collins, 1993.

Blommestein, E., Sustainable development and small island developing countries, ECLAC, Trinidad and Tobago, 1993.

Boersma, S. K. T., and T. Hoenderkamp, Simulatie een moderne methode van onderzoek, 315 pp., Academic Service, Den Haag, 1985.

Brimicombe, A. J., Flood risk assessment using spatial decision support systems, *Simulation*, 379-380, 1992.

Burrough, P. A., *Principles of Geographic Information Systems for Land Resources Assessment*, 194 pp., Clarendon Press, Oxford, 1989.

Couclelis, H., Cellular worlds: A framework for modeling micro-macro dynamics, *Environ. Plan. A*, *17*, 585-596, 1985.

Couclelis, H., Of mice and men: What rodent populations can teach us about complex spatial dynamics, *Environ. Plan. A*, *20*, 99-109, 1988.

De Meyer, R., Bananen schrijven geschiedenis, De Wereld Morgen 3, 1991.

Dendrinos, D. S., Nonlinearities, interdependent dynamics and interacting scales: Progress in urban and transportation analysis, *Chaos, Solitons and Fractals 4(4)*, 497-506, 1994.

ECLAC, Selected statistical indicators of Caribbean countries, Vol. IV, United Nations Economic Commission for Latin America and the Caribbean, Port of Spain, Trinidad and Tobago, 1991.

El-Najdawi, M. K., and A. C. Stylianou, Expert support systems: Integrating AI technologies, *Commun. of the ACM, 36*(2), 55-65, 1993.

Enchinique, M., Models: A discussion, in *Urban Space and Structures,* edited by L. Martin and L. March, pp. 164-174, Cambridge University Press, London, 1972.

Engelen, G., The theory of self-organization and modeling complex urban systems, *European J. Oper. Res., 37,* 42-57, 1988.

Engelen, G., R. White, and I. Uljee, Exploratory modeling of socio-economic impacts of climatic change, in *Climatic Change in the Intra-Americas Sea,* edited by G. A. Maul, pp. 350-368, Edward Arnold, London, 1993a.

Engelen, G., R. White, I. Uljee, and S. Wargnies, Vulnerability assessment of low-lying coastal areas and small islands to climate change and sea level rise, Final Rep. to UNEP CAR/RCU, RIKS publ. 905000/9379, Maastricht, 1993b.

Gardner, M., The fantastic combinations of John Conway's new solitaire game Life, *Scien. Amer., 223,* 120-123, 1970.

Gutowitz, H., *Cellular Automata: Theory and Experiment,* 479 pp., The MIT Press, Cambridge Massachusetts, 1991.

Han, S.-Y., and T. J. Kim, Intelligent urban information systems: Review and prospects, in *Expert Systems: Applications to Urban Planning,* edited by T. J. Kim, L. L. Wiggins, and J. R. Wright, pp. 241-264, Springer Verlag, New York, 1990.

Holtzman, S., *Intelligent Decision Systems,* 304 pp., Addison-Wesley, Reading, 1989.

Homer-Dixon, T. F., J. H. Boutwell, and G. W. Rathjens, Environmental change and violent conflict, *Scien. Amer.,* 16-23, 1993.

Huggett, R., *Systems Analysis in Geography,* 208 pp., Clarendon Press, Oxford, 1980.

Jacome, J. M., The Commenwealth Caribbean Small Island States and the Caribbean Center for Development Administration: A management development institution, in *Public Administration in Small and Island States,* edited by R. Baker, pp. 233-251, Kumarian Press, West Hartford, Connecticut, 1992.

Langton, C. G., Studying artificial life with Cellular Automata, *Physica 22D,* 120-149, 1986.

Leontief, W., *Input-Output Economics,* 436 pp., Oxford University Press, New York, 1966.

Marble, D. F., and S. E. Amundson, GIS in urban and regional planning, *Environ. Plan. B, 15,* 305-324, 1988.

Marshall, D., Migration as an agent of change in Caribbean island ecosystems, *Internat. Social Sci. J.*, *34*(3), 451-467, 1982.

Richardson, H. W., *Regional and Urban Economics*, 416 pp., Penguin Books, Harmondsworth, 1978.

Simmonds, D., Modeling contexts and purposes: Defining alternative paths of model development, in *Advances in Urban Systems Modeling*, edited by B. Hutchinson and M. Batty, pp. 75-90, North Holland, Amsterdam, 1986.

Solow, R. M., Sustainability: An economist's perspective, Eighteenth J. Seward Johnson lecture in marine policy, Marine Policy Center, Woods Hole Oceanographic Institution, Woods Hole, 1991.

The Diagram Group, *The Atlas of Central America and the Caribbean*, MacMillan, New York, 1985.

The World Bank, *World Development Report 1992: Development and the Environment*, 308 pp., Oxford University Press, New York, 1992.

Tobler, W. R., Cellular geography, in *Philosophy in Geography*, edited by S. Gale and G. Olsson, pp. 279-386, Reidel, Dordrecht, 1979.

United Nations, Agenda 21, Chapter 17: Protection of the oceans, all kind of seas, including enclosed and semi-enclosed seas, and coastal areas and the protection, rational use and development of their living resources, 1992.

van den Bergh, J. C. J. M., and P. Nijkamp, Aggregate dynamic economic-ecological models for sustainable development, *Environ. Plan. A*, 23, 1409-1428, 1991.

White, R., Dynamic central place theory: Results of a simulation approach, *Geograph. Anal.*, 9, 227-243, 1977.

White, R., and G. Engelen, Cellular automata and fractal urban form: A cellular modeling approach to the evolution of urban land use patterns, *Environ. Plan. A*, 25, 1175-1199, 1993.

White, R., and G. Engelen, Cellular dynamics and GIS: Modeling spatial complexity, *Geograph. Sys.*, 1, 237-253, 1994.

World Commission on Environment and Development (WCED), *Our Common Future*, Oxford University Press, New York, 1987.

Wright, J. R., ISIS: Toward an integrated spatial information system, in *Expert Systems: Applications to Urban Planning*, edited by T. J. Kim, L. L. Wiggins, and J. R. Wright, pp. 43-66, Springer Verlag, New York, 1989.

464

List of Contributors

Claude Augris
Dept. de Geologie et Oceanographie
Univ. Bordeaux 1
URA CNRS 197
Avenue des Facultes
33405 Talence, France

Erik Blommestein
U.N. Econ. Comm. for Latin Amer.
and the Caribbean
22-24 St. Vincent Street, 2nd Fl.
P. O. Box 1113
Port of Spain, Trinidad & Tobago

Malcolm J. Bowman
Marine Sciences Research Center
State University of New York
Stony Brook, NY 11794-5000

M. Breton
IFREMER Centre de Brest
Tech. Brest-Iroise Lab. Hydrodyn.
et Sedimentologie-DEL, B. P. 70
29280 Plouzane, France

Lance A. Busby
U. N. Econ. Comm. for Latin
Amer. and the Caribbean
22-24 St. Vincent Street, 2nd Fl.
P. O. Box 1113
Port of Spain, Trinidad & Tobago

Gillian Cambers
Sea Grant College Program
University of Puerto Rico
P.O. Box 5000
Mayaguez, Puerto Rico 00681-5000

Richard V. Cant
c/o Water and Sewage Corporation
J .F. Kennedy Drive
P. O. Box N. 3905
Nassau, Bahamas

M. P. Caprais
IFREMER Centre de Brest
Tech. Brest-Iroise Lab. Hydrodyn.
et Sedimentologie-DEL
B. P. 70
29280 Plouzane, France

Patrice Castaing
Dept. de Geologie et Oceanogr.
University de Bordeaux 1
URA CNRS 197
Avenue des Facultes
33405 Talence, France

Pierre Daniel
METEO-FRANCE
SCEM/PREVI/MAR
42, Avenue Coriolis
31 057 Toulouse Cedex, France

A. Derrien
IFREMER Centre de Brest
Tech. Brest-Iroise Lab. Hydrodyn.
et Sedimentologie-DEL
B. P. 70
29280 Plouzane, France

David. E. Dietrich
Center for Air-Sea Tech.
Mississippi State Univ.
Building 1103, Room 233
Stennis Space Center, MS 39529

Maria C. Donoso
NOAA/AOML/PhOD
4301 Rickenbacker Causeway
Miami, FL 33149-1097

E. Dubois
IFREMER Centre de Brest
Tech. Brest-Iroise Lab. Hydrodyn.
et Sedimentologie-DEL
B. P. 70
29280 Plouzane, France

E. Dupray
IFREMER Centre de Brest
Tech. Brest-Iroise Lab. Hydrodyn.
et Sedimentologie-DEL
B. P. 70
29280 Plouzane, France

Francoise Durand
Dept. de Geologie et Oceanogr.
University de Bordeaux 1
URA CNRS 197
Avenue des Facultes
33405 Talence, France

Guy Engelen
Res. Inst. for Knowledge Systems
Tongersestraat 6
P. O. Box 463
6200 Al Maastricht, The Netherlands

M. Gourmelon
IFREMER Centre de Brest
Tech. Brest-Iroise Lab. Hydrodyn.
et Sedimentologie-DEL
B. P. 70
29280 Plouzane, France

Artemio Gallegos
Inst. de Ciencias del Mar y Limnologia
Lab. de Ocenaografia Fisica
Cd. Universitaria
Apartado Postal 70-305
Coyoacan 04510
Mexico D. F., Mexico

Orman E. Granger
Department of Geography
University of California
501 Earth Science Building
Berkeley, CA 94720

Calvin R. Gray
National Meteorological Services
65 3/4 Half Way Tree Road
P. O. Box 103
Kingston 10, Jamaica

Malcolm D. Hendry
Marine Res. and Envir.
      Management Program
University of the West Indies
P.O. Box 64
Cave Hill, St. Michael
Barbados

Paul F. Holthus
University of Hawaii
East-West Center
Program on Environment
1777 East-West Road
Honolulu, HI 96848

André Klingebiel
CIBAMAR
Universite de Bordeaux 1
Avenue des Facultes
33405 Talence, France

Pascal Lazure
IFREMER Centre de Brest
Tech. Brest-Iroise Lab. Hydrodyn.
et Sedimentologie-DEL
B. P. 70
29280 Plouzane, France

466

F. Le Guyader
IFREMER Centre de Brest
Tech. Brest-Iroise Lab. Hydrodyn.
et Sedimentologie-DEL
B. P. 70
29280 Plouzane, France

Charles A. Lin
McGill University
Dept. of Atmospheric and Oceanic Sci.
Montreal, Quebec
Canada

Robin Mahon
Fisheries and Environ. Consulting
48 Sunset Crest
St. James, Barbados

Jose L. Juanes Marti
Instituto de Oceanologia
Calle Ira, No. 18406, e/184 y 186
Havana, Cuba

George A. Maul
Florida Institute of Technology
Div. of Marine and Envir. Systems
150 W. University Boulevard
Melbourne, FL 32901 6988

D. Menard
IFREMER Centre de Brest
Tech. Brest-Iroise Lab. Hydrodyn.
et Sedimentologie-DEL
B. P. 70
29280 Plouzane, France

Jean-Claude Michel
Division of Environment
P. O. Box 445
Victoria, Seychelles

Guillermo Garcia Montero
Comite Oceanografico Nacional
Calle I y 60,
Miramar, Playa
Ciudad de la Havana, Cuba

Paolo A. Pirazzoli
Lab. de Geographie Physique
Univ. de Paris l'et Paris IV
Centre Natl. de la Recherche Sci.
1, Place Artiside Briand 92195
Meudon Cedex, France

Monique Pommepuy
IFREMER Centre de Brest
Tech. Brest-Iroise Lab. Hydrodyn.
et Sedimentologie-DEL
B. P. 70
29280 Plouzane, France

J. C. Salomon
IFREMER Centre de Brest
Tech. Brest-Iroise Lab. Hydrodyn.
et Sedimentologie-DEL
B. P. 70
29280 Plouzane, France

Nirmal J. Shah
Division of Environment
P. O. Box 445
Victoria, Seychelles

Paul II. Templet
Inst. for Environmental Studies
42 Atkinson Hall
Louisiana State University
Baton Rouge, LA 70803

Georges Vernette
CIBAMAR
Universite de Bordeaux 1
Avenue des Facultes
33405 Talence, France

Vance P. Vicente
Southeast Fisheries Science Ctr.
National Marine Fishery Serv.
Suite 1108, Banco de Ponce Bldg.
Hato Rey, Puerto Rico 00918

Serge Wargnies
Res. Inst. for Knowledge Systems
Tongersestraat 6
P. O. Box 463
6200 Al Maastricht
The Netherlands

Roy A. Watlington
Eastern Caribbean Center
University of the Virgin Islands

Roger White
Memorial Univ. of Newfoundland
St. John's, Newfoundland
Canada

Judith Wolf
Church Lane, Neston
South Wirral
Cheshire L64 9UT
United Kingdom